Microplastics Pollution and its Remediation

Ajay Kumar • Vijai Singh
Editors

Microplastics Pollution and its Remediation

Springer

Editors
Ajay Kumar
Department of Biotechnology
Rama University
Kanpur, Uttar Pradesh, India

Vijai Singh
Department of Biosciences
Indrashil University
Mehsana, Gujarat, India

ISBN 978-981-97-4067-3 ISBN 978-981-97-4068-0 (eBook)
https://doi.org/10.1007/978-981-97-4068-0

This Springer imprint is published by the registered company Springer Nature Singapore Pte Ltd.
The registered company address is: 152 Beach Road, #21-01/04 Gateway East, Singapore 189721, Singapore

Foreword

I am delighted to write about "*Microplastics Pollution and Its Remediation*," a timely addition to this rapidly evolving subject. Microplastics are the most severe form of pollution contaminating the environment and ocean, as they can be easily manufactured or produced. All the garbage gets accumulated in the ocean in the form of discarded plastics exhibiting toxic effects to marine animals. Microplastics are a great threat to humans as the pathways of exposure through skin contact, oral intake, and inhalation are the common ways and the knowledge gap about the toxicity of microplastics is neglected. However, it is necessary to discuss broader aspects of the challenges, risks, and remedies to demonstrate the future directions.

Microplastics are smaller fragments that can be produced in various ways and due to their tiny size. It is believed not to pass through skin barriers, but the fact is that it has high chances of accumulating on the skin. Microplastics are believed to cause severe metabolic disorders, skin disease, intestinal issues, and also chronic disorders due to its hazardous effects.

This book contributes to all the segments of microplastics including the sustainable management and separation techniques to reduce the risk associated with the use of microplastics in cosmetic and personal care products. The everyday waste produced by different industries that has become a major concern causing pollution to ecosystem. Therefore, it is important to take actions to handle the problems coming from microplastics. Considering all the facts, this book may help to solve the problems, and it is for every individual to understand the moral duties of each to protect our environment. Additional efforts are needed such as policy actions and remediation actions on how to reduce and mitigate microplastic pollution that may require an improved waste management and reuse system, development of bioplastics and new plastic materials, as well as additional funding legally binding to cope with the increasing plastic pollution at a global scale.

I am delighted to acknowledge the commendable efforts of Prof. Ajay Kumar and Prof. Vijai Singh, who, with the support of Springer Nature, have painstakingly designed this outstanding content. This book serves as an invaluable resource to everyone including students, researchers, scientists, practitioners, policymakers, and stakeholders about the information of microplastics and its remediation.

Chemistry and Physics Department
University of Almeria,
Almería, Spain

Damià Barceló

Preface

The use of plastics coming from different sources has become a global challenge, which is faced by the environment, along with its potential health threat to animals and humans. It has been categorized into various categories, out of which the most complex tiny form with more severe toxic effects as compared to the large plastics is referred to as "microplastics." Microplastics are defined as the smaller fragments of plastics having size less than 5 mm that are produced by various industries and are disposed everyday into the water from industrial plants. Large amount of plastics are entered into the ocean; thereby, marine animals are exposed to higher risk that results in starvation, movement restriction, and reduction in nutrient uptake. The lack of knowledge about the exposure of microplastics to humans is an alarming concern, and immediate measures should be taken to understand the multiple pathways of human exposure. In order to explore the adverse toxic effects to human health, the intake of microplastics and exposure time in the human body should be deeply discussed. The exposure of microplastics to human body can not only cause liver infection, intestinal injury, and metabolic disorder but also lead to chronic diseases; therefore, the subject cannot be avoided.

Microplastics are rapidly increasing as they are used in vast amounts in cosmetic products and personal care products such as face scrubs and cleansers. Consequently, these tiny particles can hardly be detected, and they are washed out into the oceans, thereby forming clusters of microplastics that are insoluble in water. Thus, we have emphasized on the impact of microplastics on human, environment, and aquatic health and various technologies to mitigate their adverse effect. The information is also given about risks associated with the interaction of microplastics with ecosystem and the different separation techniques. Microplastics are the most neglected subject of discussion, and there is incomplete information given in literatures, so here we have discussed about the key factors of microbe interactions on animal health and how it should be tackled. The content includes the sustainable management of microplastics through the collection of clinical and epidemiological data in order to minimize the exposure to its toxic effects. This book offers the basics of microplastics, detection, separation techniques, microplastics in soils, aquatic environment, terrestrial elements, glaciers, crops, foods, cosmetics, air, interactions on animals and humans, landfill, hazardous impacts, management of microplastics, and many.

This book offers an excellent and informative text on microplastics. The content uses a rich literary text and clear figures with clarity and coverage. This book would be a primer for beginners, researchers, students, and environmentalists. It helps to establish a health risk assessment model that can help to understand broader prospects of microplastics. This book also helps for better understanding of basic knowledge from which scientific knowledge can grow, widen, and accelerate future development research on microplastics.

Kanpur, Uttar Pradesh, India — Ajay Kumar
Mehsana, Gujarat, India — Vijai Singh

Acknowledgments

Acknowledgment by Ajay Kumar

I am extremely happy to present this book, which is an outcome of rigorous studies and literature survey. I hope that this book will make significant contribution to the area of scientific studies and research.

I am especially grateful to Dr. Suraj B. S. Kushwah, Honorable Chancellor, Rama University, Kanpur, who is our great motivator and patron. I also take this opportunity to thank Dr. Prateek Singh, Director, Rama University; Dr. Pranav Singh, Director, Rama University; Dr. Hari Om Sharan, Dean, Academic and Planning; and Dr. C. S. Raghuvanshi, Dean, Faculty of Engineering and Technology, Rama University. They always encourage me to pursue academic and research activities and achieve excellence.

I am also highly indebted to my learned colleagues in the Department of Biotechnology and other members of faculty from the Faculty of Engineering and Technology, Rama University. I would especially like to mention Er. Fariya Khan and Mr. Indrajeet Singh who have always extended their helping hands in various ways.

My co-editor Dr. Vijai Singh has always offered me great personal and professional support and also inspired me to accomplish this task successfully.

I feel deeply and highly obliged to my beloved wife, Shraddha, and my loving children, Nishit and Anshika, who not only encouraged me to edit this book but also have shown patience when deprived of my much-needed attention when I was busy in editing this book. This book might not have seen the light of the day without the support of my family.

I seek the blessing of my elder brothers, Er. Prem Chandra and Dr. Ram Kumar, who have always guided me in my passion. I am also thankful to the Almighty God who has blessed me with the wisdom to edit this book.

Finally, I dedicate this book to my parents for their countless blessings that helped me accomplish the target.

Acknowledgment by Vijai Singh

I am delighted to express my sincere gratitude and deep appreciation to Dr. J. S. Yadav, Director (Research), Indrashil University, India, for extending his outstanding support and motivation to complete this book. I would like to give many thanks to co-editor Dr. Ajay Kumar of this book who gave me outstanding personal and professional support as well as inspiration to finish this book.

I am delighted to thank all the authors for their excellent contributions to this book. I would like to thank Mrs. Swati Sharma (Associate Editor—Biomedicine) and Mrs. Aishwarya Thyagarajan (Production Editor) from Springer for their excellent management of this project.

I would like to thank Prof. Rakesh Rawal, Prof. Bharat Maitreya, Prof. Pawan K. Dhar, Dr. Poonam Bhargava, Dr. Madhvi Joshi, Dr. Bhabatosh Das, Dr. Pablo Carbonell, Dr. Rupesh Maurya, Dr. Satya Prakash, Dr. Vimal C. Pandey, Dr. Suresh Ramakrishna, Dr. Dinh-Toi Chu, Dr. Mueksh Kumar Awasthi, and those whose names do not feature here but have directly or indirectly contributed to shaping this project.

I greatly appreciate the support of my students Dr. Nisarg Gohil, Mr. Khushal Khambhati, and Dr. Gargi Bhattacharjee, whose discussion and comments helped to shape this book.

I wish to express my gratitude to my beloved wife Pritee Singh for her endless support, patience, and inspiration. I also wish to express lots of affection for my kids, Aaradhya and Ayush, who missed me during this project. I would like to warmly thank the faculty and staff of Indrashil University for providing a great working environment.

I am aware that even despite our best efforts, the first version always comes with some error that may have crept in the compilation. I would be delighted to receive feedback from readers to further improve the future book. Last but not least, my sincere thanks to God for his supreme power for endowing me to live with joy and victory in the shape of this book.

Contents

Editors and Contributors

About the Editors

Ajay Kumar is a Professor and Head of the Department of Biotechnology at the Faculty of Engineering and Technology at Rama University, Kanpur, Uttar Pradesh. He has a proven experience in genomics and proteomics, bioprocess engineering, bioinformatics, industrial microbiology, environmental biotechnology, and food biotechnology. He has held several key positions in well-renowned universities and engineering institutes. Dr. Kumar received his M. Tech. (Biotechnology) from the Institute of Engineering and Technology, Lucknow, India, and Ph.D. from the ICAR-Central Institute for Research on Goats, Mathura, India. His expertise lies in research that includes computational vaccine and drug development, cancer biology genomics and proteomics, and fermentation technology. He has published 100 articles, 20 book chapters, 4 books, and 7 patents. He has also served as associate editor, editorial board member, and reviewer of several peer-reviewed journals. Besides all these achievements, he is also a member of the Board of Study and Academic Council of Rama University, Kanpur, and Regional Food Research Analysis Center, Department of Horticulture and Food Processing, Lucknow, Uttar Pradesh. Being a member of a professional body such as the International Association of Engineers (IAENG) and INSA, he has rendered consultancy services in vaccine research.

Vijai Singh is a Professor and Head of the Department of Biosciences, School of Science, at Indrashil University, Rajpur, Mehsana, Gujarat, India. He has worked as Associate Professor in the Department of Biosciences, School of Science, at Indrashil University, Rajpur, Mehsana, India. He was an Assistant Professor in the Department of Biological Sciences and Biotechnology at the Institute of Advanced Research, Gandhinagar, India, and also an Assistant Professor in the Department of Biotechnology at the Invertis University, Bareilly, India. Prior to that, he was a Postdoctoral Fellow at the Institute of Systems and Synthetic Biology, Paris, France, and School of Energy and Chemical Engineering at the Ulsan National Institute of Science and Technology, Ulsan, South Korea. He received his Ph.D. in Biotechnology (2009) from the National Bureau of Fish Genetic Resources, Uttar Pradesh Technical University, Lucknow, India, with a research focus on the development of molecular and immunoassays for the diagnosis of *Aeromonas hydrophila*. His research interests are focused on building novel biosynthetic pathways for production of medically and industrially important biomolecules. Additionally, his laboratory is working on CRISPR-Cas9 tools for genome editing. He has more than 12 years of research and teaching experience in synthetic biology, metabolic engineering, bioinformatics, microbiology, and industrial microbiology. He has published 107 articles, 80 book chapters, 25 books, and 3 patents. He serves as an associate editor, editorial board member, and reviewer of several peer-reviewed journals. He is also a member of the Board of Study and Academic Council of Indrashil University and is the Member Secretary of the Institutional Biosafety Committee (IBSC) at the same university.

Contributors

Nahid Akhtar School of Bioengineering and Biosciences, Lovely Professional University, Jalandhar, Punjab, India

Mahbub Alam Department of Environmental Science, Bangladesh University of Professionals (BUP), Dhaka, Bangladesh

Center for Environmental Nanoscience and Risk, Department of Environmental Health Sciences, Arnold School of Public Health, University of South Carolina, Columbia, SC, USA

Amit Hasan Anik Department of Environmental Science, Bangladesh University of Professionals (BUP), Dhaka, Bangladesh

Priyancka Arora Department of Biotechnology, Amity University, Mohali, Punjab, India

Reet Arora Department of Microbiology, Institute of Home Economics, University of Delhi, New Delhi, India

Shruti Chaudhari Department of Environmental Studies, Faculty of Science, The Maharaja Sayajirao University of Baroda, Vadodara, Gujarat, India

Debolina Department of Microbiology, Institute of Home Economics, University of Delhi, New Delhi, India

Sapna Devi Department of Food Science and Technology, Khalsa College of Amritsar, Amritsar, Punjab, India

Hardik Giri Gosai Department of Environmental Studies, Faculty of Science, The Maharaja Sayajirao University of Baroda, Vadodara, Gujarat, India

MD Tanvir Hasnine Environmental Engineering Program, Department of Civil Engineering, University of Manitoba, Winnipeg, MB, Canada

Shabiha Hossain Department of Environmental Science, Bangladesh University of Professionals (BUP), Dhaka, Bangladesh

Department of Geography and Sustainability, University of Tennessee, Knoxville, TN, USA

Jagriti Department of Microbiology, Institute of Home Economics, University of Delhi, New Delhi, India

Amit Joshi Department of Biochemistry, Kalinga University, Naya Raipur, Chhattisgarh, India

Alamgir Kabir Department of Environmental Science, Bangladesh University of Professionals (BUP), Dhaka, Bangladesh

Kirandeep Department of Microbiology, Institute of Home Economics, University of Delhi, New Delhi, India

Ajay Kumar Department of Biotechnology, Faculty of Engineering and Technology, Rama University, Kanpur, Uttar Pradesh, India

Rajendra Kumar Department of Applied Sciences and Humanities, Faculty of Engineering and Technology, Rama University, Kanpur, Uttar Pradesh, India

Indra Mani Department of Microbiology, Gargi College, University of Delhi, New Delhi, India

Mantasha Department of Microbiology, Institute of Home Economics, University of Delhi, New Delhi, India

Himakshi Parmar Department of Environmental Studies, Faculty of Science, The Maharaja Sayajirao University of Baroda, Vadodara, Gujarat, India

Md Mostafizur Rahman Department of Environmental Science, Bangladesh University of Professionals (BUP), Dhaka, Bangladesh

Laboratory of Environmental Health and Ecotoxicology, Department of Environmental Sciences, Jahangirnagar University, Dhaka, Bangladesh

Centre for Environmental Research and Development (cERD), Dhaka, Bangladesh

D. Rajnandini Department of Microbiology, Institute of Home Economics, University of Delhi, New Delhi, India

Deeksha Ranjan Department of Applied Sciences and Humanities, Faculty of Engineering and Technology, Rama University, Kanpur, Uttar Pradesh, India

Md Rayhanul Islam Rayhan Laboratory of Environmental Health and Ecotoxicology, Department of Environmental Sciences, Jahangirnagar University, Dhaka, Bangladesh

Centre for Environmental Research and Development (cERD), Dhaka, Bangladesh

Vanshika Sahani Department of Microbiology, Institute of Home Economics, University of Delhi, New Delhi, India

Ashok Saini Department of Microbiology, Institute of Home Economics, University of Delhi, New Delhi, India

Prakash Samnani Department of Chemistry, Faculty of Science, The Maharaja Sayajirao University of Baroda, Vadodara, Gujarat, India

Himanshi Sharma Department of Biotechnology, Amity University, Mohali, Punjab, India

Aakash Singh Department of Biotechnology, Faculty of Engineering and Technology, Rama University, Kanpur, Uttar Pradesh, India

Indrajeet Singh Department of Biotechnology, Faculty of Engineering and Technology, Rama University, Kanpur, Uttar Pradesh, India

Laxmi Singh Department of Biotechnology, Faculty of Engineering and Technology, Rama University, Kanpur, Uttar Pradesh, India

Department of Applied Sciences and Humanities, Faculty of Engineering and Technology, Rama University, Kanpur, Uttar Pradesh, India

Department of Biotechnology and Bioinformatics, School of Life Sciences, University of Hyderabad, Hyderabad, Telangana, India

Maisha Binte Sultan Department of Environmental Science, Bangladesh University of Professionals (BUP), Dhaka, Bangladesh

Mohammad Toha Department of Environmental Science, Bangladesh University of Professionals (BUP), Dhaka, Bangladesh

Laboratory of Environmental Health and Ecotoxicology, Department of Environmental Sciences, Jahangirnagar University, Dhaka, Bangladesh

Jaya Verma Department of Microbiology, Faculty of Science, Guru Nanak Dev University, Amritsar, Punjab, India

Monika Yadav Department of Applied Sciences and Humanities (Environmental Science), Faculty of Engineering and Technology, Rama University, Kanpur, Uttar Pradesh, India

Qiuyan Yuan Environmental Engineering Program, Department of Civil Engineering, University of Manitoba, Winnipeg, MB, Canada

Navigating Microplastic Challenges: Separation and Detection Strategies in Wastewater Treatment

1

M. D. Tanvir Hasnine, Amit Hasan Anik, Mahbub Alam, and Qiuyan Yuan

Abstract

Wastewater treatment plants (WWTPs) function as primary receptors of microplastics (MPs), playing a key role in managing these pollutants within wastewater streams. Despite their essential function, WWTPs face limitations in effectively removing MPs from wastewater, indicating to the release of millions of MPs into the surrounding ecosystem. This chapter explores the origins of MPs within WWTPs and characterizing their properties, including polymer composition, morphology, size, and color, across influents, effluents, and sludges. Additionally, it explores the classification of MPs based on various forms such as fragments, films, pellets, beads, and foams, alongside their occurrence within WWTP systems. Subsequent sections discuss the methodologies employed for sampling, extraction, and analysis of MPs, encompassing both physical (visual identification and scanning electron microscopy) and chemical (Fourier-transform infrared spectroscopy, Raman spectroscopy, and thermal analysis) techniques, highlighting their respective strengths and limitations. However, the chapter emphasizes the lack of standardized protocols for isolating and charac-

M. D. T. Hasnine · Q. Yuan (✉)
Environmental Engineering Program, Department of Civil Engineering, University of Manitoba, Winnipeg, MB, Canada
e-mail: qiuyan.yuan@umanitoba.ca

A. H. Anik
Department of Environmental Science, Bangladesh University of Professionals (BUP), Dhaka, Bangladesh

M. Alam
Department of Environmental Science, Bangladesh University of Professionals (BUP), Dhaka, Bangladesh

Center for Environmental Nanoscience and Risk, Department of Environmental Health Sciences, Arnold School of Public Health, University of South Carolina, Columbia, SC, USA

A. Kumar, V. Singh (eds.), *Microplastics Pollution and its Remediation*,
https://doi.org/10.1007/978-981-97-4068-0_1

terizing MPs within WWTP streams. Consequently, while current methodologies have demonstrated efficacy in environmental MP analysis, further research is needed to establish standardized, reliable, and user-friendly techniques accessible for managing MPs within WWTPs.

Keywords

Microplastics · Wastewater treatment plants · Sources · Sampling and extraction · Characterization · Identification techniques

1.1 Introduction

Since the 1950s, the global production of plastic has experienced a significant rise, surpassing 348 million tonnes annually, and is projected to double by the year 2035 (Walker et al. 2020). The United Nations Environment Program reports that over 80% of marine litter is plastic, leading to concerns about oceans potentially having more plastics than fish (Kershaw 2016). Plastics find applications across various sectors, including personal care, toys, textiles, automobiles, and packaging (Shen et al. 2019). However, a significant amount of plastic materials have been released into the environment as waste (Monira et al. 2021). These plastics are referred to as primary plastics, and they are compelled to a variety of processes such as biodegradation, photodegradation (Li et al. 2018a), mechanical abrasion, photo-oxidative degradation (Wan et al. 2019), and hydrolysis (Li et al. 2020), all of which result in the release of microplastics (MPs) under 5 mm in size into the environment. Furthermore, MPs may discharge into drainage systems and subsequently flow into rivers and soils (Edo et al. 2020). MPs have a deleterious impact on living organisms, preventing them from growing, developing, or reproducing (Strungaru et al. 2019). Ingestion, inhalation, and digestion are all ways for microplastics to reach the human food chain (Revel et al. 2018). They can bioaccumulate upward in the food chain once they enter it, posing a concern to human health.

In recent investigations, attention has been directed toward the occurrence of MPs within freshwater, marine, and terrestrial ecosystems (Akdogan and Guven 2019). Recent research has observed a notable increase in the documentation of MP occurrence within both effluent and sludge released from WWTPs. This underscores the role of WWTPs as substantial sources of MPs, which therefore contribute significantly to the presence of MPs in nearby freshwater and soil environments (Acarer 2023; Liu et al. 2022). Herein, WWTPs are the sinks of MPs that receive urban sewage, industrial wastewater, and rainfall. The organic and inorganic particles found in wastewater treatment plants (WWTPs) function as carrier particles. Furthermore, given the massive quantities of wastewater released by WWTPs, even a tiny percentage of MPs present in the effluent might have a significant environmental impact. Although typical WWTPs could remove >90% of MPs from sewage, studies have shown that wastewater treatment facilities played an essential role in releasing MPs to receiving water environments (Bayo et al. 2020). Daily, millions of MPs are disposed of into water bodies worldwide following secondary treatment processes, highlighting the inefficiency of current methods in completely eliminating MPs (Gies et al. 2018; Gündoğdu et al. 2018; Mason et al. 2016). MPs had a deleterious effect on activated sludge (AS), disrupting microbial populations and inhibiting sludge activities including hydrogen generation and anaerobic digestion (AD) (Wei et al. 2019). On the other hand, the retention of 50–85% of MPs in sewage sludge, which is frequently used as a biofertilizer, suggested significant soil contamination (Akarsu et al. 2020). These microplastics can enter the terrestrial food chain, ultimately impacting human health (He et al. 2018). For instance, Ahmed et al. (2024) found that the predominant portion (around 90%) of MPs present in WWTPs consist of polymeric polyester (PES) originating from products like facial or body cleansers and water bottles. Moreover, their study identified 37 different polymers in seafood, including ethylene vinyl alcohol (EVOH), polyacetylene (PAc), and polyacrylamide (PAMs) that may pose health hazards to consumers and adversely affecting aquatic environments (Ahmed et al. 2024).

However, analyzing MPs in wastewater and wastewater sludge presents greater complexity compared to samples taken from freshwater and seawater, primarily due to the intricate composition and higher diversity of constituents within these samples (Liu et al. 2022). Additionally, environmental interactions and sample matrix effects, such as those from sludge or wastewater, MPs require thorough purification and fitted analytical approaches (Hale et al. 2022). Current detection techniques include microscopy, spectroscopy, and chromatography, each with limitations, particularly in dealing with complex matrices. Numerous research endeavors have explored the separation and characterization of MPs within wastewater and sludge samples, shedding light on practical hurdles and analytical challenges encountered in WWTPs (Okoffo et al. 2019; Sun et al. 2019). Moreover, recent advancements in quantifying MPs through techniques such as Fourier-transform infrared spectroscopy (FTIR) and Raman spectroscopy warrant further examination and updating (Liu et al. 2022). Furthermore, hyphenated techniques, which combine separation methods with sophisticated detectors, offer enhanced accuracy but are more complex (Wang and Wang 2018). To date, WWTPs have not employed a suitable method for the separation, detection, and removal of MPs from wastewater and sludge. For complete understanding of the environmental impact of MPs, it also requires a robust analytical method for their quantification, considering their chemical diversity and factors like size, chemical composition, shape, and concentration (Hasan Anik et al. 2021; Song et al. 2015). The objective of this chapter is to investigate the sources and characteristics of MPs and assess the different techniques for their separation, identification, and detection within wastewater and sludge contexts. Key characteristics such as fragments, films, pellets, beads, and foams are detailed and comparatively analyzed to explain their distribution and influence within wastewater treatment contexts.

1.2 Overview of Microplastics

1.2.1 Definition and Characteristics of Microplastics

MPs are small plastic particles less than 5 mm in diameter and those greater than 5 mm are known as macroplastics (Hasan Anik et al. 2021; Song et al. 2015). MPs are classified as primary or secondary. Primary MPs are intentionally manufactured small particles like pellets, powders, and beads, used in various applications or as raw materials for producing plastic products, known for their use as abrasives (Kershaw 2016). Secondary MPs form unintentionally from the degradation of larger plastic items into smaller fragments (Koelmans et al. 2019). Nanoplastics, a specific category of MPs, measure 1–100 nm in one dimension and primarily result from the degradation of larger plastic pieces (Rist and Hartmann 2018). This breakdown into smaller microplastics, occurs through photolytic, mechanical, and biological degradation, particularly in marine environments (Andrady 2015). MPs are categorized based on their morphology into fibers, fragments, microbeads, and sheets, and they are primarily composed of polymers such as polyethylene (PE), polypropylene (PP), polyamide (PA), polyvinyl chloride (PVC), polystyrene (PS), and polyethylene terephthalate (PET) (Rezania et al. 2018).

MPs, characterized by their lightweight and insolubility, are widespread and easily transported within aquatic ecosystems (Holland et al. 2016). The bioavailability and environmental fate of MPs are largely governed by processes of partitioning and sorption (Elizalde-Velázquez et al. 2020). Smaller plastic particles are more bioavailable (Gregory 2009), and have a higher tendency to absorb and release toxic substances due to their larger surface area (Lee et al. 2013). Partitioning, a dynamic process reaching equilibrium between phases, and sorption, the transfer of molecules across phases involving adsorption on surfaces and absorption into materials (Piccin et al. 2017), are key to understanding these interactions. The study of sorption kinetics through empirical equations is crucial to explain these processes (Piccin et al. 2017).

1.2.2 Sources of Microplastics

Land-based sources include cosmetics and personal care products, synthetic textiles, terrestrial transport such as tire wear, and the accidental loss of resin pellets from plastic producers and fabricators (Kershaw 2016; Koelmans et al. 2019). Synthetic textiles and clothing are particularly significant, with microfibers released during wear, laundering, and from fishing gear like nets (Environment and Climate Change Canada 2019). Sea-based contributors include the fisheries, aquaculture, and shipping industries, which may lose plastics through accidents, abandonment, or deliberate disposal (GESAMP Joint Group of Experts on the Scientific Aspects of Marine Environmental Protection 2015). These microplastics can then enter marine environments via multiple pathways, including wastewater, rivers, and

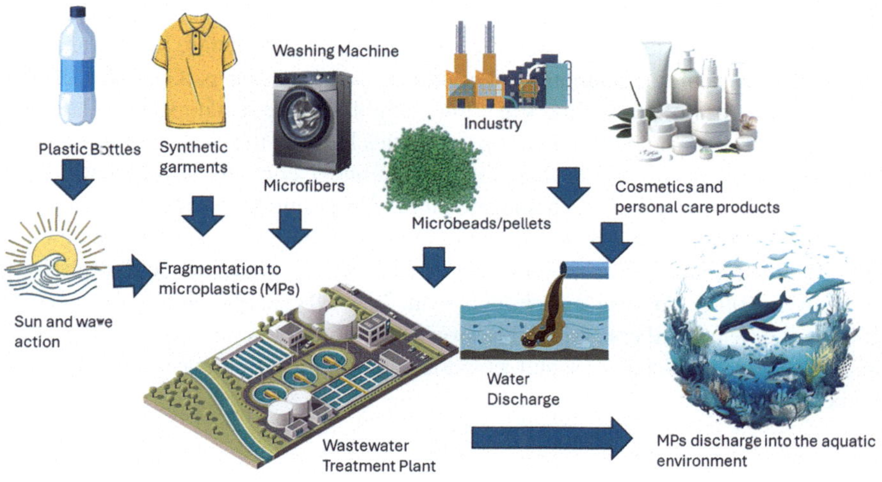

Fig. 1.1 Sources of MPs in wastewater treatment plants

coastal regions, varying by geographical area (Kershaw 2016). Figure 1.1 represents various sources of MPs entering WWTPs.

Non-domestic sources of MPs in wastewater include a variety of industrial activities. These sources include plastic particles from air blasting used in cleaning paints (Gregory 2009); pre-production pellets lost during transport or manufacturing (Sheavly and Register 2007); lost Styrofoam from shipping; and fibers from the synthetic textile industry. These diverse sources emphasize the importance of comprehensive management strategies.

1.3 Classifications of Microplastics in Wastewater Treatment Plant

Based on visual characteristics, MPs were classified into five types such as fragments (hard, jagged pieces), films (thin, fragile sheets from disintegrated plastic products), pellets (round, hard particles for plastic production), beads (small, manufactured polyethylene particles in beauty products), and foams (soft, porous polystyrene plastic) (Ben-David et al. 2021).

1.3.1 Fibers

Fibers, which originate from textiles like clothing and upholstery, and pellets, often deriving from industrial sources, represent significant sources of MPs (Rochman et al. 2014). Notably, synthetic textiles account for approximately 35% of marine MPs, with washing processes releasing millions of fibers per load—polyester being a major contributor with over six million fibers, in contrast to acrylic's 700,000

(Boucher and Friot 2017; De Falco et al. 2019). In one research they highlight that in wastewater, fibers predominate (65.6%), followed by fragments (28.1%) and pellets (5.4%), with minimal amounts of granules, foam, and sheets (Gies et al. 2018). The high presence of plastic fibers in effluent highlights the real impact that laundry, washing, cleaning, and textile handling activities have on the release of microplastics (McCormick et al. 2014). This distribution underscores the significant role of domestic and industrial textile management in MPs pollution. Moreover, blue fibers as well as red and black fibers are predominantly found in aquatic environments, although their visibility in analyses may be enhanced by their distinct morphology compared to other MP forms (Hamidian et al. 2021). Typically ranging from 100 to 1000 mm in length, fibers are notably larger than other MPs, highlighting their distinct physical characteristics and potential environmental impact (Dubaish and Liebezeit 2013).

1.3.2 Fragments

The wastewater samples mainly consist of secondary MPs, predominantly in the form of fibers and fragments. While fragments are commonly found, some studies have observed their dominance in effluents (McCormick et al. 2016). Factors like combined sewers and stormwater runoff contribute to the increased presence of fragments and foam particles in the effluent (Mason et al. 2016). This study investigates that smaller particles are distributed evenly between fragments and fibers, but larger particles are mainly fibers (80%). Kay et al. (2018) emphasized that secondary MPs, given their significant presence as fragments and flakes, are key pollutants, thus reducing their discharge should be a priority.

1.3.3 Microbeads or Pellets

Microbeads, primarily sourced from cosmetic and personal care products like scrubs, constitute about 6–7% of these products. These cosmetics may include microbeads, which are primary microplastics, in concentrations ranging from 0.5 to 5%. These microbeads typically have an average diameter of 250 μm (Hamidian et al. 2021). Microbeads in personal care products become more concentrated as they decrease in size, ranging from 8 to less than 100 μm (Cheung and Fok 2017; Kalčíková et al. 2017). These particles, often irregular with some spherical forms (Godoy et al. 2019) can avoid wastewater filtration and degrade slowly because of their low surface-to-volume ratio (Chang 2015). They vary in size based on whether they originate from industrial or municipal effluents and tend to accumulate in freshwater ecosystems, particularly under increased flow conditions (Kalčíková et al. 2017; Moriarty et al. 2018), while their smooth surface reduces adsorption compared to irregular MPs (Lambert et al. 2017). Despite their widespread presence, the environmental impact of these microbeads is still underexplored.

1.3.4 Foams and Films

Foams and films, often resulting from plastic bag and packaging corrosion, contribute minimally to MPs, constituting about 5–10% of total MPs, unlike pellets and fragments found more abundantly, ranging from 25 to 70% (Bui et al. 2020; Harrison et al. 2012). A research by Murphy et al. (2016) detailed the distribution of MP types in wastewater, revealing fragments as the majority (67.3%), followed by fibers (18.5%), films (9.9%), microbeads (3.0%), and foams (1.3%). Foams, mainly expanded polystyrene (PS), are prevalent in beach environments due to their higher buoyancy and are widely used in construction and the food industry for insulation (Faure et al. 2015).

1.4 Approaches for Sampling and Separating Microplastic from Wastewater

The separation of MPs from wastewater is complex, with various techniques being employed to address this issue. It involves a combination of physical methods (like centrifugation, filtration, and flotation), chemical treatments (such as oxidation and enzymatic digestion), and advanced technologies (including magnetic separation and various degradation methods). Figure 1.2 represents various steps of analysis for MP separation and identification in wastewater treatment systems.

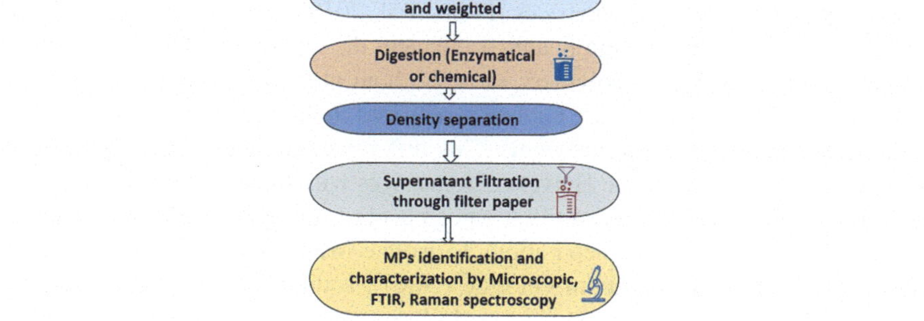

Fig. 1.2 Steps of microplastic analysis in wastewater treatment

1.4.1 Sample Collection and Filtration

MPs collection from WWTP effluents employs various sampling methods. These methods involve either integrating the separation step during collection, such as through pumping with filtration and surface filtration or using auto-samplers, or they need collecting samples in containers that are later separated in laboratories. Auto sampling, described by Gündoğdu et al. (2018), involves collecting 24-h composite samples from WWTP influents and effluents. Talvitie et al. (2017) noted discrete sampling at 15-min intervals over a day. Separating samples in situ at the sampling location enables larger volumes processing; however, this approach carries a greater risk of contamination compared to processing in the controlled conditions of a laboratory. Separation typically uses sieves with 38–4750 μm mesh sizes, tailored to the MPs' sizes, facilitating size categorization (Hidalgo-Ruz et al. 2012; Wang and Wang 2018). Post-separation, the material is washed and stored for analysis (Ziajahromi et al. 2017). Sieve blockage due to high organic loads and the influence of microparticle morphology, like microfibers, on separation efficiency are noted challenges (Michielssen et al. 2016; Ziajahromi et al. 2017). Further separation may include visual sorting, density separation, and filtration.

However, the sampling method affects microparticle morphology, with factors like organic load and microfiber characteristics influencing separation efficiency. Standardized, precise sampling procedures are vital for reliable, error-minimized results. The choice of sampling technique should align with study goals, considering factors like sample volume and daily MP concentration for selecting between filtration sampling and automated sampling for efficiency and accuracy. The absence of standardized separation protocols complicates data comparison across studies, underscoring the need for standardized methodologies to ensure consistency and reliability in MPs quantification.

1.4.2 Density Separation

Density separation is a key method used for isolating microplastics from wastewater and complex environmental matrices (Prata et al. 2019). The density, an essential characteristic of polymers, significantly influences the separation of MPs from environmental samples. Virgin polymers, those without manufacturing additives, have densities ranging from 0.90 to 1.6 g/cm^3, compared to the higher typical density of soil and sediments at about 2.6 g/cm^3. This difference facilitates MPs' separation using density-based methods. However, in matrices with high organic content, pre-separation steps like sieving and filtration can lead to equipment saturation, hindering MPs' identification and separation. To isolate MPs, samples are treated with dense solutions such as sodium chloride (NaCl), sodium iodide (NaI), and zinc chloride ($ZnCl_2$), followed by stirring and filtration to collect the MPs for further analysis, which is mainly depending on the specific characteristics of the environmental matrices. For instance, matrices rich in clay or organic matter, where MPs are closely bound within soil particles, necessitate the use of mechanical stirrers or centrifuges to ensure well separation (Liu et al. 2019). NaCl solution, with a density of 1.2 g/cm^3, is commonly used for extracting low-density polymers like PE, PS, and PP due to its cost-effectiveness and eco-friendliness. In contrast, solutions with higher densities, such as NaI (1.8 g/cm^3) and $ZnCl_2$ (1.5–1.7 g/cm^3), are necessary for removing heavier MPs like PET and PVC. Additional techniques such as aeration, stirring, and centrifugation play crucial roles in enhancing the separation process of MPs from other contaminants (Mohamed Nor and Obbard 2014). During the aeration phase, an air pump introduces so many bubbles into the extraction solutions, facilitating the separation of MPs that are initially adherent to soil or sediment particles. This bubble-induced movement can transport the MPs upward, a process known as froth flotation (Zobkov and Esiukova 2017). After flotation, a vacuum pump is used to improve the filtration of the supernatant, generating concentrated MPs. Key innovations include elutriation columns and sediment separators, achieving up to 100% recovery rates (Imhof et al. 2012) and semiautomatic flotation devices for both general and specific uses, circular extraction devices for soil MPs, ensuring over 90% recovery and environmentally friendly results (Liu et al. 2019). These developments demonstrate that combining appropriate flotation reagents with specialized devices can significantly improve the effectiveness of density-based separation processes for MPs. Table 1.1 represents density separation methods for MPs from wastewater and biosolids across different countries (Bretas Alvim et al. 2020).

1.4.3 Supernatant Filtration

After separation processes, the supernatant is commonly filtered using paper with pore sizes ranging from 1 to 2 μm. The particles captured on these filters are then individually extracted and examined for identification using tweezers (Rocha-Santos 2018). Different types of filter media, such as glass fiber, nitrocellulose, and polycarbonate, are utilized. However, filtration efficiency can be hampered by the clogging of filter pores with particulate matter (Wang and Wang 2018). To mitigate this issue, either the volume of the filtered sample is reduced, or coagulants are added to form solid particles, thus enhancing filtration efficiency.

1.4.4 Electrostatic Separation

The electrostatic separation method offers significant benefits such as automation, ease of use, and high precision. Simulation tests on environmental samples, which vary in organic matter content and microplastic sizes, have demonstrated that this method can effectively isolate components. It successfully removes up to 99% of the initial sample mass while retaining the microplastics (Felsing et al. 2018). It achieves a high recovery rate of over 99%, even for MPs as small as 63 μm (Bretas Alvim et al. 2020; Felsing et al. 2018). This process is effective across samples with diverse compositions, successfully collecting all MPs in the nonconductive fraction,

Table 1.1 Techniques for isolating microplastics in sewage and biosolids

Location	Sample types	Separation techniques	References
Germany	Sewage	Sieve; separated density using ($ZnCl_2$)	Mintenig et al. (2017)
Australia	Sewage	Sieving; separate density with (NaI)	Ziajahromi et al. (2017)
Italy	Sewage	Sieving; density separation ($ZnCl_2$); 8 μm cellulose nitrate membrane filters	Mason et al. (2016)
Turkey	Sewage	Sieving; density separation using (NaI)	Gündoğdu et al. (2018)
Scotland	Sewage	Sieving; 11 μm filter paper	Murphy et al. (2016)
Norway	Soils and biosolids	Glass fiber filter, density separation (NaI)	Hurley et al. (2018)
China	Sewage sludge	Sieving; density separation with (NaCl); glass fiber filter	Li et al. (2018b)
Ireland	Biosolids	Sieving; elutriation; density separation ($ZnCl_2$); 1.2 μm glass fiber filter	Mahon et al. (2017)

regardless of high organic or biomass content. The advantages of electrostatic separation include straightforward operations, rapid processing times (usually just a few hours), and the capability for automated operation. However, the primary limitation is the relatively large size of the electrostatic equipment, which necessitates downsizing for practical MP separation tasks in analytical settings.

1.4.5 Magnetic Separation

Magnetic separation is a technique used to extract microplastics from environmental samples by attaching hydrophobic iron nanoparticles to them, thereby making them magnetic (Grbic et al. 2019). This technique allows for the efficient and rapid extraction of MPs, including in situ separation of specific nanoparticle-bound microorganisms (Xu et al. 2011). It was found that magnetic filtration effectively removed fine MP particles from dilute suspensions (Rhein et al. 2019). Research on MPs of different polymer types found within sediments has shown recovery rates ranging from 49 to 90% for medium-sized MPs, measuring between 0.2 and 1 mm and up to 90% for finer MPs (<20 μm), indicating size-dependent separation efficiency (Grbic et al. 2019). The efficiency of separating MPs is influenced by several factors including the aging of MPs, the presence of pollutants, the pH of the environmental sample, and the characteristics of the magnetic nanoparticles employed (Rhein et al. 2019).

1.4.6 Sample Processing: Digestion Methods

The primary objective of the sample processing digestion is to eliminate biological material and organic matter from samples derived through separation techniques since such matter can cause confusion and lead to an overestimation of polymers in the MPs analysis (He et al. 2018; Prata et al. 2019). The following sections provide a concise overview of various digestion methods.

1.4.6.1 Hydrolytic Digestion with Acidic Substances

Acids such as nitric acid (HNO_3), sulfuric acid (H_2SO_4), and hydrochloric acid (HCl) are employed in the chemical digestion process, which can lead to the degradation or damage of polymers in MPs (He et al. 2018). To mitigate this effect, a lower concentration of acids is recommended during the removal of biological material (Naidu 2019). However, there is evidence that acid digestion can still destroy certain polymers, potentially causing the breakdown of some plastic materials (He et al. 2018). This degradation can lead to inaccuracies in the quantification of MP samples. In the study of chemical digestion on polymers, polyamide (PA), polyurethane (PU), and black tire rubber elastomer experienced full degradation (Enders et al. 2017). Other polymers, including polycarbonate (PC) and expanded solid polystyrene (EPS, PS); (PET), showed partial degradation, evidenced by color loss. However, (PP, high- and low-density polyethylene (HDPE, LDPE), ethylene-vinyl acetate (EVA), and polytetrafluoroethylene (PTFE) remained unaffected by the chemical treatment (Enders et al. 2017). Research indicates that while nitric acid and KOH can harm biopolymers and cellulose, H_2O_2 demonstrates enhanced capability in degrading organic substances without impacting MPs (Tagg et al. 2015). Consequently, 30% H_2O_2 is preferred for removing organic matter before analyzing samples using FTIR and Raman spectroscopy (Leslie et al. 2017). A short heating period (30 min) has been shown to enhance the reaction and facilitate easier sample filtration. However, excessive heating should be avoided as it may lead to the melting of some MPs (Munno et al. 2018).

1.4.6.2 Hydrolytic Digestion with Alkaline Substances

Alkaline solutions, like acidic substances, can alter the morphology of MPs, potentially causing sample loss and underestimation of particle counts, which can lead to inaccurate results. For instance, using 10 M NaOH at 60 °C has been shown to be too severe for certain polymers, causing the partial degradation of nylon fibers and the melting of polyethylene fragments (Cole et al. 2013). Hurley et al. (2018) investigated six chemical digestions to determine their efficacy in removing organic matter and found that alkaline solutions (1 M and 10 M NaOH and 10% KOH at 60 °C) were ineffective, unable to remove 70% of organic matter from sludge. In contrast, oxidative methods showed higher success in digestion. Similarly, Nuelle et al. (2014) found that 30% and 35% H_2O_2 solutions were more effective at removing organic matter from sediments than NaOH and HCl solutions at various concentrations.

1.4.6.3 Digestion by Oxidation

Oxidation with hydrogen peroxide (H_2O_2) is used for treating wastewater, with the process parameters such as contact time and temperature varying based on the organic content. Digestion involves heating until H_2O_2 evaporates. Studies have utilized different concentrations of H_2O_2 for digesting various wastewater samples, employing methods like sieving, chemical digestion, and filtration (Ziajahromi et al. 2017). In a study it was found that using Fenton's reagent (Fe^{2+}) as a catalyst in the oxidation and separation processes of MPs not only accelerates the reaction, with an optimal pH of around 3 being crucial for its effectiveness, but also results in an 86% reduction of organic matter, compared to a 67% removal rate when using potassium hydroxide (KOH) and sodium hydroxide (NaOH) (Tagg et al. 2017). The literature suggests that maintaining appropriate conditions during the oxidation process, including the use of catalysts and temperature control, is essential for effective treatment and MPs identification without altering the chemical composition of materials like PP, PE, PVC, and nylon. Recent research indicates that temperatures exceeding 60 °C can adversely affect the digestion process, compromising the accurate assessment of MPs in samples (Bretas Alvim et al. 2020). Carr et al. (2016) highlighted this negative impact on digestion, while Munno et al. (2018) discovered that high temperatures can cause microbeads to melt, leading to an underestimation of their quantities in analyses. Similarly, Napper and Thompson (2016) found that the application of heat might result in a significant underrepresentation of microbeads in cosmetics. To mitigate the detrimental effects of elevated temperatures on MPs analysis, Hurley et al. (2018) recommend a chemical digestion method utilizing the Fenton reaction, maintaining a temperature below 40 °C to ensure accurate quantification of MPs.

1.4.6.4 Enzymatic Digestion

Enzymatic digestion is considered a potential method for extracting organic matter from MP samples due to its nondestructive nature (Bretas Alvim et al. 2020). Cole et al. (2014) achieved 88% digestion efficiency with Proteinase-K, which increased to over 97% with adjusted conditions, without damaging the samples. However, the cost and complexity of using Proteinase-K led to exploring cheaper alternatives like trypsin, collagenase, and papain (Courtene-Jones et al. 2017). von Friesen et al. (2019) found that pancreatic enzymes did not significantly alter the FTIR spectral profiles of ten polymers after digestion. However, enzymatic digestion's effectiveness and method consistency in treating WWTP samples remain under-researched. Simon et al. (2018) employed cellulase for breaking down cellulose fibers in wastewater, with subsequent oxidation via the Fenton reaction. Mintenig et al. (2017) achieved good results in removing organic material and MPs from WWTP effluents using protease, lipase, and cellulase, but the process was time-consuming (over 10 days) and prone to contamination and sample loss. Therefore, more research is needed to evaluate enzymatic digestion's efficacy and practicality in WWTP samples compared to other techniques.

1.4.6.5 Various Techniques for Detecting Microplastics in Wastewater

The challenges in accurately identifying MPs arise from the identical characteristics between certain nonplastics that appear like plastics and vice versa. Microplastic analysis typically follows a two-step procedure: first, initial physical identification using microscopy to categorize potential plastic particles based on size, shape (such as fiber, film, foam, pellets, or fragments), and color, followed by chemical verification using spectroscopy to confirm the presence of plastics.

1.4.7 Physical Identification

1.4.7.1 Visual Identification

Visual identification, primarily using the naked eye or optical microscopes, is the very common method for detecting MPs, focusing on shape and color to determine the presence of MP particles (Crawford and Quinn 2016). To enhance accuracy, certain criteria must be met, such as uniformity in color and thickness for fibers, and clear coloration for particles, with more scrutiny required for transparent and white particles under higher magnification (Hidalgo-Ruz et al. 2012). While cost-effective for large samples, visual methods are subject to bias, influenced by the observer's subjectivity, the sample's context, and particle characteristics. Misidentification risks increase with smaller particle sizes, and weathering can alter plastic morphology, complicating visual assessment (Crawford and Quinn 2016). Thus, confirming suspected microplastics with spectroscopic or other analytical methods is advised, especially for smaller items. However, for analyzing smaller microplastics, optical microscopes are employed to capture images that reveal the particles' shape and quantity. Specifically, polarized light microscopy is utilized to generate stacked images at varying heights within the sample, facilitating the estimation of microplastic counts in synthetic water samples (Rodríguez Chialanza et al. 2018). Certain microplastics, like polypropylene fibers, exhibit distinct colors, aiding identification, but similarities between plastics and materials like perlite can lead to errors, necessitating careful evaluation of physical characteristics (Heo et al. 2013; Song et al. 2015).

1.4.7.2 Scanning Electron Microscopy

The scanning electron microscope (SEM) serves as an effective instrument for nanoscale resolution imaging, employing a high-intensity electron beam to scan surfaces and identify MPs in samples. This method achieves less than 0.5 nm resolution, enabling the differentiation of microplastics from other substances through detailed surface morphology analysis (Crawford and Quinn 2016). Furthermore, SEM helps assess the weathering of MPs by detecting surface alterations like cracks and pits (Zhang et al. 2016). Integrating SEM with energy dispersive X-ray spectroscopy (SEM-EDS) offers comprehensive insights into the elemental composition and inorganic constituents of MPs, facilitating the distinction between natural materials and MPs and reducing the number of samples required for detailed spectroscopic analysis (Fries et al. 2013). However, despite its efficacy in detailed surface examination, SEM's extensive sample preparation requirements limit its suitability for processing large sample volumes (Wang and Wang 2018).

1.4.8 Chemical Identification

1.4.8.1 Fourier Transform Infrared (FTIR) Spectroscopy

FTIR spectroscopy is a key method for characterizing MPs, as it generates unique spectra based on chemical bonds, enabling identification of materials by comparing these spectra with known references. FTIR's reliability has established it as a prevalent tool for analyzing microplastics in environmental samples (Hale et al. 2022). It's utilized either for scanning suspected MP particles or confirming visual identifications through subsample analyses (Browne et al. 2011; Wang et al. 2017). However, FTIR's effectiveness is size-dependent, typically limited to particles larger than 10–20 mm, and the analysis can be time-consuming and technically demanding (Huppertsberg and Knepper 2018). Advanced FTIR techniques like micro-FTIR, ATR-FTIR, and FPA-FTIR have enhanced microplastic research, allowing for the detection of smaller or irregularly shaped particles with higher throughput and resolution (Tagg et al. 2015). Despite being slower and more expensive, FPA-FTIR excels in rapid, detailed analysis compared to standard FTIR. Both FTIR and its variant techniques provide not only compositional information but also insights into the physicochemical weathering of MPs, such as oxidation levels (Corcoran et al. 2009; Wang and Wang 2018). Four primary techniques are employed in FTIR analysis to obtain accurate infrared spectra, depending on the sample properties: transmission (Käppler et al. 2018), true specular reflectance/reflection-adsorption (Roodenko et al. 2018), and attenuated total reflection (ATR) (Tobin et al. 2014). The selection of the appropriate mode is determined by the characteristics of the samples under examination. Table 1.2 represents the benefits and suitable sample types for various operational modes in FTIR for MP identification (Fu et al. 2020).

1.4.8.2 Raman Spectroscopy

Raman spectroscopy, recognized for its advanced applications in wastewater systems, effectively identifies MPs in both solid and liquid samples (Araujo et al. 2018). The process involves irradiating the sample with a monochromatic laser beam, causing a Raman shift reflected in the light's frequency, uniquely characterizing each polymer (Crawford and Quinn 2017). This nondestructive method offers high reliability and precise chemical characterization, essential for subsequent analyses (Shim et al. 2017). Raman spectroscopy boasts advantages over FTIR, including higher spatial resolution, wider spectral range, and minimal water interference,

Table 1.2 Benefits and suitable samples for various operating modes in FTIR

Modes	Benefits	Suitable samples	Suitable forms of MPs
Attenuated total reflection (ATR)	Sample preparation is minimal to nonexistent, nondestructive, ideal for samples that are thick or have a high IR absorbency	Plastics, laminates, and rubbers, as well as samples that are highly absorbent or thick	Fibers of various thickness, powders, thick and thin films, irregular shaped
Diffuse reflectance	Sample preparation is minimal to nonexistent	Sample could be fine to soft powder or mixed, tablets, rigid polymers	Applicable for thin films (<1 μm) and irregular-shaped MPs
Transmission	Spectral information is great and good for quantitative work	Samples that can be passed via an infrared beam, thermoplastic granules, thin polymer films, soluble polymers	Thin fibers and films, powdered MPs
Micro-FTIR	Detect the distinctive bands of samples with diameters as tiny as 10 μm and provide chemical mapping	Applies to all types of samples	Most of the MPs are applicable for micro-ATR mode, regular-sized, thick, and opaque (MPs) are suitable for micro-reflectance mode, while transparent MPs are suited for micro-transmittance mode

with micro-Raman enabling the identification of MPs as small as 1 mm (Araujo et al. 2018). Techniques like micro-Raman coupled with spectral imaging provide detailed spatial chemical imaging, while its integration with confocal laser scanning microscopy helps locate MPs within biological tissues (Löder and Gerdts 2015). Despite these benefits, the technique's sensitivity to additives, pigments, and chemicals on MPs can impede accuracy, and its low signal-to-noise ratio complicates spectrum analysis (Araujo et al. 2018).

1.4.8.3 Thermal Analysis

Thermal degradation and analysis methods, including pyrolysis gas chromatography-mass spectrometry (Pyr-GC/MS), thermal extraction desorption gas chromatography-mass spectrometry (TED-GC/MS), and thermogravimetric analysis coupled with differential scanning calorimetry (TGA-DSC), have been utilized to quantify and analyze the composition of MPs in wastewater treatment plant samples (Dümichen et al. 2017; Okoffo et al. 2019). Pyr-GC/MS, a destructive technique, identifies polymer types by analyzing thermal degradation products and comparing them with reference pyrograms (Fries et al. 2013; Löder and Gerdts 2015). It offers advantages in chemical analysis of polymers and additives with minimal sample preparation and is not affected by the particle's shape, size, or contaminants, though it is limited by the small sample size and longer analysis time, handling only larger particles (>100 μm) (Löder and Gerdts 2015; Vianello et al. 2013). TED-GC-MS, an advanced method, overcomes some limitations by integrating thermal extraction and thermogravimetric analysis for rapid and efficient environmental MPs analysis (Dümichen et al. 2017). Despite the efficacy in chemical characterization, these thermal analytical methods are destructive and do not provide morphological details like particle size distribution, necessitating their use alongside spectroscopic techniques for comprehensive MP analysis (Käppler et al. 2018).

1.5 Conclusion

In conclusion, the issue of MPs in wastewater is a significant environmental concern that requires urgent and comprehensive action. The complexity of MP characteristics, such as their size, shape, and chemical composition, complicates their effective removal and detection in wastewater treatment processes. It also highlights the absence of standardized methods for separating, quantifying, and identifying MPs, which hinders the ability to compare results across different studies. Therefore, there is an urgent need for the development and implementation of efficient, quick, and uniform protocols that cover all aspects of MP analysis. This includes determining sizes, conducting chemical digestion, performing density and visual separations, and using analytical techniques to identify the polymers. Establishing such standardized methods is crucial for a unified and effective strategy for managing MP contamination in wastewater on a global scale. Future research should focus on developing effective methods for microplastic removal in wastewater treatment facilities, understanding the long-term ecological and health impacts of MPs, and developing policies and practices that reduce microplastic pollution at the source. By addressing these key areas, it is possible to mitigate the impact of MPs on the environment and public health, ensuring the preservation of water quality for future generations.

Competing Interest The author declares no conflict of interest.

References

Acarer S (2023) Microplastics in wastewater treatment plants: sources, properties, removal efficiency, removal mechanisms, and interactions with pollutants. Water Sci Technol 87(3):685. https://doi.org/10.2166/wst.2023.022

Ahmed SF, Islam N, Tasannum N, Mehjabin A, Momtahin A, Chowdhury AA, Almomani F, Mofijur M (2024) Microplastic removal and management strategies for wastewater treatment plants. Chemosphere 347:140648. https://doi.org/10.1016/j.chemosphere.2023.140648

Akarsu C, Kumbur H, Gökdağ K, Kıdeyş AE, Sanchez-Vidal A (2020) Microplastics composition and load from three wastewater treatment plants discharging into Mersin Bay, north eastern Mediterranean Sea. Mar Pollut Bull 150:110776. https://doi.org/10.1016/j.marpolbul.2019.110776

Akdogan Z, Guven B (2019) Microplastics in the environment: a critical review of current understanding and identification of future research needs. Environ Pollut 254:113011. https://doi.org/10.1016/j.envpol.2019.113011

Andrady AL (2015) Persistence of plastic litter in the oceans. In: Marine anthropogenic litter, pp 29–56. https://doi.org/10.1007/978-3-319-16510-3_3

Araujo CF, Nolasco MM, Ribeiro AMP, Ribeiro-Claro PJA (2018) Identification of microplastics using Raman spectroscopy: latest developments and future prospects. Water Res 142:426–440. https://doi.org/10.1016/j.watres.2018.05.060

Bayo J, Olmos S, López-Castellanos J (2020) Microplastics in an urban wastewater treatment plant: the influence of physicochemical parameters and environmental factors. Chemosphere 238:124593. https://doi.org/10.1016/j.chemosphere.2019.124593

Ben-David EA, Habibi M, Haddad E, Hasanin M, Angel DL, Booth AM, Sabbah I (2021) Microplastic distributions in a domestic wastewater treatment plant: removal efficiency, seasonal variation and influence of sampling technique. Sci Total Environ 752:141880. https://doi.org/10.1016/j.scitotenv.2020.141880

Boucher J, Friot D (2017) Primary microplastics in the oceans: a global evaluation of sources. In: Primary microplastics in the oceans: a global evaluation of sources. https://doi.org/10.2305/iucn.ch.2017.01.en

Bretas Alvim C, Mendoza-Roca JA, Bes-Piá A (2020) Wastewater treatment plant as microplastics release source—Quantification and identification techniques. J Environ Manag 255:109739. https://doi.org/10.1016/j.jenvman.2019.109739

Browne MA, Crump P, Niven SJ, Teuten E, Tonkin A, Galloway T, Thompson R (2011) Accumulation of microplastic on shorelines woldwide: sources and sinks. Environ Sci Technol 45(21):9175–9179. https://doi.org/10.1021/es201811s

Bui XT, Vo TDH, Nguyen PT, Nguyen VT, Dao TS, Nguyen PD (2020) Microplastics pollution in wastewater: characteristics, occurrence and removal technologies. Environ Technol Innov 19:101013. https://doi.org/10.1016/j.eti.2020.101013

Carr SA, Liu J, Tesoro AG (2016) Transport and fate of microplastic particles in wastewater treatment plants. Water Res 91:174–182. https://doi.org/10.1016/j.watres.2016.01.002

Chang M (2015) Reducing microplastics from facial exfoliating cleansers in wastewater through treatment versus consumer product decisions. Mar Pollut Bull 101(1):330–333. https://doi.org/10.1016/j.marpolbul.2015.10.074

Cheung PK, Fok L (2017) Characterisation of plastic microbeads in facial scrubs and their estimated emissions in Mainland China. Water Res 122:53–61. https://doi.org/10.1016/j.watres.2017.05.053

Cole M, Lindeque P, Fileman E, Halsband C, Goodhead R, Moger J, Galloway TS (2013) Microplastic ingestion by zooplankton. Environ Sci Technol 47(12):6646–6655. https://doi.org/10.1021/es400663f

Cole M, Webb H, Lindeque PK, Fileman ES, Halsband C, Galloway TS (2014) Isolation of microplastics in biota-rich seawater samples and marine organisms. Sci Rep 4:4528. https://doi.org/10.1038/srep04528

Corcoran PL, Biesinger MC, Grifi M (2009) Plastics and beaches: a degrading relationship. Mar Pollut Bull 58(1):80–84. https://doi.org/10.1016/j.marpolbul.2008.08.022

Courtene-Jones W, Quinn B, Murphy F, Gary SF, Narayanaswamy BE (2017) Optimisation of enzymatic digestion and validation of specimen preservation methods for the analysis of ingested microplastics. Anal Methods 9(9):1437. https://doi.org/10.1039/c6ay02343f

Crawford CB, Quinn B (2016) Microplastic pollutants. In: Microplastic pollutants. https://doi.org/10.1016/c2015-0-04315-5

Crawford CB, Quinn B (2017) The emergence of plastics. In: Microplastic pollutants, pp 1–17. https://doi.org/10.1016/b978-0-12-809406-8.00001-3

De Falco F, Di Pace E, Cocca M, Avella M (2019) The contribution of washing processes of synthetic clothes to microplastic pollution. Sci Rep 9(1):6633. https://doi.org/10.1038/s41598-019-43023-x

Dubaish F, Liebezeit G (2013) Suspended microplastics and black carbon particles in the Jade system, southern North Sea. Water Air Soil Pollut 224(2). https://doi.org/10.1007/s11270-012-1352-9

Dümichen E, Eisentraut P, Bannick CG, Barthel AK, Senz R, Braun U (2017) Fast identification of microplastics in complex environmental samples by a thermal degradation method. Chemosphere 174:572–584. https://doi.org/10.1016/j.chemosphere.2017.02.010

Edo C, González-Pleiter M, Leganés F, Fernández-Piñas F, Rosal R (2020) Fate of microplastics in wastewater treatment plants and their environmental dispersion with effluent and sludge. Environ Pollut 259:113837. https://doi.org/10.1016/j.envpol.2019.113837

Elizalde-Velázquez A, Subbiah S, Anderson TA, Green MJ, Zhao X, Cañas-Carrell JE (2020) Sorption of three common nonsteroidal anti-inflammatory drugs (NSAIDs) to microplastics. Sci Total Environ 715:136974. https://doi.org/10.1016/j.scitotenv.2020.136974

Enders K, Lenz R, Beer S, Stedmon CA (2017) Extraction of microplastic from biota: recommended acidic digestion destroys common plastic polymers. ICES J Mar Sci 74(1):326. https://doi.org/10.1093/icesjms/fsw173

Environment and Climate Change Canada (2019) Canadian plastic industry, markets and waste. Environment and Climate Change Canada

Faure F, Demars C, Wieser O, Kunz M, De Alencastro LF (2015) Plastic pollution in Swiss surface waters: nature and concentrations, interaction with pollutants. Environ Chem 12(5):582–591. https://doi.org/10.1071/EN14218

Felsing S, Kochleus C, Buchinger S, Brennholt N, Stock F, Reifferscheid G (2018) A new approach in separating microplastics from environmental samples based on their electrostatic behavior. Environ Pollut 234:20. https://doi.org/10.1016/j.envpol.2017.11.013

Fries E, Dekiff JH, Willmeyer J, Nuelle MT, Ebert M, Remy D (2013) Identification of polymer types and additives in marine microplastic particles using pyrolysis-GC/MS and scanning electron microscopy. Environ Sci Process Impacts 15(10):1949–1956. https://doi.org/10.1039/c3em00214d

Fu W, Min J, Jiang W, Li Y, Zhang W (2020) Separation, characterization and identification of microplastics and nanoplastics in the environment. Sci Total Environ 721:137561. https://doi.org/10.1016/j.scitotenv.2020.137561

GESAMP Joint Group of Experts on the Scientific Aspects of Marine Environmental Protection (2015) Sources, fate and effects of microplastics in the marine environment: a global assessment. In: Reports and studies GESAMP, vol 90. issn:1020–4873. http://ec.europa.eu/environment/marine/good-environmental-status/descriptor-10/pdf/GESAMP_microplastics full study.pdf

Gies EA, LeNoble JL, Noël M, Etemadifar A, Bishay F, Hall ER, Ross PS (2018) Retention of microplastics in a major secondary wastewater treatment plant in Vancouver, Canada. Mar Pollut Bull 133:553–561. https://doi.org/10.1016/j.marpolbul.2018.06.006

Godoy V, Martín-Lara MA, Calero M, Blázquez G (2019) Physical-chemical characterization of microplastics present in some exfoliating products from Spain. Mar Pollut Bull 139:91–99. https://doi.org/10.1016/j.marpolbul.2018.12.026

Grbic J, Nguyen B, Guo E, You JB, Sinton D, Rochman CM (2019) Magnetic extraction of microplastics from environmental samples. Environ Sci Technol Lett 6(2):68. https://doi.org/10.1021/acs.estlett.8b00671

Gregory MR (2009) Environmental implications of plastic debris in marine settings—entanglement, ingestion, smothering, hangers-on, hitch-hiking and alien invasions. Philos Trans R Soc B Biol Sci 364(1526):2013–2025. https://doi.org/10.1098/rstb.2008.0265

Gündoğdu S, Çevik C, Güzel E, Kilercioğlu S (2018) Microplastics in municipal wastewater treatment plants in Turkey: a comparison of the influent and secondary effluent concentrations. Environ Monit Assess 190(11):626. https://doi.org/10.1007/s10661-018-7010-y

Hale RC, Seeley ME, King AE, Yu LH (2022) Analytical chemistry of plastic debris: sampling, methods, and instrumentation, pp 17–67. https://doi.org/10.1007/978-3-030-78627-4_2

Hamidian AH, Ozumchelouei EJ, Feizi F, Wu C, Zhang Y, Yang M (2021) A review on the characteristics of microplastics in wastewater treatment plants: a source for toxic chemicals. J Clean Prod 295:126480. https://doi.org/10.1016/j.jclepro.2021.126480

Harrison JP, Ojeda JJ, Romero-González ME (2012) The applicability of reflectance micro-Fourier-transform infrared spectroscopy for the detection of synthetic microplastics in marine sediments. Sci Total Environ 416:455–463. https://doi.org/10.1016/j.scitotenv.2011.11.078

Hasan Anik A, Hossain S, Alam M, Binte Sultan M, Hasnine MT, Rahman MM (2021) Microplastics pollution: a comprehensive review on the sources, fates, effects, and potential remediation. Environ Nanotechnol Monit Manag 16:100530. https://doi.org/10.1016/j.enmm.2021.100530

He D, Luo Y, Lu S, Liu M, Song Y, Lei L (2018) Microplastics in soils: analytical methods, pollution characteristics and ecological risks. TrAC Trends Anal Chem 109:163. https://doi.org/10.1016/j.trac.2018.10.006

Heo NW, Hong SH, Han GM, Hong S, Lee J, Song YK, Jang M, Shim WJ (2013) Distribution of small plastic debris in cross-section and high strandline on Heungnam beach, South Korea. Ocean Sci J 48(2):225–233. https://doi.org/10.1007/s12601-013-0019-9

Hidalgo-Ruz V, Gutow L, Thompson RC, Thiel M (2012) Microplastics in the marine environment: a review of the methods used for identification and quantification. Environ Sci Technol 46(6):3060–3075. https://doi.org/10.1021/es2031505

Holland ER, Mallory ML, Shutler D (2016) Plastics and other anthropogenic debris in freshwater birds from Canada. Sci Total Environ 571:251–258. https://doi.org/10.1016/j.scitotenv.2016.07.158

Huppertsberg S, Knepper TP (2018) Instrumental analysis of microplastics—benefits and challenges. Anal Bioanal Chem 410(25):6343–6352. https://doi.org/10.1007/s00216-018-1210-8

Hurley RR, Lusher AL, Olsen M, Nizzetto L (2018) Validation of a method for extracting microplastics from complex, organic-rich, environmental matrices. Environ Sci Technol 52(13):7409. https://doi.org/10.1021/acs.est.8b01517

Imhof HK, Schmid J, Niessner R, Ivleva NP, Laforsch C (2012) A novel, highly efficient method for the separation and quantification of plastic particles in sediments of aquatic environments. Limnol Oceanogr Methods 10:524. https://doi.org/10.4319/lom.2012.10.524

Kalčíková G, Alič B, Skalar T, Bundschuh M, Gotvajn AŽ (2017) Wastewater treatment plant effluents as source of cosmetic polyethylene microbeads to freshwater. Chemosphere 188:25–31. https://doi.org/10.1016/j.chemosphere.2017.08.131

Käppler A, Fischer M, Scholz-Böttcher BM, Oberbeckmann S, Labrenz M, Fischer D, Eichhorn KJ, Voit B (2018) Comparison of μ-ATR-FTIR spectroscopy and py-GCMS as identification tools for microplastic particles and fibers isolated from river sediments. Anal Bioanal Chem 410(21):5313–5327. https://doi.org/10.1007/s00216-018-1185-5

Kay P, Hiscoe R, Moberley I, Bajic L, McKenna N (2018) Wastewater treatment plants as a source of microplastics in river catchments. Environ Sci Pollut Res 25(20):20264–20267. https://doi.org/10.1007/s11356-018-2070-7

Kershaw PJ (2016) Marine plastic debris and microplastics global lessons and research to inspire action and guide policy change. In: Unep

Koelmans B, Pahl S, Backhaus T, Bessa F, van Calster G, Contzen N, et al; SAPEA, science advice for policy by European academies (2019) A scientific perspective on microplastics in nature and society. In: Evidence review report (4)

Lambert S, Scherer C, Wagner M (2017) Ecotoxicity testing of microplastics: considering the heterogeneity of physicochemical properties. Integr Environ Assess Manag 13(3):470–475. https://doi.org/10.1002/ieam.1901

Lee KW, Shim WJ, Kwon OY, Kang JH (2013) Size-dependent effects of micro polystyrene particles in the marine copepod Tigriopus japonicus. Environ Sci Technol 47(19):11278–11283. https://doi.org/10.1021/es401932b

Leslie HA, Brandsma SH, van Velzen MJM, Vethaak AD (2017) Microplastics en route: field measurements in the Dutch river delta and Amsterdam canals, wastewater treatment plants, North Sea sediments and biota. Environ Int 101:133–142. https://doi.org/10.1016/j.envint.2017.01.018

Li J, Liu H, Paul Chen J (2018a) Microplastics in freshwater systems: a review on occurrence, environmental effects, and methods for microplastics detection. Water Res 137:362–374. https://doi.org/10.1016/j.watres.2017.12.056

Li X, Chen L, Mei Q, Dong B, Dai X, Ding G, Zeng EY (2018b) Microplastics in sewage sludge from the wastewater treatment plants in China. Water Res 142:75–85. https://doi.org/10.1016/j.watres.2018.05.034

Li L, Geng S, Li Z, Song K (2020) Effect of microplastic on anaerobic digestion of wasted activated sludge. Chemosphere 247:125874. https://doi.org/10.1016/j.chemosphere.2020.125874

Liu X, Yuan W, Di M, Li Z, Wang J (2019) Transfer and fate of microplastics during the conventional activated sludge process in one wastewater treatment plant of China. Chem Eng J 362:176–182. https://doi.org/10.1016/j.cej.2019.01.033

Liu Y, Mendoza-Perilla P, Clavier KA, Tolaymat TM, Bowden JA, Solo-Gabriele HM, Townsend TG (2022) Municipal solid waste incineration (MSWI) ash co-disposal: influence on per- and polyfluoroalkyl substances (PFAS) concentration in landfill leachate. Waste Manag 144:49. https://doi.org/10.1016/j.wasman.2022.03.009

Löder MGJ, Gerdts G (2015) Methodology used for the detection and identification of microplastics—a critical appraisal. In: Marine anthropogenic litter, pp 201–227. https://doi.org/10.1007/978-3-319-16510-3_8

Mahon AM, O'Connell B, Healy MG, O'Connor I, Officer R, Nash R, Morrison L (2017) Microplastics in sewage sludge: effects of treatment. Environ Sci Technol 51(2):810–818. https://doi.org/10.1021/acs.est.6b04048

Mason SA, Garneau D, Sutton R, Chu Y, Ehmann K, Barnes J, Fink P, Papazissimos D, Rogers DL (2016) Microplastic pollution is widely detected in US municipal wastewater treatment plant effluent. Environ Pollut 218:1045–1054. https://doi.org/10.1016/j.envpol.2016.08.056

McCormick A, Hoellein TJ, Mason SA, Schluep J, Kelly JJ (2014) Microplastic is an abundant and distinct microbial habitat in an urban river. Environ Sci Technol 48(20):11863–11871. https://doi.org/10.1021/es503610r

McCormick AR, Hoellein TJ, London MG, Hittie J, Scott JW, Kelly JJ (2016) Microplastic in surface waters of urban rivers: concentration, sources, and associated bacterial assemblages. Ecosphere 7(11):e01556. https://doi.org/10.1002/ecs2.1556

Michielssen MR, Michielssen ER, Ni J, Duhaime MB (2016) Fate of microplastics and other small anthropogenic litter (SAL) in wastewater treatment plants depends on unit processes employed. Environ Sci Water Res Technol 2(6):1064–1073. https://doi.org/10.1039/c6ew00207b

Mintenig SM, Int-Veen I, Löder MGJ, Primpke S, Gerdts G (2017) Identification of microplastic in effluents of waste water treatment plants using focal plane array-based micro-Fourier-transform infrared imaging. Water Res 108:365–372. https://doi.org/10.1016/j.watres.2016.11.015
Mohamed Nor NH, Obbard JP (2014) Microplastics in Singapore's coastal mangrove ecosystems. Mar Pollut Bull 79(1–2):278. https://doi.org/10.1016/j.marpolbul.2013.11.025
Monira S, Bhuiyan MA, Haque N, Shah K, Roychand R, Hai FI, Pramanik BK (2021) Understanding the fate and control of road dust-associated microplastics in stormwater. Process Saf Environ Protect 152:47. https://doi.org/10.1016/j.psep.2021.05.033
Moriarty M, Lee E, Brown S, Buksa B, Niekrewicz T, Barnes J, Chaskey E (2018) Microplastic pollution: a survey of wastewater effluent in the Lake Champlain Basin. Digital Commons: Center for Earth and Environmental Science Student Posters, 12901. https://digitalcommons.plattsburgh.edu/cees_student_posters/42
Munno K, Helm PA, Jackson DA, Rochman C, Sims A (2018) Impacts of temperature and selected chemical digestion methods on microplastic particles. Environ Toxicol Chem 37(1):91. https://doi.org/10.1002/etc.3935
Murphy F, Ewins C, Carbonnier F, Quinn B (2016) Wastewater treatment works (WwTW) as a source of microplastics in the aquatic environment. Environ Sci Technol 50(11):5800–5808. https://doi.org/10.1021/acs.est.5b05416
Naidu SA (2019) Preliminary study and first evidence of presence of microplastics and colorants in green mussel, Perna viridis (Linnaeus, 1758), from southeast coast of India. Mar Pollut Bull 140:416. https://doi.org/10.1016/j.marpolbul.2019.01.024
Napper IE, Thompson RC (2016) Release of synthetic microplastic plastic fibres from domestic washing machines: effects of fabric type and washing conditions. Mar Pollut Bull 112(1–2):39–45. https://doi.org/10.1016/j.marpolbul.2016.09.025
Nuelle MT, Dekiff JH, Remy D, Fries E (2014) A new analytical approach for monitoring microplastics in marine sediments. Environ Pollut 184:161. https://doi.org/10.1016/j.envpol.2013.07.027
Okoffo ED, O'Brien S, O'Brien JW, Tscharke BJ, Thomas KV (2019) Wastewater treatment plants as a source of plastics in the environment: a review of occurrence, methods for identification, quantification and fate. Environ Sci Water Res Technol 5(11):1908–1931. https://doi.org/10.1039/c9ew00428a
Piccin JS, Cadaval TRSA, De Pinto LAA, Dotto GL (2017) Adsorption isotherms in liquid phase: experimental, modeling, and interpretations. In: Adsorption processes for water treatment and purification, pp 19–51. https://doi.org/10.1007/978-3-319-58136-1_2
Prata JC, da Costa JP, Duarte AC, Rocha-Santos T (2019) Methods for sampling and detection of microplastics in water and sediment: a critical review. TrAC Trends Anal Chem 110:150. https://doi.org/10.1016/j.trac.2018.10.029
Revel M, Châtel A, Mouneyrac C (2018) Micro(nano)plastics: a threat to human health? Curr Opin Environ Sci Health 1:17. https://doi.org/10.1016/j.coesh.2017.10.003
Rezania S, Park J, Md Din MF, Mat Taib S, Talaiekhozani A, Kumar Yadav K, Kamyab H (2018) Microplastics pollution in different aquatic environments and biota: a review of recent studies. Mar Pollut Bull 133:191–208. https://doi.org/10.1016/j.marpolbul.2018.05.022
Rhein F, Scholl F, Nirschl H (2019) Magnetic seeded filtration for the separation of fine polymer particles from dilute suspensions: microplastics. Chem Eng Sci 207:1278. https://doi.org/10.1016/j.ces.2019.07.052
Rist S, Hartmann NB (2018) Aquatic ecotoxicity of microplastics and nanoplastics: lessons learned from engineered nanomaterials. In: Handbook of environmental chemistry, vol 58, pp 25–49. https://doi.org/10.1007/978-3-319-61615-5_2
Rocha-Santos TAP (2018) Editorial overview: micro and nano-plastics. Curr Opin Environ Sci Health 1:52. https://doi.org/10.1016/j.coesh.2018.01.003
Rochman CM, Kurobe T, Flores I, Teh SJ (2014) Early warning signs of endocrine disruption in adult fish from the ingestion of polyethylene with and without sorbed chemical pollutants from the marine environment. Sci Total Environ 493:656–661. https://doi.org/10.1016/j.scitotenv.2014.06.051

Rodríguez Chialanza M, Sierra I, Pérez Parada A, Fornaro L (2018) Identification and quantitation of semi-crystalline microplastics using image analysis and differential scanning calorimetry. Environ Sci Pollut Res 25(17):16767–16775. https://doi.org/10.1007/s11356-018-1846-0

Roodenko K, Aureau D, Yang F, Thissen P, Rappich J (2018) Characterization of thin organic films with surface-sensitive FTIR spectroscopy. In: Springer series in surface sciences, vol 52, pp 483–503. https://doi.org/10.1007/978-3-319-75895-4_21

Sheavly SB, Register KM (2007) Marine debris & plastics: environmental concerns, sources, impacts and solutions. J Polym Environ 15(4):301–305. https://doi.org/10.1007/s10924-007-0074-3

Shen M, Zhang Y, Zhu Y, Song B, Zeng G, Hu D, Wen X, Ren X (2019) Recent advances in toxicological research of nanoplastics in the environment: a review. Environ Pollut 252:511–521. https://doi.org/10.1016/j.envpol.2019.05.102

Shim WJ, Hong SH, Eo SE (2017) Identification methods in microplastic analysis: a review. Anal Methods 9(9):1384–1391. https://doi.org/10.1039/c6ay02558g

Simon M, van Alst N, Vollertsen J (2018) Quantification of microplastic mass and removal rates at wastewater treatment plants applying Focal Plane Array (FPA)-based Fourier Transform Infrared (FT-IR) imaging. Water Res 142:1. https://doi.org/10.1016/j.watres.2018.05.019

Song YK, Hong SH, Jang M, Han GM, Rani M, Lee J, Shim WJ (2015) A comparison of microscopic and spectroscopic identification methods for analysis of microplastics in environmental samples. Mar Pollut Bull 93(1–2):202–209. https://doi.org/10.1016/j.marpolbul.2015.01.015

Strungaru SA, Jijie R, Nicoara M, Plavan G, Faggio C (2019) Micro- (nano) plastics in freshwater ecosystems: abundance, toxicological impact and quantification methodology. TrAC Trends Anal Chem 110:116. https://doi.org/10.1016/j.trac.2018.10.025

Sun J, Dai X, Wang Q, van Loosdrecht MCM, Ni BJ (2019) Microplastics in wastewater treatment plants: detection, occurrence and removal. Water Res 152:21–37. https://doi.org/10.1016/j.watres.2018.12.050

Tagg AS, Sapp M, Harrison JP, Ojeda JJ (2015) Identification and quantification of microplastics in wastewater using focal plane array-based reflectance micro-FT-IR imaging. Anal Chem 87(12):6032–6040. https://doi.org/10.1021/acs.analchem.5b00495

Tagg AS, Harrison JP, Ju-Nam Y, Sapp M, Bradley EL, Sinclair CJ, Ojeda JJ (2017) Fenton's reagent for the rapid and efficient isolation of microplastics from wastewater. Chem Commun 53(2):372. https://doi.org/10.1039/c6cc08798a

Talvitie J, Mikola A, Koistinen A, Setälä O (2017) Solutions to microplastic pollution—Removal of microplastics from wastewater effluent with advanced wastewater treatment technologies. Water Res 123:401–407. https://doi.org/10.1016/j.watres.2017.07.005

Tobin MJ, Bambery KR, Martin DE, Puskar L, Beattie DA, Ivanova EP, Nguyen SH, Webb HK, Vongsvivut J (2014) Attenuated total reflection FTIR microspectroscopy at the Australian synchrotron. Optics InfoBase Conference Papers, 3. https://doi.org/10.1364/FTS.2016.FTu2E.5

Vianello A, Boldrin A, Guerriero P, Moschino V, Rella R, Sturaro A, Da Ros L (2013) Microplastic particles in sediments of Lagoon of Venice, Italy: first observations on occurrence, spatial patterns and identification. Estuar Coast Shelf Sci 130:54–61. https://doi.org/10.1016/j.ecss.2013.03.022

von Friesen LW, Granberg ME, Hassellöv M, Gabrielsen GW, Magnusson K (2019) An efficient and gentle enzymatic digestion protocol for the extraction of microplastics from bivalve tissue. Mar Pollut Bull 142:129. https://doi.org/10.1016/j.marpolbul.2019.03.016

Walker T, Gramlich D, Dumont-Bergeron A (2020) The case for a plastic tax: a review of its benefits and disadvantages within a circular economy. Sustainability 4:185–211. https://doi.org/10.1108/S2514-175920200000004010

Wan Z, Wang C, Zhou J, Shen M, Wang X, Fu Z, Jin Y (2019) Effects of polystyrene microplastics on the composition of the microbiome and metabolism in larval zebrafish. Chemosphere 217:646. https://doi.org/10.1016/j.chemosphere.2018.11.070

Wang W, Wang J (2018) Investigation of microplastics in aquatic environments: an overview of the methods used, from field sampling to laboratory analysis. TrAC Trends Anal Chem 108:195–202. https://doi.org/10.1016/j.trac.2018.08.026

Wang W, Ndungu AW, Li Z, Wang J (2017) Microplastics pollution in inland freshwaters of China: a case study in urban surface waters of Wuhan, China. Sci Total Environ 575:1369–1374. https://doi.org/10.1016/j.scitotenv.2016.09.213

Wei Z, Xu T, Zhao D (2019) Treatment of per- and polyfluoroalkyl substances in landfill leachate: status, chemistry and prospects. Environ Sci Water Res Technol 5(11):1814–1835. https://doi.org/10.1039/c9ew00645a

Xu L, Guo C, Wang F, Zheng S, Liu CZ (2011) A simple and rapid harvesting method for microalgae by in situ magnetic separation. Bioresour Technol 102(21):10047. https://doi.org/10.1016/j.biortech.2011.08.021

Zhang K, Su J, Xiong X, Wu X, Wu C, Liu J (2016) Microplastic pollution of lakeshore sediments from remote lakes in Tibet plateau, China. Environ Pollut 219:450–455. https://doi.org/10.1016/j.envpol.2016.05.048

Ziajahromi S, Neale PA, Rintoul L, Leusch FDL (2017) Wastewater treatment plants as a pathway for microplastics: development of a new approach to sample wastewater-based microplastics. Water Res 112:93–99. https://doi.org/10.1016/j.watres.2017.01.042

Zobkov MB, Esiukova EE (2017) Evaluation of the Munich plastic sediment separator efficiency in extraction of microplastics from natural marine bottom sediments. Limnol Oceanogr Methods 15(11):967. https://doi.org/10.1002/lom3.10217

Microplastics Detection Techniques

2

Amit Joshi, Nahid Akhtar, and Ajay Kumar

Abstract

Microplastics, defined as small plastic particles typically measuring less than 5 mm, have become widespread environmental pollutants, posing significant risks to ecosystems and human health. The detection and characterization of microplastics in diverse environmental settings are crucial for understanding their distribution, fate, and potential consequences. This chapter offers an extensive overview of the various techniques utilized for the detection and analysis of microplastics. Traditional approaches such as visual inspection and screening have laid the foundation for more sophisticated methodologies, including spectroscopic, microscopic, and chromatographic techniques. Spectroscopic methods such as Fourier-transform infrared spectroscopy (FTIR) and Raman spectroscopy provide precise identification based on molecular signatures, whereas microscopy-based techniques such as scanning electron microscopy (SEM) and micro-Fourier-transform infrared spectroscopy (μ-FTIR) offer high-resolution imaging and chemical analysis capabilities. Chromatographic methods like gas chromatography-mass spectrometry (GC-MS) and liquid chromatography-mass spectrometry (LC-MS) facilitate the identification and quantification of microplastic additives and degradation products. Additionally, emerging technologies like hyperspectral imaging, microfluidic devices, and

A. Joshi
Department of Biochemistry, Kalinga University, Naya Raipur, Chhattisgarh, India

N. Akhtar
School of Bioengineering and Biosciences, Lovely Professional University, Jalandhar, Punjab, India

A. Kumar (✉)
Department of Biotechnology, Faculty of Engineering and Technology, Rama University, Kanpur, Uttar Pradesh, India
e-mail: drajay.fet@ramauniversity.ac.in

A. Kumar, V. Singh (eds.), *Microplastics Pollution and its Remediation*,
https://doi.org/10.1007/978-981-97-4068-0_2

artificial intelligence (AI) algorithms show promise in enhancing detection efficiency and accuracy. Nonetheless, challenges persist, including sample preparation, standardization, and differentiation of microplastics from natural particles, emphasizing the ongoing necessity for research and development in this field.

Keywords

Microplastics · Detection techniques · Spectroscopic analysis · Chromatographic methods · Environmental monitoring

2.1 Introduction

Microplastics, a term coined to describe minute plastic particles measuring less than 5 mm in size, have become a ubiquitous and concerning environmental issue in recent years (Wirnkor et al. 2019). These tiny plastic fragments originate from various sources, including the breakdown of larger plastic items, microbeads from personal care products, and industrial processes (Duis and Coors 2016). Due to their small size, microplastics pose unique challenges for environmental management, as they can easily infiltrate ecosystems, impacting both aquatic and terrestrial environments (Roy et al. 2022). The proliferation of microplastics is a consequence of widespread plastic usage and insufficient waste management practices. Plastics, known for their durability and versatility, have become integral to modern life (Evode et al. 2021). However, the persistent nature of plastic materials also means that they can persist in the environment for extended periods, breaking down into smaller particles over time.

The environmental implications of microplastics are far-reaching. These particles have been found in oceans, rivers, soil, and even in the air, highlighting the extent of their dispersal. Once in the environment, microplastics can absorb and accumulate harmful chemicals, posing potential threats to aquatic and terrestrial organisms (Thacharodi et al. 2024). The ingestion of microplastics by marine life, for example, can lead to bioaccumulation and adversely affect the food chain, with potential repercussions for human health. The detection and quantification of microplastics have become critical components of environmental monitoring and research efforts (Ghanadi et al. 2024; Vitali et al. 2024). Understanding the distribution, abundance, and characteristics of microplastics is essential for devising effective mitigation and management strategies. Various techniques and methods have been developed to identify and measure microplastics in different environmental matrices, ranging from water and sediment to organisms.

This comprehensive exploration of microplastics detection techniques aims to shed light on the diverse methodologies employed in analyzing these tiny particles. From visual inspection and counting methods to advanced spectroscopic and chromatographic techniques, researchers employ a range of tools to unravel the intricate dimensions of the microplastics conundrum (see Table 2.1). This knowledge is vital not only for scientific understanding but also for informing policies and practices aimed at minimizing the impact of microplastics on ecosystems and human well-being. In the subsequent sections, we will delve into the challenges associated with microplastics detection, provide an overview of the various detection techniques, and explore the implications of microplastics on the environment and public health. Through this exploration, we aim to contribute to a deeper understanding of the complexities surrounding microplastics and highlight the importance of proactive measures to address this global environmental challenge.

2.2 Importance of Detecting Microplastics

The pervasive presence of microplastics in the environment has raised significant concerns about their potential impact on ecosystems, wildlife, and human health. Detecting microplastics has become a crucial aspect of environmental research and management due to the following key reasons.

2.2.1 Ecological Impact

Microplastics, being minute plastic particles, have the potential to infiltrate ecosystems and adversely affect biodiversity (Mingming et al. 2024). The ingestion of microplastics by aquatic organisms, such as fish and marine invertebrates, can lead to bioaccumulation, disrupting food webs and posing threats to various species (Mondal et al. 2024). Understanding the extent of microplastic pollution is essential for assessing its ecological impact and implementing conservation measures.

2.2.2 Human Health Concerns

As microplastics enter the food chain through the consumption of contaminated organisms, there are growing concerns about the potential health impacts on humans. The ability of microplastics to absorb and accumulate toxic substances raises questions about the safety of consuming seafood and other products affected by microplastic contamination (Saeedi 2024). Detecting and quantifying microplastics in various food sources is crucial for assessing potential health risks and implementing measures to ensure food safety (Lin et al. 2024).

2.2.3 Environmental Monitoring

Microplastics are found in diverse environments, including oceans, rivers, soil, and even the air. Monitoring the presence and concentration of microplastics in different matrices is essential for understanding their distribution and transport mechanisms (De Arbeloa and Marzadri 2024). This information is vital for developing effective strategies to mitigate microplastic pollution and protect ecosystems.

Table 2.1 Types of microparticles: Based on size and detection method

Microparticle type	Size (range)	Detection method(s)	Advantages	Disadvantages	Environmental/Health impacts
Nanoparticles	1–100 nm	Dynamic light scattering (DLS), AFM, TEM, SEM, fluorescence spectroscopy	Small size allows for easy penetration and targeting. Versatile applications in drug delivery, imaging, and diagnostics	Risk of environmental accumulation and potential toxicity. Difficulty in monitoring and regulation of exposure	Potential environmental contamination, toxicity to aquatic life, respiratory health concerns for humans (e.g., inhalation)
Microspheres	1–100 μm	Flow cytometry, optical microscopy, laser diffraction, microfluidic techniques	Uniform size distribution, easy surface modification, high stability. Suitable for drug delivery, diagnostics, and cell sorting	Non-biodegradable, may persist in the environment. Potential risk of ingestion by wildlife	Accumulation in ecosystems, ingestion by marine animals and birds leading to physical harm or obstruction of digestive systems
Liposomes	50–500 nm	Fluorescence spectroscopy, DLS, Cryo-EM	Biocompatible, versatile drug delivery systems. Encapsulation of hydrophilic and hydrophobic drugs	Limited stability and shelf life. Issues with reproducibility in large-scale production	Potential discharge into aquatic environments, biodegradability may vary depending on lipid composition
Quantum Dots	2–10 nm	Fluorescence spectroscopy, confocal microscopy, TEM	Bright and stable fluorescence, tunable emission properties. Used in cellular imaging, biosensing, and photovoltaic devices	Toxicity concerns due to heavy metal content (e.g., cadmium, lead). Potential environmental persistence	Hazardous waste generation during production and disposal, risk of leaching heavy metals into soil and water
Polystyrene beads	0.1–10 μm	Flow cytometry, optical microscopy, laser diffraction, microfluidic techniques	Uniform size, inert material, easy functionalization. Suitable for immunoassays, cell separation, and standards in microscopy	Non-biodegradable, potential ingestion by marine organisms and wildlife	Accumulation in marine environments, ingestion by marine animals causing harm or death, potential introduction into food chain

Magnetic beads	1–100 μm	MRI, magnetoresistive sensors, SQUIDs, MPI	Easy manipulation and separation using magnetic fields. Widely used in diagnostics, biomolecule separation, and drug delivery	Magnetic field interference with sensitive equipment. Potential bioaccumulation of magnetic materials in organisms	Potential impact on magnetic-sensitive organisms, concerns about leaching of magnetic materials into the environment
Gold nanoparticles	1–100 nm	SPR, colorimetric assays, TEM, SEM	Unique optical properties, facile surface modification. Applied in biosensing, imaging, and therapeutic applications	Costly production and purification processes. Potential environmental persistence	Risk of environmental contamination during production and disposal, potential toxicity to aquatic organisms

2.2.4 Source Identification and Mitigation

Identifying the sources of microplastic pollution is crucial for implementing targeted mitigation measures. Detection techniques help researchers trace the origins of microplastics, whether from plastic manufacturing processes, improper waste disposal, or other sources (Fältström and Carlsson 2024). By understanding the sources, authorities can implement policies and practices to reduce the introduction of microplastics into the environment.

2.2.5 Scientific Understanding and Research

Research on microplastics provides valuable insights into their characteristics, behavior, and interactions with the environment (Kong et al. 2024). Detecting and characterizing microplastics using advanced analytical techniques contribute to the scientific understanding of this complex issue. Such knowledge is essential for developing effective management strategies and technological innovations to address the challenges posed by microplastics.

2.2.6 Public Awareness and Education

Detecting and publicizing the presence of microplastics in the environment raise awareness among the general public. Educating communities about the impact of microplastics fosters a sense of responsibility and encourages sustainable practices (Rangel-Buitrago and Gracia 2024). Informed citizens are more likely to support initiatives aimed at reducing plastic use and improving waste management.

The importance of detecting microplastics cannot be overstated. It is a multidimensional challenge that requires a comprehensive understanding of their distribution, effects on ecosystems, and potential risks to human health. Through continued research and vigilant monitoring, stakeholders can work together to develop strategies that mitigate the impact of microplastics on the environment and safeguard the well-being of both ecosystems and human populations.

2.3 Challenges in Microplastics Detection

The detection of microplastics presents a complex set of challenges owing to the minute size, diverse sources, and widespread distribution of these plastic particles in the environment. Addressing these challenges is essential for advancing our understanding of microplastic pollution and developing effective mitigation strategies. The key challenges in microplastics detection include the following.

2.3.1 Size Variability

Microplastics encompass a wide range of sizes, from microscopic particles to larger fragments. The diversity in size makes detection challenging, as conventional methods may overlook smaller particles. The need for specialized equipment and techniques to capture and analyze these tiny plastic fragments is a significant challenge in the field of microplastics research (Steiner et al. 2024).

2.3.2 Method Standardization

The absence of standardized methods for microplastics detection poses challenges in comparing data across studies. Different researchers and laboratories may employ varied techniques, making it difficult to establish consistent baseline information. Standardizing detection methods is crucial for ensuring the reliability and accuracy of microplastics data (Jones et al. 2024), allowing for meaningful comparisons between different regions and ecosystems.

2.3.3 Sample Collection and Preparation

Obtaining representative samples for analysis is a critical step in microplastics detection. However, the heterogeneous nature of environmental samples, such as water, sediment, or soil, complicates the collection process (Dris et al. 2024). Additionally, preparing samples for analysis without introducing contamination is challenging. Developing standardized protocols for sample collection and preparation is essential to enhance the reliability of microplastics detection results.

2.3.4 Lack of Reference Materials

The absence of universally accepted reference materials for microplastics hinders accurate quantification and identification (Fakayode et al. 2024). Establishing a set of reference materials with known properties and concentrations would aid in calibrating detection instruments and validating measurement results. This would contribute to the accuracy and reproducibility of microplastics research.

2.3.5 Identification of Microplastic Types

Distinguishing between different types of microplastics is a complex task, especially when dealing with fragmented particles with diverse polymer compositions. Advanced techniques, such as spectroscopy and microscopy, are required for accurate identification (Fakayode et al. 2024). However, these methods often demand specialized expertise and equipment, adding to the overall complexity of microplastics detection.

2.3.6 Dynamic Environmental Conditions

Environmental conditions, such as temperature, pH, and salinity, can influence the behavior and distribution of microplastics. Detecting microplastics in dynamic and changing environments poses a challenge, as their transport and fate can vary based on these factors. Understanding the impact of environmental conditions on microplastics detection is crucial for interpreting research findings accurately (Binda et al. 2024).

2.3.7 Cost and Accessibility of Technology

Some advanced detection technologies come with high costs and may not be readily accessible, particularly for researchers in resource-constrained settings. Developing cost-effective detection methods that maintain accuracy is crucial for promoting widespread participation in microplastics research and monitoring (Gayathri et al. 2024).

2.3.8 Data Interpretation and Integration

Analyzing and interpreting microplastics data is complex due to the variability in detection methods and the dynamic nature of environmental systems. Integrating data from different studies and sources requires robust frameworks and methodologies to ensure accurate assessments of the overall microplastics landscape (Norling et al. 2024).

Addressing these challenges is essential for advancing the field of microplastics detection and understanding the full extent of their impact on ecosystems and human health. Collaborative efforts among researchers, standardization organizations, and policymakers are necessary to overcome these obstacles and develop effective strategies for mitigating the pervasive issue of microplastic pollution.

2.4 Overview of Microplastics Detection Techniques

The identification and quantification of microplastics in the environment have become pivotal in addressing the escalating concerns surrounding plastic pollution. Researchers employ a variety of techniques to detect and analyze these minuscule plastic particles, each method offering unique advantages and limitations. This overview explores some of the key microplastics detection techniques, shedding light on their principles and applications:

2.4.1 Visual Inspection and Counting

Visual inspection involves manually examining environmental samples, such as water or sediment, under a microscope to identify and count microplastics (Ghanadi et al. 2024). While this method is straightforward and cost-effective, it is limited by the potential for human error and may not capture the full spectrum of microplastics, especially smaller particles.

2.4.2 Spectroscopic Techniques

Spectroscopic techniques, including Fourier-transform infrared spectroscopy (FTIR) and Raman spectroscopy, are employed for the identification of microplastic polymers. These methods rely on the unique spectral fingerprints of different polymers, enabling researchers to categorize and quantify microplastics in samples (Circelli et al. 2024). Spectroscopy provides valuable information about the composition of microplastics but may require specialized equipment and expertise.

2.4.3 Microscopy-Based Approaches

Microscopy techniques, such as scanning electron microscopy (SEM) and transmission electron microscopy (TEM), offer high-resolution imaging of microplastics (Hu et al. 2024). These methods allow for detailed morphological analysis, aiding in the characterization of particle shapes and sizes. Microscopy is particularly useful for studying individual particles but may be time-consuming for large-scale sample analysis.

2.4.4 Chromatographic Methods

Chromatographic techniques, including gas chromatography (GC) and liquid chromatography (LC), are employed to separate and analyze microplastic components (Martín-Gómez et al. 2024). These methods are effective in detecting additives and plasticizers associated with microplastics. Chromatography provides detailed chemical information but may require extensive sample preparation and sophisticated instrumentation.

2.4.5 Fluorescence Techniques

Fluorescence spectroscopy and microscopy leverage the natural fluorescence of certain polymers or fluorescent dyes to detect and visualize microplastics (Zhang et al. 2024). These techniques offer rapid screening capabilities and can be applied to larger sample volumes. However, challenges include potential interference from natural organic matter and the need for proper calibration.

2.4.6 Density Separation Methods

Density separation techniques involve isolating microplastics based on their buoyancy in different density solutions. Common methods include density gradient centrifugation and flotation (Zhu et al. 2024). These techniques are effective in concentrating microplastics from complex environmental samples, facilitating subsequent analysis.

2.4.7 Emerging Technologies

Continuous advancements in technology have led to the development of innovative approaches for microplastics detection (Ribeiro et al. 2024). This includes the use of artificial intelligence and machine learning algorithms to automate detection processes, improving efficiency and accuracy. Nanotechnology-based sensors and microfluidic devices are also being explored for their potential in enhancing detection sensitivity.

2.4.8 Quantitative Polymerase Chain Reaction (qPCR)

Quantitative polymerase chain reaction (qPCR) is a molecular biology technique that targets specific DNA associated with microorganisms adhered to microplastics (Akarsu et al. 2023). This method allows for the quantification of microplastics indirectly through the measurement of microbial DNA. While not applicable to all microplastics, qPCR provides insights into the microbial colonization of plastic surfaces.

This illustrates the diverse array of techniques employed in the detection of microplastics, highlighting the need for a multifaceted approach to comprehensively understand the prevalence and characteristics of microplastic pollution (Niu et al. 2023). Researchers often combine multiple methods to cross-verify results and gain a more comprehensive understanding of the impact of microplastics on the environment. As the field continues to evolve, ongoing research and technological innovations will further refine and expand the toolkit for microplastics detection.

2.5 Visual Inspection and Counting Methods

Visual inspection and counting methods represent a fundamental and widely used approach in the detection and analysis of microplastics. As microscopic plastic particles, often measuring less than 5 mm, continue to pose environmental challenges, these methods serve as essential tools for researchers seeking to understand the prevalence and distribution of microplastics in various ecosystems.

2.5.1 Principles of Visual Inspection

Visual inspection involves the manual examination of environmental samples, such as water, sediment, or soil, under a microscope. The primary objective is to visually identify and classify microplastics based on their size, shape, and color (Niu et al. 2023; Bydalek et al. 2023). Researchers scrutinize samples to distinguish plastic particles from natural components, such as organic matter or minerals.

2.5.2 Sampling Techniques

Collecting representative samples is a crucial aspect of visual inspection. Researchers employ various sampling methods, including plankton nets, sediment corers, and water filtration, to gather environmental matrices containing microplastics. These samples are then processed and prepared for microscopic examination.

2.5.3 Microscopic Examination

Microscopic examination is conducted using light microscopes or stereomicroscopes, providing the necessary magnification for visualizing microplastics. Researchers carefully scan the samples, identifying particles that exhibit characteristics indicative of plastic origin (Macairan et al. 2023). Manual counting is then employed to quantify the number of microplastics within a given sample.

2.5.4 Challenges and Limitations

Despite its widespread use, visual inspection and counting have inherent challenges. The subjectivity introduced by human interpretation may lead to variations in results among different analysts. Moreover, smaller microplastics may go unnoticed, potentially resulting in underestimations. To address these limitations, researchers often complement visual inspection with additional analytical techniques for a more comprehensive analysis.

2.5.5 Enhancements Through Staining

To improve the visibility of microplastics under the microscope, staining techniques are sometimes employed. Fluorescent dyes or stains that selectively bind to plastic polymers enhance the contrast between microplastics and background materials.

This staining approach facilitates a more accurate identification and enumeration of microplastics during visual inspection.

2.5.6 Quality Assurance and Standardization

Recognizing the potential for variability in visual inspection results, efforts are underway to establish quality assurance protocols and standardize methodologies. Standardization involves developing consistent criteria for identifying and categorizing microplastics, reducing the likelihood of discrepancies between different studies and laboratories.

2.5.7 Integration with Other Techniques

Visual inspection is often used in conjunction with other detection methods to overcome its limitations. For example, researchers may combine visual counting with spectroscopic techniques or microscopy for a more comprehensive analysis (Singh and Kumar 2023). This integrative approach helps validate results and provides a more nuanced understanding of microplastics in the studied environment.

2.5.8 Educational and Citizen Science Applications

Visual inspection and counting methods are not limited to professional researchers; they also find application in educational programs and citizen science initiatives. By involving the public in microplastics monitoring efforts, these methods contribute to raising awareness about plastic pollution and fostering a sense of environmental responsibility.

Visual inspection and counting methods play a vital role in the initial identification and quantification of microplastics. While subject to certain limitations, ongoing advancements in standardization, staining techniques, and integration with other methodologies continue to enhance the reliability and accuracy of visual inspection (Ho et al. 2023). As part of a broader toolkit for microplastics detection, visual methods contribute valuable data to the collective understanding of the extent and impact of microplastic pollution in our ecosystems.

2.6 Spectroscopic Techniques for Microplastics Analysis

Spectroscopic techniques are instrumental in the analysis of microplastics, providing researchers with valuable insights into the composition and characteristics of these minute plastic particles (Fang et al. 2023). As a diverse group of analytical methods, spectroscopy enables the identification and quantification of microplastics by exploiting their unique spectral features. Several spectroscopic techniques are employed in microplastics analysis, each offering distinct advantages in unraveling the complexities of plastic pollution.

2.6.1 Fourier-Transform Infrared Spectroscopy (FTIR)

Fourier-transform infrared spectroscopy (FTIR) is a widely used spectroscopic technique in microplastics analysis. It operates on the principle that different polymers exhibit unique infrared absorption spectra (Mikulec et al. 2023). Microplastic samples are exposed to infrared radiation, and the resulting absorption spectra provide a fingerprint for identifying the polymer composition. FTIR is particularly effective in distinguishing between various plastic types, aiding in the precise characterization of microplastics.

2.6.2 Raman Spectroscopy

Raman spectroscopy is another powerful tool for microplastics analysis. It relies on inelastic scattering of monochromatic light, producing a unique Raman spectrum for each polymer (Yasri and Wiwanitkit 2023). This technique allows for nondestructive identification of microplastics, offering advantages in differentiating polymers with similar FTIR spectra. Raman spectroscopy is valuable in providing complementary information to FTIR, enhancing the overall accuracy of microplastics characterization.

2.6.3 Near-Infrared Spectroscopy (NIRS)

Near-infrared spectroscopy (NIRS) involves the measurement of light absorption in the near-infrared region. While less commonly used than FTIR and Raman spectroscopy, NIRS has shown promise in rapidly screening large sample volumes for microplastics (Liu et al. 2023). Its nondestructive nature and potential for high-throughput analysis make it suitable for certain applications, particularly in preliminary assessments of microplastic contamination.

2.6.4 Ultraviolet-Visible (UV-Vis) Spectroscopy

Ultraviolet-visible (UV-Vis) spectroscopy is employed for the analysis of microplastics containing colorants or additives that absorb in the ultraviolet or visible regions (Shen et al. 2024). By measuring the absorption of light, this technique can provide information about the presence of specific compounds within microplastics. UV-Vis spectroscopy is often used in conjunction with other spectroscopic methods for a more comprehensive analysis.

2.6.5 Fluorescence Spectroscopy

Fluorescence spectroscopy exploits the natural fluorescence of certain polymers or fluorescent dyes bound to microplastics (Morgana et al. 2024). This technique is sensitive and can be used for rapid screening of microplastics in environmental samples. By leveraging the fluorescence properties of plastics, researchers can enhance the detection limits and efficiency of microplastics analysis.

While spectroscopic techniques offer advanced capabilities, they are not without challenges. The need for specialized equipment and expertise can be a barrier, particularly in resource-limited settings. Additionally, the presence of interfering substances in complex environmental matrices may affect the accuracy of results. Careful sample preparation and calibration are crucial to mitigating these challenges. Ongoing advancements in spectroscopic technologies, including miniaturization and automation, are facilitating the broader application of these techniques in microplastics research. Integration with other analytical methods and the development of portable devices are areas of active exploration, aiming to enhance the accessibility and efficiency of microplastics analysis. Spectroscopic techniques play a pivotal role in advancing our understanding of microplastics. Their ability to provide detailed information about polymer composition and characteristics contributes to the comprehensive analysis of microplastics in environmental samples. As technology continues to evolve, spectroscopy remains at the forefront of innovative approaches for accurate and efficient microplastics analysis, furthering our efforts to address the global challenge of plastic pollution.

2.7 Microscopy-Based Approaches

Microscopy-based approaches are pivotal in the detailed examination and characterization of microplastics, offering researchers the ability to visualize and analyze these minute plastic particles with high precision (Kaushik et al. 2024). As microplastic pollution becomes an escalating environmental concern, microscopy-based techniques play a crucial role in identifying, quantifying, and understanding the morphological features of these particles. Several microscopy methods are employed in microplastics analysis, each contributing unique insights to the broader understanding of plastic pollution.

2.7.1 Scanning Electron Microscopy (SEM)

Scanning electron microscopy (SEM) is a powerful microscopy technique that provides high-resolution, three-dimensional images of microplastics. It utilizes electron beams to scan the surface of samples, offering detailed insights into particle morphology, size, and surface characteristics (Wei et al. 2024). SEM is particularly effective in studying individual microplastics, enabling researchers to distinguish between different types based on their distinctive surface features.

2.7.2 Transmission Electron Microscopy (TEM)

Transmission electron microscopy (TEM) goes beyond SEM by allowing researchers to explore the internal structures of microplastics at the nanoscale (Surana et al. 2024). This technique involves transmitting electrons through thin sections of samples, producing highly detailed images. TEM is valuable for investigating the internal composition, crystallinity, and defects within microplastic particles, providing a deeper understanding of their properties.

2.7.3 Light Microscopy

Light microscopy, including both compound and stereomicroscopy, is commonly employed for the initial identification and counting of microplastics (Gheorghe et al. 2024). While it may lack the resolution of electron microscopy, light microscopy is efficient for rapid screening of samples. This method allows researchers to observe the general characteristics of microplastics and estimate their abundance in environmental matrices.

2.7.4 Confocal Microscopy

Confocal microscopy utilizes lasers and pinhole apertures to eliminate out-of-focus light, resulting in optical sections that can be reconstructed into three-dimensional images. This technique is advantageous for visualizing the three-dimensional structure of microplastics and assessing their interactions with other environmental components (Tötzke et al. 2024). Confocal microscopy is particularly useful in ecological studies examining microplastics in biological tissues.

2.7.5 Infrared Microscopy

Infrared microscopy combines the principles of microscopy with infrared spectroscopy, enabling the simultaneous visualization and identification of microplastics (Fakayode et al. 2024). This technique provides spatially resolved chemical information, allowing researchers to correlate morphological features with polymer composition. Infrared microscopy enhances the comprehensive analysis of microplastics by integrating both structural and compositional data.

2.7.6 Staining Techniques

Staining methods, using dyes or fluorescent markers, enhance the visibility of microplastics under various microscopy techniques (Mallah et al. 2024). Staining aids in differentiating microplastics from background materials and organic matter.

Fluorescent staining, in particular, is valuable for improving the contrast between microplastics and their surroundings, facilitating more accurate identification and quantification.

2.7.7 Automated Image Analysis

With the advancement of technology, automated image analysis is increasingly employed in microscopy-based approaches. This involves the use of software algorithms to process and analyze large sets of microscopic images, improving the efficiency and consistency of microplastics identification and quantification (Fritz et al. 2024). Automated approaches complement manual methods and contribute to large-scale studies.

Despite their versatility, microscopy-based approaches face challenges such as time-consuming sample preparation, potential artifacts introduced during processing, and limitations in analyzing large sample volumes. Researchers must carefully consider these factors to ensure accurate and reliable results. Microscopy-based approaches are indispensable tools in the realm of microplastics analysis. Their ability to provide detailed morphological and structural information contributes significantly to our understanding of the distribution and impact of microplastics in the environment. As technology continues to advance, microscopy-based techniques will likely play an increasingly crucial role in advancing research and addressing the complex challenges associated with microplastic pollution.

2.8 Chromatographic Methods in Microplastics Detection

Chromatographic methods represent a sophisticated and powerful set of techniques utilized in the detection and analysis of microplastics (Kaur et al. 2024). These methods, which are rooted in the separation of components within a mixture based on their specific chemical and physical properties, have proven instrumental in providing detailed insights into the composition and characteristics of microplastic particles. Several chromatographic techniques are employed for microplastics detection, each contributing to a more comprehensive understanding of the extent and impact of plastic pollution.

2.8.1 Gas Chromatography (GC)

Gas chromatography (GC) is a widely employed technique in microplastics analysis, particularly for identifying and quantifying plastic additives and monomers (Jiménez-Skrzypek et al. 2024). GC separates volatile components by their vaporization and subsequent movement through a chromatographic column. By analyzing the elution times of specific compounds, researchers can identify and quantify the presence of plastic-related chemicals associated with microplastics.

2.8.2 Liquid Chromatography (LC)

Liquid Chromatography (LC), including techniques such as high-performance liquid chromatography (HPLC) and ultra-high-performance liquid chromatography (UHPLC), is employed for the separation and analysis of nonvolatile and semi-volatile components associated with microplastics (Nguyen et al. 2024). Liquid chromatography allows for the identification of plasticizers, dyes, and other chemical additives present in microplastic samples.

2.8.3 Solid-Phase Microextraction (SPME)

Solid-Phase Microextraction (SPME) is a sample preparation technique often coupled with gas chromatography (Murtada and Pawliszyn 2024). SPME involves the extraction of analytes from the sample matrix onto a solid-phase sorbent. This pre-concentration step enhances the sensitivity of GC analysis for microplastics, enabling the detection of trace levels of plastic-related compounds.

2.8.4 Pyrolysis-Gas Chromatography-Mass Spectrometry (Py-GC-MS)

Pyrolysis-gas chromatography-mass spectrometry (Py-GC-MS) is a specialized technique that combines pyrolysis, gas chromatography, and mass spectrometry to analyze the polymer composition of microplastics (Barber et al. 2024). Pyrolysis involves the thermal degradation of polymers into smaller fragments, and the resulting compounds are then separated and identified by GC-MS. This technique offers detailed information about the molecular structure of microplastics.

2.8.5 Size-Exclusion Chromatography (SEC)

Size-exclusion chromatography (SEC) is employed to separate microplastics based on their size and molecular weight (Subair et al. 2024). This technique utilizes a porous stationary phase to selectively allow smaller molecules, including microplastics, to enter the pores, while larger molecules are excluded. SEC aids in characterizing the size distribution of microplastic particles in a sample.

Chromatographic methods, while powerful, come with certain challenges. Sample preparation for chromatography requires meticulous attention to avoid contamination, and the complexity of environmental matrices can impact the accuracy of results. Additionally, the availability of standards for microplastic-related compounds is crucial for calibration and reliable quantification. Ongoing advancements in chromatographic technologies aim to improve the sensitivity, speed, and efficiency of microplastics analysis. The development of multidimensional chromatography and the integration of chromatographic techniques with mass spectrometry contribute to enhanced capabilities in identifying and quantifying diverse components associated with microplastics.

Chromatographic methods play a pivotal role in the detection and analysis of microplastics, offering a detailed examination of their chemical composition. These techniques provide critical information for understanding the sources, types, and environmental fate of microplastics, contributing to the broader efforts aimed at mitigating the impact of plastic pollution on ecosystems. As technology continues to evolve, chromatographic methods will likely remain at the forefront of innovative approaches in microplastics research.

2.9 Emerging Technologies for Microplastics Detection

As the issue of microplastic pollution continues to gain global attention, the development of innovative and advanced technologies for microplastics detection has become a key focus in environmental research (Yanuar et al. 2024; Pramanik et al. 2024). These emerging technologies aim to overcome existing challenges, enhance detection sensitivity, and provide more efficient and reliable methods for identifying and quantifying microplastics in diverse environmental matrices. Several cutting-edge approaches are shaping the future of microplastics detection:

2.9.1 Nanotechnology-Based Sensors

Nanotechnology plays a pivotal role in the development of sensors designed specifically for detecting microplastics (Mustapha et al. 2024). Nanoscale sensors can exhibit high sensitivity, enabling the identification of even minute concentrations of microplastics in environmental samples. These sensors often utilize unique interactions between nanomaterials and microplastics, offering rapid and selective detection capabilities.

2.9.2 Microfluidic Devices

Microfluidic devices are miniaturized systems that manipulate small volumes of fluids within channels, allowing for precise control and analysis (Suqi et al. 2024). In microplastics detection, these devices can be designed to efficiently concentrate, separate, and detect microplastics from complex environmental samples. Microfluidic platforms offer advantages such as reduced sample volumes, faster analysis times, and increased portability.

2.9.3 Hyperspectral Imaging

Hyperspectral imaging involves the acquisition of images across a wide range of wavelengths, providing spectral information for each pixel. In microplastics detection, hyperspectral imaging enables the identification of different polymers based on their unique spectral signatures (Hsu et al. 2024). This nondestructive technique offers the potential for rapid, remote sensing of microplastics in various environments.

2.9.4 Machine Learning and Artificial Intelligence

Machine learning and artificial intelligence (AI) algorithms are increasingly being integrated into microplastics detection processes (Kida et al. 2024). These technologies can automate the analysis of large datasets generated by microscopic or spectroscopic techniques, improving the efficiency and accuracy of microplastics identification. AI applications also enable the classification of microplastics based on their shape, size, and other characteristics.

2.9.5 DNA Barcoding

DNA barcoding involves using genetic markers to identify species. In the context of microplastics detection, researchers are exploring the use of DNA barcoding to trace the origin of plastic particles and understand their sources (Rishan et al. 2024). This method relies on the unique genetic signatures associated with different types of microplastics.

2.9.6 Advanced Spectroscopic Techniques

Continued advancements in spectroscopy, such as Terahertz spectroscopy and Raman spectroscopy, contribute to the refinement of microplastics detection (Withana et al. 2024). Terahertz spectroscopy offers nondestructive, subsurface imaging capabilities, while Raman spectroscopy provides enhanced molecular identification. These techniques contribute to a more comprehensive analysis of microplastics in various environmental samples.

2.9.7 Bioinformatics and Metagenomics

Bioinformatics and metagenomics approaches involve analyzing genetic information from environmental samples to understand microbial communities. These methods can be applied to study the interactions between microplastics and microbial life (Parida et al. 2024; Joshi et al. 2024; Joshi and Kaushik 2021). Understanding the microbial degradation of plastics provides insights into potential biodegradation strategies.

2.9.8 Autonomous Underwater Vehicles (AUVs) and Drones

Autonomous underwater vehicles (AUVs) and drones equipped with specialized sensors and imaging technologies are being employed for the remote and autonomous detection of microplastics in aquatic environments (Kwon et al. 2023). These technologies enhance survey capabilities and allow for efficient monitoring of large water bodies.

While these emerging technologies show promise in advancing microplastics detection, ongoing research is essential to validate their reliability, optimize their performance, and address potential limitations (Kieu and Law 2021). As the field continues to evolve, the integration of these innovative approaches holds the potential to revolutionize our ability to monitor and mitigate the impact of microplastic pollution on ecosystems worldwide.

2.10 Case Studies: Application of Detection Techniques

Microplastics, tiny plastic particles less than 5 mm in size, have become a pervasive environmental concern, posing risks to ecosystems and human health. Here case studies showcasing the application of various detection techniques to identify and quantify microplastic pollution in different environmental matrices are presented. Each case study highlights the specific challenges addressed, the detection methods employed, and the insights gained. By examining these real-world examples, we gain a deeper understanding of the efficacy and versatility of microplastic detection techniques in environmental monitoring and research.

2.10.1 Water Bodies: Monitoring Microplastic Pollution in Rivers

Rivers serve as conduits for the transport of microplastics from urban and industrial areas to oceans, making them crucial sites for monitoring plastic pollution. In this case study, researchers focused on a major river system impacted by urban runoff and industrial discharges. The research team utilized a combination of filtration methods and spectroscopic analysis to isolate and identify microplastic particles in water samples (Luo et al. 2019). Fourier-transform infrared spectroscopy (FTIR) and Raman spectroscopy were employed for polymer identification, while microscopy techniques aided in particle characterization. The detection techniques revealed the widespread presence of microplastics in the river, with varying concentrations along different stretches (Fu et al. 2020). The analysis also identified the predominant types of polymers present, providing insights into potential pollution sources and informing targeted mitigation strategies.

2.10.2 Soil: Assessing Microplastic Contamination in Agricultural Land

Agricultural activities can contribute to the accumulation of microplastics in soil, posing risks to ecosystem health and food safety. In this case study, researchers investigated the extent of microplastic contamination in agricultural soils used for crop cultivation. The research employed a combination of density separation techniques and microscopy analysis to extract and quantify microplastic particles from soil samples. Density separation methods, such as flotation and sedimentation, helped isolate microplastics from soil matrix components, followed by microscopy for particle identification and enumeration (Corradini et al. 2019). The study revealed the presence of microplastics in agricultural soils, with concentrations varying depending on factors such as land use practices and proximity to urban areas. By understanding the distribution of microplastics in soil, researchers can assess potential risks to crop health and explore mitigation strategies to reduce contamination.

2.10.3 Marine Environment: Mapping Microplastic Accumulation in Coastal Sediments

Coastal sediments act as sinks for microplastics transported by ocean currents, posing threats to marine ecosystems and biodiversity. This case study focused on assessing microplastic accumulation patterns in coastal sediments influenced by anthropogenic activities and natural processes (Qiu et al. 2015). The research team employed sediment coring techniques combined with microplastic extraction and characterization methods. Sediment cores were collected from multiple locations along the coastline and analyzed using microscopy and spectroscopic techniques to quantify microplastic abundance and identify polymer types. The study revealed significant accumulations of microplastics in coastal sediments, particularly in areas near urban centers and industrial facilities. By mapping microplastic distribution patterns, researchers gained insights into the sources and transport pathways of plastic pollution in marine environments, informing conservation efforts and policy interventions.

2.11 Future Trends in Microplastics Detection

Microplastics, defined as plastic particles less than 5 mm in size, have emerged as a global environmental concern due to their widespread presence in various ecosystems and potential impacts on wildlife and human health (Kamel et al. 2024). As research into microplastics detection continues to evolve, several future trends are emerging, driven by advancements in technology and analytical techniques. This chapter explores some of the key trends shaping the future of microplastics detection and monitoring.

2.11.1 Development of High-Throughput Screening Methods

Future trends in microplastics detection involve the development of high-throughput screening methods capable of analyzing large volumes of samples quickly and efficiently. These methods leverage automation, miniaturization, and robotics to streamline the detection process, enabling researchers to assess microplastic contamination levels across broader geographic scales and environmental matrices.

2.11.2 Integration of Spectroscopic and Imaging Techniques

Spectroscopic and imaging techniques, such as Fourier-transform infrared spectroscopy (FTIR), Raman spectroscopy, and hyperspectral imaging, are increasingly being integrated into microplastics detection workflows. These techniques offer enhanced capabilities for polymer identification, particle characterization, and spatial mapping of microplastic distribution in environmental samples. Integration of spectroscopic and imaging data allows for more comprehensive analysis and interpretation of microplastics data, facilitating a deeper understanding of their sources, fate, and impacts.

2.11.3 Advancements in Nanotechnology-Based Sensors

Nanotechnology-based sensors hold promise for the development of highly sensitive and selective detection platforms for microplastics. Functionalized nanomaterials, such as carbon nanotubes, graphene, and quantum dots, can be engineered to selectively bind to microplastic particles, enabling their rapid detection and quantification. Nanotechnology-based sensors offer advantages such as high sensitivity, low detection limits, and real-time monitoring capabilities, making them valuable tools for environmental surveillance and early warning systems.

2.11.4 Implementation of Machine Learning and Artificial Intelligence

Machine learning (ML) and artificial intelligence (AI) algorithms are increasingly being applied to microplastics detection and classification tasks, leveraging their ability to analyze complex datasets and identify patterns. ML and AI algorithms can assist in automated image analysis, pattern recognition, and data interpretation, accelerating the analysis process and enhancing the accuracy of microplastics detection results. Integration of ML and AI into microplastics detection workflows enables researchers to extract meaningful insights from large datasets and make data-driven decisions regarding microplastics management and mitigation strategies.

2.11.5 Multi-omics Approaches for Source Tracking

Multi-omics approaches, combining genomics, transcriptomics, proteomics, and metabolomics, hold promise for tracking the sources and pathways of microplastics in the environment. By analyzing the genetic, biochemical, and physiological responses of organisms exposed to microplastics, researchers can identify biomarkers indicative of exposure and develop molecular tools for source attribution. Multi-omics approaches offer a holistic understanding of the interactions between microplastics and biota, shedding light on their ecological impacts and informing strategies for pollution prevention and remediation.

The future of microplastics detection is characterized by the convergence of advanced technologies, interdisciplinary approaches, and data-driven methodologies. By embracing high-throughput screening methods, spectroscopic and imaging techniques, nanotechnology-based sensors, machine learning, and multi-omics approaches, researchers can enhance their ability to detect, characterize, and understand microplastics in the environment. These future trends hold the potential to drive innovation, improve analytical capabilities, and inform evidence-based strategies for mitigating the impacts of microplastic pollution on ecosystems and human health.

2.12 Environmental Implications and Policy Considerations

As the issue of environmental degradation continues to gain global attention, it is essential to understand the implications of human activities on the environment and formulate effective policy measures to address these challenges (Yusuf et al. 2022; Khan et al. 2022). This chapter explores the environmental implications of various human activities and discusses key policy considerations for promoting sustainable development and conservation efforts.

2.12.1 Climate Change and Global Warming

Climate change, driven primarily by greenhouse gas emissions from human activities such as burning fossil fuels and deforestation, poses significant threats to ecosystems, biodiversity, and human well-being. Rising temperatures, changing precipitation patterns, and extreme weather events are leading to shifts in habitats, species distributions, and ecosystem functions, impacting agriculture, water resources, and coastal communities. Policy responses to climate change include mitigation measures to reduce greenhouse gas emissions, such as transitioning to renewable energy sources, improving energy efficiency, and implementing carbon pricing mechanisms. Additionally, adaptation strategies are necessary to enhance resilience to climate impacts, including investments in infrastructure, ecosystem restoration, and disaster preparedness.

2.12.2 Biodiversity Loss and Habitat Destruction

Biodiversity loss and habitat destruction, resulting from activities such as deforestation, urbanization, and industrial agriculture, threaten the stability and resilience of ecosystems. Loss of habitat leads to species extinction, disrupts ecosystem services, and diminishes genetic diversity, compromising ecosystem functions and resilience to environmental change. Policy interventions to address biodiversity loss and habitat destruction include protected area management, habitat restoration, and sustainable land use planning. Conservation strategies should prioritize the preservation of critical habitats, restoration of degraded ecosystems, and promotion of biodiversity-friendly agricultural practices. Integrating biodiversity considerations into development planning and decision-making processes is essential for achieving sustainable development goals.

2.12.3 Pollution and Environmental Contamination

Pollution from industrial activities, transportation, agriculture, and waste disposal contaminates air, water, and soil, posing risks to human health and ecosystem integrity. Air pollution contributes to respiratory diseases, cardiovascular problems, and premature mortality, while water pollution affects aquatic ecosystems, wildlife, and drinking water quality. Policy measures to address pollution and environmental contamination include regulatory frameworks, pollution control technologies, and pollution prevention strategies. Implementing pollution control laws, enforcing emission standards, and promoting cleaner production practices are essential for reducing pollution levels and protecting environmental health. Integrated approaches that consider the interconnectedness of environmental media and address pollution sources holistically are critical for achieving long-term environmental sustainability.

2.12.4 Resource Depletion and Overexploitation

Overexploitation of natural resources, including freshwater, forests, fisheries, and minerals, threatens the availability and quality of essential ecosystem services. Unsustainable resource extraction leads to habitat degradation, loss of biodiversity, and ecosystem collapse, exacerbating environmental degradation and undermining the resilience of ecosystems and human societies. Policy responses to resource depletion and overexploitation involve sustainable resource management, conservation strategies, and circular economy initiatives. Implementing sustainable resource extraction practices, promoting resource efficiency, and fostering resource recycling and reuse are essential for reducing environmental pressures and promoting long-term resource security. Strengthening governance frameworks, enhancing stakeholder engagement, and fostering international cooperation are critical for addressing transboundary resource management challenges.

Addressing environmental implications requires concerted efforts at local, national, and global levels, guided by effective policy measures and informed by scientific evidence and stakeholder engagement. By integrating environmental considerations into decision-making processes, promoting sustainable development practices, and fostering collaboration among governments, businesses, civil society, and academia, we can safeguard the environment, protect human health, and ensure a sustainable future for generations to come.

2.13 Conclusion

Microplastic pollution represents a complex and pervasive environmental challenge with far-reaching implications for ecosystems, wildlife, and human health. The development and application of effective detection techniques are crucial for accurately assessing the extent and impacts of microplastic contamination in various environmental matrices. In conclusion, the advancements in microplastic detection techniques hold promise for enhancing our understanding of microplastic pollution and informing evidence-based strategies for mitigation and management. The future of microplastic detection is characterized by the convergence of innovative technologies, interdisciplinary approaches, and data-driven methodologies. High-throughput screening methods, spectroscopic and imaging techniques, nanotechnology-based sensors, machine learning, and multi-omics approaches offer valuable tools for detecting, characterizing, and tracking microplastics in the environment. These techniques enable researchers to analyze complex samples, identify microplastic particles, and elucidate their sources, distribution patterns, and ecological impacts. Effective microplastic detection requires collaboration and coordination among scientists, policymakers, industries, and civil society stakeholders. By fostering partnerships and sharing knowledge and resources, we can accelerate the development and adoption of standardized protocols and quality assurance measures for microplastic detection. Moreover, investing in research and innovation is essential for advancing detection technologies, improving analytical capabilities, and addressing knowledge gaps related to microplastic pollution. Policy considerations play a critical role in supporting microplastic detection efforts and driving proactive measures to reduce plastic pollution. Policymakers need to prioritize the development and implementation of regulations and incentives aimed at curbing plastic production, promoting sustainable consumption and waste management practices, and fostering the transition to a circular economy. Additionally, integrating microplastic monitoring and reporting requirements into environmental monitoring programs can enhance transparency and accountability in addressing plastic pollution at local, national, and global scales.

References

Akarsu C, Özdemir S, Ozay Y, Acer Ö, Dizge N (2023) Investigation of two different size microplastic degradation ability of thermophilic bacteria using polyethylene polymers. Environ Technol 44(24):3710–3720

Barber TR, Claes S, Ribeiro F, Dillon AE, More SL, Thornton S et al (2024) Abundance and distribution of tire and road wear particles in the Seine River, France. Sci Total Environ 913:169633

Binda G, Kalčíková G, Allan IJ, Hurley R, Rødland E, Spanu D, Nizzetto L (2024) Microplastic aging processes: environmental relevance and analytical implications. TrAC Trends Anal Chem 172:117566

Bydalek F, Webster G, Barden R, Weightman AJ, Kasprzyk-Hordern B, Wenk J (2023) Microplastic biofilm, associated pathogen and antimicrobial resistance dynamics through a wastewater treatment process incorporating a constructed wetland. Water Res 235:119936

Circelli L, Cheng Z, Garwood E, Yuksel K, Di Iorio E, Angelico R, Colombo C (2024) Comparison of ATR-FTIR and NIR spectroscopy for identification of microplastics in biosolids. Sci Total Environ 916:170215

Corradini F, Meza P, Eguiluz R, Casado F, Huerta-Lwanga E, Geissen V (2019) Evidence of microplastic accumulation in agricultural soils from sewage sludge disposal. Sci Total Environ 671:411–420

De Arbeloa NP, Marzadri A (2024) Modeling the transport of microplastics along river networks. Sci Total Environ 911:168227

Dris R, Beaurepaire M, Bouzid N, Stratmann C, Nguyen MT, Bordignon F et al (2024) Sampling and analyzing microplastics in rivers: what methods are being used after a decade of research? In: Microplastic contamination in aquatic environments. Elsevier, pp 65–91

Duis K, Coors A (2016) Microplastics in the aquatic and terrestrial environment: sources (with a specific focus on personal care products), fate and effects. Environ Sci Eur 28(1):1–25

Evode N, Qamar SA, Bilal M, Barceló D, Iqbal HM (2021) Plastic waste and its management strategies for environmental sustainability. Case Stud Chem Environ Eng 4:100142

Fakayode SO, Mehari TF, Fernand Narcisse VE, Grant C, Taylor ME, Baker GA et al (2024) Microplastics: challenges, toxicity, spectroscopic and real-time detection methods. Appl Spectrosc Rev, 1–95

Fältström E, Carlsson A (2024) Exploring how municipalities address microplastics pollution in stormwater—a case study in a Swedish municipality. J Environ Plan Manag, 1–19

Fang C, Luo Y, Naidu R (2023) Microplastics and nanoplastics analysis: options, imaging, advancements and challenges. TrAC Trends Anal Chem 166:117158

Fritz M, Deutsch LF, Wijaya KP, Götz T, Fischer CB (2024) An image-processing tool for size and shape analysis of manufactured irregular polyethylene microparticles. Microplastics 3(1):124–146

Fu W, Min J, Jiang W, Li Y, Zhang W (2020) Separation, characterization and identification of microplastics and nanoplastics in the environment. Sci Total Environ 721:137561

Gayathri NP, Priya V, Prabhakaran V, Prasad G (2024) Understanding the impact of microplastic contamination on soil quality and eco-toxicological risks in horticulture: a comprehensive review. Case Stud Chem Environ Eng 9:100633

Ghanadi M, Joshi I, Dharmasiri N, Jaeger JE, Burke M, Bebelman C et al (2024) Quantification and characterization of microplastics in coastal environments: insights from laser direct infrared imaging. Sci Total Environ 912:168835

Gheorghe S, Stoica C, Harabagiu AM, Neidoni DG, Mighiu ED, Bumbac C et al (2024) Laboratory assessment for determining microplastics in freshwater systems—characterization and identification along the Somesul Mic River. Water 16(2):233

Ho D, Liu S, Wei H, Karthikeyan KG (2023) The glowing potential of Nile red for microplastics identification: science and mechanism of fluorescence staining. Microchem J 197:109708

Hsu YJ, Huang C, Lee M (2024) Unveiling microplastic spectral signatures under weathering and digestive environments through shortwave infrared hyperspectral sensing. Environ Pollut 342:123106

Hu JZ, Qiao L, Zhao X, Liu CJ, Hu G (2024) Visualizing mitochondrial budding in intact BS-C-1 cells by in situ cryo-electron tomography. Biophys J 123(3):419A

Jiménez-Skrzypek G, Lusiardi R, González-Sálamo J, Vega-Moreno D, Hernández-Borges J (2024) Insights into emerging organic pollutants extraction from polypropylene, polystyrene, and polyethylene microplastics. Anal Chim Acta 1287:342071

Jones NR, de Jersey AM, Lavers JL, Rodemann T, Rivers-Auty J (2024) Identifying laboratory sources of microplastic and nanoplastic contamination from the air, water, and consumables. J Hazard Mater 465:133276

Joshi A, Kaushik V (2021) Big data and its analytics in agriculture. In: Bioinformatics for agriculture: high-throughput approaches, pp 71–83

Joshi A, Kumar A, Kaushik V (2024) Functional genomics and network biology. In: Advances in bioinformatics. Springer Nature Singapore, Singapore, pp 71–96

Kamel AH, Hefnawy A, Hazeem LJ, Rashdan SA, Abd-Rabboh HS (2024) Current perspectives, challenges, and future directions in the electrochemical detection of microplastics. RSC Adv 14(3):2134–2158

Kaur K, Singh R, Kaur G (2024) A comparative study of ethylene detection methods in fruit supply chains: a review. Food Anal Methods 17(1):14–32

Kaushik A, Gupta P, Kumar A, Saha M, Varghese E, Shukla G et al (2024) Identification and physico-chemical characterization of microplastics in marine aerosols over the northeast Arabian Sea. Sci Total Environ 912:168705

Khan NA, Khan AH, López-Maldonado EA, Alam SS, López JRL, Herrera PFM et al (2022) Microplastics: occurrences, treatment methods, regulations and foreseen environmental impacts. Environ Res 215:114224

Kida M, Pochwat K, Ziembowicz S (2024) Assessment of machine learning-based methods predictive suitability for migration pollutants from microplastics degradation. J Hazard Mater 461:132565

Kieu HT, Law AWK (2021) Remote sensing of coastal hydro-environment with portable unmanned aerial vehicles (pUAVs) a state-of-the-art review. J Hydro Environ Res 37:32–45

Kong X, Zhou A, Chen X, Cheng X, Lai Y, Li C et al (2024) Insight into the adsorption behaviors and bioaccessibility of three altered microplastics through three types of advanced oxidation processes. Sci Total Environ 917:170420

Kwon DY, Kim J, Park S, Hong S (2023) Advancements of remote data acquisition and processing in unmanned vehicle technologies for water quality monitoring: an extensive review. Chemosphere 343:140198

Lin B, Wang L, Chen Q, Liu Z, Liu B, Wen S et al (2024) Health assessment based on exposure to microplastics in tropical agricultural soil. J Hazard Mater 465:133372

Liu Z, Wang W, Liu X (2023) Automated characterization and identification of microplastics through spectroscopy and chemical imaging in combination with chemometric: latest developments and future prospects. TrAC Trends Anal Chem 160:116956

Luo W, Su L, Craig NJ, Du F, Wu C, Shi H (2019) Comparison of microplastic pollution in different water bodies from urban creeks to coastal waters. Environ Pollut 246:174–182

Macairan JR, Nguyen B, Li F, Tufenkji N (2023) Tissue clearing to localize microplastics via three-dimensional imaging of whole organisms. Environ Sci Technol 57:8476

Mallah MA, Yu Z, Qu Z, Ali M, Liu N (2024) Recognition and detection technology for microplastic, its source and health effects. Environ Sci Pollut Res 31(8):11428–11452

Martín-Gómez B, Elmore JS, Valverde S, Ares AM, Bernal J (2024) Recent applications of chromatography for determining microplastics and related compounds (bisphenols and phthalate esters) in food. Microchem J 197:109903

Mikulec V, Adamović P, Cvetković Ž, Ivešić M, Gajdoš Kljusurić J (2023) Green techniques for detecting microplastics in marine with emphasis on FTIR and NIR spectroscopy—short review. Processes 11(8):2360

Mingming H, Huaqiang L, Zhu T, Chenxi Z, Zihan Z, Jiang Q et al (2024) Toxic effects of micro (nano)-plastics on terrestrial ecosystems and human health. TrAC Trends Anal Chem 172:117517

Mondal C, Mondal N, Moon U, Biswas SJ, Hossain A (2024) Contamination of microplastics in the marine food web with special reference to seafood: present status and future concern. In: Spatial modeling of environmental pollution and ecological risk. Woodhead Publishing, Sawston, pp 175–207

Morgana S, Casentini B, Tirelli V, Grasso F, Amalfitano S (2024) Fluorescence-based detection: a review of current and emerging techniques to unveil micro/nanoplastics in environmental samples. TrAC Trends Anal Chem 172:117559

Murtada K, Pawliszyn J (2024) Solid-phase microextraction. In: Microextraction techniques. Springer, Cham, pp 85–116

Mustapha S, Tijani JO, Elabor R, Salau RB, Egbosiuba TC, Amigun AT et al (2024) Technological approaches for removal of microplastics and nanoplastics in the environment. J Environ Chem Eng 12:112084

Nguyen TT, Nguyen DV, Tran QH, Pham MD, Nguyen VM, Nguyen TT et al (2024) Choline chloride based natural deep eutectic solvents coupling with ultra-high performance liquid chromatography-tandem mass spectroscopy for effective extraction and rapid detection of adenosine and cordycepin in Cordyceps millitaris. J Mol Liquids 397:124107

Niu L, Chen Y, Li Y, Wang Y, Shen J, Wang L et al (2023) Diversity, abundance and distribution characteristics of potential polyethylene and polypropylene microplastic degradation bacterial communities in the urban river. Water Res 232:119704

Norling M, Hurley R, Schell T, Futter MN, Rico A, Vighi M et al (2024) Retention efficiency for microplastic in a landscape estimated from empirically validated dynamic model predictions. J Hazard Mater 464:132993

Parida D, Katare K, Ganguly A, Chakraborty D, Konar O, Nogueira R, Bala K (2024) Molecular docking and metagenomics assisted mitigation of microplastic pollution. Chemosphere 351:141271

Pramanik DD, Kay P, Goycoolea FM (2024) A rapid and portable fluorescence spectroscopy staining method for the detection of plastic microfibers in water. Sci Total Environ 908:168144

Qiu Q, Peng J, Yu X, Chen F, Wang J, Dong F (2015) Occurrence of microplastics in the coastal marine environment: first observation on sediment of China. Mar Pollut Bull 98(1–2):274–280

Rangel-Buitrago N, Gracia A (2024) From the closet to the shore: fashion waste pollution on Colombian Central Caribbean beaches. Mar Pollut Bull 199:115976

Ribeiro F, Duarte AC, da Costa JP (2024) Staining methodologies for microplastics screening. TrAC Trends Anal Chem 172:117555

Rishan ST, Kline RJ, Rahman MS (2024) Exploitation of environmental DNA (eDNA) for ecotoxicological research: a critical review on eDNA metabarcoding in assessing marine pollution. Chemosphere 351:141238

Roy P, Mohanty AK, Misra M (2022) Microplastics in ecosystems: their implications and mitigation pathways. Environ Sci Adv 1(1):9–29

Saeedi M (2024) How microplastics interact with food chain: a short overview of fate and impacts. J Food Sci Technol 61(3):403–413

Shen T, Ma H, Xing B (2024) Interfacial interactions of polyethylene terephthalate microplastics and malachite green, tetracycline in aqueous environments. Mar Pollut Bull 200:116093

Singh B, Kumar A (2023) Advances in microplastics detection: a comprehensive review of methodologies and their effectiveness. TrAC Trends Anal Chem 170:117440

Steiner T, Leitner LC, Zhang Y, Möller JN, Löder MG, Greiner A et al (2024) Detection and specific chemical identification of submillimeter plastic fragments in complex matrices such as compost. Sci Rep 14(1):2282

Subair A, Krishnamoorthy Lakshmi P, Chellappan S, Chinghakham C (2024) Removal of polystyrene microplastics using biochar-based continuous flow fixed-bed column. Environ Sci Pollut Res Int 31(9):13753–13765

Suqi W, Rong CLJ, Abdulbari HA, Mahmood WK (2024) Microfluidic approaches in water quality monitoring: an insight and a comprehensive review. ChemBioEng Rev 11:215

Surana M, Pattanayak DS, Yadav V, Singh VK, Pal D (2024) An insight decipher on photocatalytic degradation of microplastics: mechanism, limitations, and future outlook. Environ Res 247:118268

Thacharodi A, Meenatchi R, Hassan S, Hussain N, Bhat MA, Arockiaraj J et al (2024) Microplastics in the environment: a critical overview on its fate, toxicity, implications, management, and bioremediation strategies. J Environ Manag 349:119433

Tötzke C, Kozhuharova B, Kardjilov N, Lenoir N, Manke I, Oswald SE (2024) Non-invasive 3D analysis of microplastic particles in sandy soil—exploring feasible options and capabilities. Sci Total Environ 907:167927

Vitali C, Peters RJ, Janssen HG, Undas AK, Munniks S, Ruggeri FS, Nielen MW (2024) Quantitative image analysis of microplastics in bottled water using artificial intelligence. Talanta 266:124965

Wei M, Zhang Y, Wang Y, Liu X, Li X, Zheng X (2024) Employing atomic force microscopy (AFM) for microscale investigation of interfaces and interactions in membrane fouling processes: new perspectives and prospects. Membranes 14(2):35

Wirnkor VA, Ebere EC, Ngozi VE (2019) Microplastics, an emerging concern: a review of analytical techniques for detecting and quantifying microplatics. Anal Methods Environ Chem J 2:13–30

Withana PA, Senadheera SS, He D, Bank MS, Gu C, Hwang SY, Ok YS (2024) Soil microplastic analysis: a harmonized methodology. Crit Rev Environ Sci Technol, 1–26

Yanuar AT, Pramudia Z, Susanti YAD, Kurniawan A (2024) Analysis of microplastics in spring water. Emerg Contam 10(1):100277

Yasri S, Wiwanitkit V (2023) Scattering (X-ray, light, neutron) studies of PVC composites and nanocomposites. In: Poly (vinyl chloride) based composites and nanocomposites. Springer International Publishing, Cham, pp 275–291

Yusuf A, Sodiq A, Giwa A, Eke J, Pikuda O, Eniola JO et al (2022) Updated review on microplastics in water, their occurrence, detection, measurement, environmental pollution, and the need for regulatory standards. Environ Pollut 292:118421

Zhang J, Wang Y, Feng H, Xue Y, Sha X, Fu D, Wang L (2024) Analysis and detection methods of microplastics in the environment. In: Microplastic contamination in aquatic environments. Elsevier, pp 33–63

Zhu J, Wakisaka M, Omura T, Yang Z, Yin Y, Fang W (2024) Advances in industrial harvesting techniques for edible microalgae: recent insights into sustainable, efficient methods and future directions. J Clean Prod 436:140626

3 Microplastics in Soil: Unveiling the Threat Within the Earth's Crucial Ecosystem

Monika Yadav and Rajendra Kumar

Abstract

Plastic particles smaller than 5 mm in size are known as microplastics, which are considered as major global environmental contaminants affecting many different ecosystems. The sources, impacts, and possible remediation techniques of microplastics in solid environments are examined in this chapter. Microplastics in solid environments come from a variety of human activities, including the decomposition of bigger plastic trash, industrial processes, and plastic littering. Microplastics have a variety of negative effects on solid environments, including harm to wildlife, ecosystems, and human health. Microplastics have the potential to have cascade effects on ecosystem functioning by altering nutrient cycling, upsetting soil structure, and impairing soil biota. Microplastic consumption can lead to bioaccumulation and biomagnification across the food chain, endangering higher trophic level species including humans. Numerous remediation techniques, including biological techniques like phytoremediation and microbial degradation, as well as physical techniques like filtration, sediment dredging, and soil amendments, reduce the effects of microplastics in solid environment. Addressing the underlying causes of microplastic contamination requires significant efforts to decrease the amount of plastic trash that is introduced into the environment through better waste management techniques and legislative changes.

Keywords

Health effects · Microplastics · Remediation · Sources

M. Yadav (✉) · R. Kumar
Department of Applied Sciences and Humanities, Faculty of Engineering and Technology, Rama University, Kanpur, UP, India

A. Kumar, V. Singh (eds.), *Microplastics Pollution and its Remediation*,
https://doi.org/10.1007/978-981-97-4068-0_3

3.1 Introduction

In the contemporary era, the presence of plastic pollution has emerged as a crucial global concern, transcending geographical boundaries and infiltrating various ecosystems. While much attention has been devoted to the detrimental impacts of plastic waste in oceans and water bodies, a relatively understudied yet equally critical dimension of this crisis unfolds in the soil—the very foundation of terrestrial life (Surendran et al. 2023). The insidious infiltration of microplastics into soil ecosystems has become an alarming environmental challenge, demanding urgent scientific scrutiny and concerted efforts for mitigation (Alimi et al. 2018).

Microplastics, defined as plastic particles with a diameter less than 5 mm, are derived from the fragmentation of larger plastic debris or are intentionally manufactured at microscopic scales for various industrial purposes (Zobkov and Esiukova 2018). Their presence in soil represents a relatively recent but growing area of concern, raising questions about the consequences for soil health, biodiversity, and the intricate web of life supported by this essential component of the Earth's ecosystem (Loder and Gerdts 2015). This introduction aims to shed light on the origins, distribution, and potential ecological ramifications of microplastics in soil, emphasizing the need for a comprehensive understanding and proactive measures to address this hidden menace along with their different sources and transport (Schell et al. 2020).

The current study is comprehensive in its approach, aiming to provide insights into the existence of microplastics in soil and their multifaceted impacts on humans, plants, soil composition, and ecosystems. This type of research is crucial for developing informed strategies to address and mitigate the potential negative consequences of microplastic pollution in the environment.

3.1.1 Origins of Microplastics in Soil

The introduction of microplastics into soil can be attributed to a complex interplay of anthropogenic activities, atmospheric deposition, and the degradation of larger plastic items. Agricultural practices, industrial activities, and improper waste management contribute significantly to the introduction of microplastics into the soil matrix (Campanale et al. 2020). For instance, using plastic mulching in agriculture is a popular technique to increase crop output, but during deterioration, microplastics are unintentionally released into the soil. Investigations on plastic contamination of agricultural soils have just lately begun, despite the fact that it accounts for a sizeable portion (14%) of the world's plastic pollution (Butbunchu and Pathom-Aree 2019). Additionally, the breakdown of plastic waste in landfills, coupled with the fragmentation of plastic debris, further amplifies the presence of microplastics in soil. Moreover, atmospheric deposition serves as a substantial pathway for microplastics to infiltrate soil ecosystems (Brennecke et al. 2016). Airborne microplastic particles, transported over long distances, settle on the land surface through precipitation or dry deposition. This introduces microplastics into diverse terrestrial ecosystems, ranging from urban areas to remote wilderness. As a result, even regions far removed from direct sources of plastic pollution exhibit traces of microplastics in their soils (de Souza Machado et al. 2018). Once introduced, these minute particles exhibit remarkable staying power due to their resistance to biodegradation. Unlike larger plastic items that eventually break down into smaller components, microplastics persist in the environment for extended periods, interacting with soil particles, organic matter, and microbial communities. The distribution of microplastics within soil profiles is not uniform, with concentrations varying based on factors such as land use, proximity to pollution sources, and soil characteristics (Fu et al. 2022; Saccá et al. 2017).

The infiltration of microplastics into soil ecosystems poses multifaceted threats to soil health, biodiversity, and the delicate balance of ecological processes. As microplastics intermingle with soil particles, they can alter the physical structure of the soil, affecting water retention, aeration, and nutrient availability (Bostan et al. 2023). This disruption in soil properties may have cascading effects on plant growth, microbial communities, and the overall functionality of terrestrial ecosystems. Furthermore, the impact of microplastics extends beyond the abiotic components of soil to encompass the diverse array of organisms inhabiting this environment. Soil-dwelling organisms, including earthworms, insects, and microorganisms, may inadvertently ingest or come into contact with microplastics (Crichton 2023; Muhammad et al. 2020). The consequences of such interactions are still being unraveled, but emerging evidence suggests potential adverse effects on reproduction, behavior, and overall population dynamics.

The intricate web of connections between soil, microplastics, and human activities becomes particularly pronounced when considering the potential implications for food security. Microplastics present in soil may be taken up by plants, entering the food chain and ultimately reaching human consumers. The pathways through which microplastics translocate within plants, and the potential impacts on crop quality and safety, are areas of active research (Bostan et al. 2023; Akdogan and Guven 2019). As a worldwide issue, developing practical solutions to address the problems associated with microplastic pollution is imperative. Significant initiatives have been made, such as the creation of bio-based feedstock, enhanced recycling capability, and material reduction. A variety of agreements and rules have also been put into place as tools to deal with the problem of microplastics. The European Union has launched a comprehensive plastic plan to control the flow of microplastics into streams and minimize microplastic contamination from water sources (Browne et al. 2011). Identifying the entire range of microplastic types, shapes, morphology, and surface chemistry presents the first challenge in eliminating these pollutants. It might be possible to develop techniques for locating, identifying, and characterizing microplastics in environmental matrices in order to give precise information on its properties, concentrations, and toxicity (Goh et al. 2022; Huang et al. 2023).

3.2 Sources of Microplastics

Microplastics in soil originate from a wide range of indirect environmental processes as well as direct human activity. There are several direct and indirect sources of these microplastics. Plastic pollution from land-based sources is one of the main causes of microplastics in soil. Illegal dumping and other improper waste disposal practices lead to the accumulation of macroplastics in the soil, which then decompose into microplastics. The application of plastic mulching in agriculture is another cause of microplastics in soil. Plastic films are used in agriculture for weed control and soil moisture retention, among other uses. These plastic films break down into tiny plastic particles over time, which are then released into the soil (Chae and An 2018; Auta et al. 2017). Microplastics are released and redistributed in soil by maritime activities including shipping and fishing. Through processes like erosion and sediment deposition, microplastics that are introduced into the marine environment by these activities may eventually find their way to coastal areas and soil. Moreover, there are other sources of microplastic contamination in soil than marine ones (Wanner 2021; Cole et al. 2011).

There is increasing evidence suggesting that freshwater and terrestrial systems are significant contributors to microplastic pollution in soil. While there is limited data on the exact sources, transport routes, and transformations of microplastics in freshwater and terrestrial systems, it is believed that most of the nano- and microplastic pollution in these environments also originates from land-based sources.

There are some prominent sources also shown in Fig. 3.1. Ineffective waste management techniques, such as illegal dumping, littering, and inadequately managed landfills, significantly raise the amount of macroplastics in the soil, which then decompose into microplastics. Microplastics are released into the soil as a consequence of the long-term fragmentation of plastic films used in agriculture, such as mulching to suppress weeds and preserve soil moisture (Fu et al. 2022). Long distances can be covered by microplastics traveling through the atmosphere. Microplastic particles are brought to the ground by atmospheric deposition, both wet and dry, and this helps explain their existence in soil that is remote from direct sources of pollution (Mao et al. 2022). Urban areas play a major role in the soil's microplastic contamination. Microplastics from roadways and urban surfaces can be carried into adjacent soil. Microplastics can be found in wastewater treatment plant effluents, and they can also be incorporated into soil through the use of sewage sludge as fertilizer in agricultural practices. Microplastics are introduced into the soil when biosolids or treated sewage sludge are used as fertilizer in agriculture. Microplastics from urban wastewater may be present in biosolids (Corradini et al. 2019). Larger plastic objects, such bottles and packaging, are subject to mechanical deterioration and weathering over time as a result of exposure to sunshine, temperature changes, and external forces. Microplastics produced by this process have the potential to seep into the soil. While washing synthetic fabrics like nylon and polyester, microplastics are released. Through the application of biosolids or other routes, these microplastics have the potential to enter wastewater systems and ultimately make their way into soil. Microplastics are directly present in soil due to

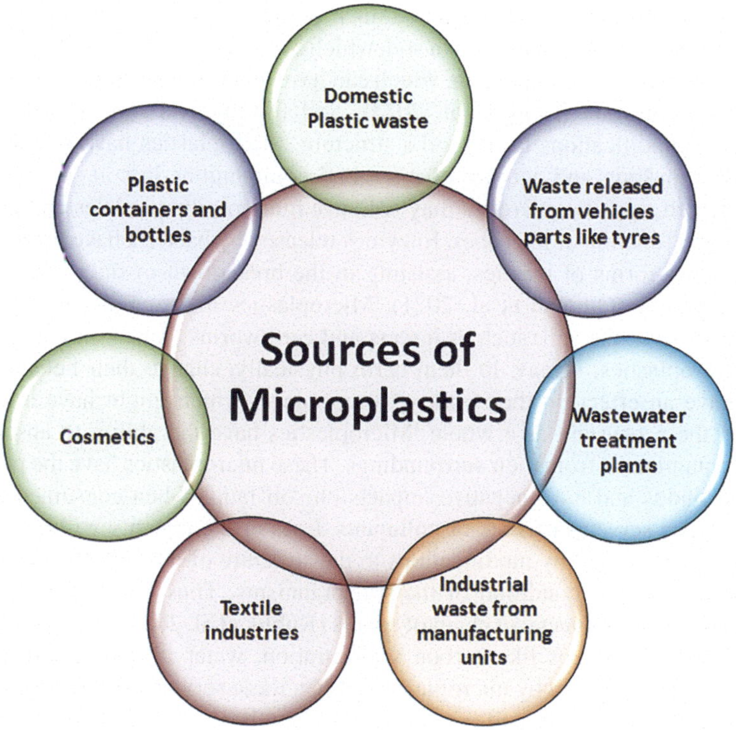

Fig. 3.1 Various sources of microplastics in environment

inappropriate disposal of a variety of plastic-containing products, including packaging materials, disposable objects, and broken plastic items (Hansen et al. 2013). Microplastics from fields may wash into adjacent soils as a result of agricultural practices, including the use of plastic-based agrochemical packaging, particularly during periods of heavy rainfall. To minimize the introduction of microplastics into soil ecosystems, focused methods must take into account the multitude of sources (Gan et al. 2023). Microplastics possess mainly two types of chemicals including raw materials of polymers and additives from plastics and the chemicals absorbed from surrounding environment (Campanale et al. 2020; Koelmans et al. 2016). Efforts should focus on improved waste management, sustainable agricultural practices, and regulatory measures to mitigate the impact of microplastics on soil health and biodiversity.

3.3 Effects of Microplastics

Microplastics in soil have a complex impact on many elements of ecosystem health, soil health, and possibly even human health. Our knowledge of these consequences is continually growing because research in this area is still developing. The following are some significant ways that microplastics affect soil: Microplastics can change the physical characteristics of soil, which can affect its structure. Microplastic buildup can cause soil compaction, which can have an impact on drainage, porosity, and water infiltration (Deng et al. 2023). Soil fertility and root growth can be affected by modifications to the soil's structure. Microplastics have the ability to change the makeup and activity of microbial communities in soil. According to some research, specific microbes may colonize microplastic particles and alter the diversity and function of microbes. Enzymes released by bacteria have the ability to degrade some forms of plastics, assisting in the breakdown of microplastics into smaller particles (Othman et al. 2021). Microplastics may interact with invertebrates that live in the soil, such as insects and earthworms. When these organisms ingest microplastics, it may do them harm physically, change their behavior, and maybe have an effect on their ability to reproduce, which might have a domino effect on the ecosystem as a whole. Microplastics have the ability to absorb and gather contaminants from their surroundings. These microplastics have the potential to be poisonous and have negative impacts on soil fauna when consumed by soil organisms and serve as carriers of pollutants. Loss of biodiversity within soil ecosystems can be caused by modifications to the structure of the soil, the availability of nutrients, and the condition of the soil organisms. Thus, the larger terrestrial ecosystem may be subject to a chain of events (Kublik et al. 2022). The provision of vital ecosystem services like carbon sequestration, water filtration, and nutrient cycling may be impacted by microplastics. When these services are interrupted, the well-being of people may suffer greatly. Plant uptake of microplastics in the soil may allow them to enter the food chain. There is still more research to be done on the effects on human health, particularly when it comes to eating infected crops. Concerns over long-term soil deposition are raised by the persistent nature of microplastics in the environment (Chang et al. 2022). As a result, the effects of continuous exposure to microplastics over extended periods are not fully understood. The effects of microplastics on soil are complex and interconnected, involving physical, chemical, and biological processes (Fig. 3.2). Addressing the challenges posed by microplastics in soil requires a holistic understanding of their interactions with the soil environment and the implementation of sustainable practices to mitigate their impact on soil health and ecosystem functioning.

3.3.1 Effect on Humans

The health impacts of microplastics in soil are an emerging area of concern, and research in this field is ongoing. While the direct effects on human health are not yet fully understood, there are potential pathways through which microplastics in soil may pose risks to ecosystems and, subsequently, to human well-being (Schirinzi et al. 2017). Here are key considerations regarding the potential health impacts:

Food Chain Contamination: Plants have the ability to absorb microplastics from the soil, which can result in their accumulation in edible sections. There is a chance that eating of these contaminated plants could lead to the entry of

Fig. 3.2 Effects of microplastics in environment

microplastic particles into the human body through the food chain. A number of activity, including construction, agriculture, and wind erosion, can release microplastics into the atmosphere. Dust particles containing microplastics may be inhaled as a result, possibly resulting in respiratory exposure. Microplastics possess the ability to absorb and gather diverse chemical contaminants. Microplastics may act as a vector for the spread of toxins in the environment if these chemicals are persistent organic pollutants (POPs) or other dangerous substances. According to investigation 159 worldwide sources of tap water testing shows 81% microplastic contamination with less than 5 mm measurement (Yee et al. 2021). Earthworms and other soil-dwelling creatures may consume microplastics as they consume the organic materials in the soil. The possibility of microplastic transfer to higher trophic levels, including humans, arises if these species are a component of the food chain. The potential toxicity of microplastics may be influenced by their small size and variety of forms (Avio et al. 2015). According to certain research, particles that are smaller in size might possess a greater surface area in comparison to their mass, which could improve their capacity to absorb and move pollutants. It has been suggested that microplastics may cause inflammatory reactions in living things

(Cverenkárová et al. 2021). Although the reasons and implications of inflammatory reactions are still being studied, they may have an impact on immune function and general health. **Biomagnification**: There is concern regarding biomagnification, which occurs when microplastic concentrations rise up the trophic levels, if microplastics enter into the food chain. Microplastic levels may rise as a result in the tissues of higher trophic level species, such as humans. Within an organism in a food web, the concentration of microplastics increases in relation to its prey (Miller and Kroon 2020). Depending on future predictions of microplastics, species-dependent food chains, and geographic scales, the biomagnification process of microplastics may vary. According to Karbalaei et al. (2018), humans may be affected by biomagnification as a result of microplastics generated by exfoliants, medication delivery particles in medicine, cosmetics, etc. **Gastrointestinal Absorption**: Although there is not much of research on the matter, there exists assumption on the possibility that microplastics consumed through food could enter the bloodstream and be absorbed in the gastrointestinal tract (Wright and Kelly 2017). We still don't fully understand what such a translocation will mean down the road. It's important to remember that knowledge about how microplastics in soil affect human health is still developing, and further study is required to determine the full scope of the risks (Glausiusz 2014). In order to reduce possible harm to ecosystems and human health, efforts should be directed at minimizing the input of microplastics into soil and adopting sustainable waste management procedures.

3.3.2 Effect on Plants

Microplastics in the soil can impact a variety of aspects of a plant's physiological functions and general health, hence influencing plant growth in multiple ways. Plant growth is the process of a plant gaining additional volume or mass, regardless of the development of new structures like organs, cells, or tissues. This process is connected to physical cell specialization, reproduction, and physiological processes; growth conditions, however, have a big influence on this process. Any variation from ideal growth circumstances hinders the growth and development of plants because it causes alterations in these systems. Microplastic (MP) stress restricts plant growth and development by imposing (1) direct implications connected to the physical barrier and (2) indirect impacts associated to a reduction in soil production (Awet et al. 2018). The following are a few possible impacts of microplastics on plant growth: **Compaction**: The buildup of microplastics in the soil can cause porosity to decrease and water infiltration to be hampered. The total growth of plants can be impacted by this changed soil structure, which can also restrict root development and water and nutrient availability. **Adsorption of Nutrients**: Microplastics possess the ability to absorb and retain nutrients, hence modifying their accessibility for plants. Nutrient imbalances could result from this, which would affect the plant's capacity to absorb nutrients needed for growth and development. **Root Growth Inhibition**: Microplastics in the soil matrix have the potential to inhibit root growth, which could have an impact on the morphology and architecture of the roots. The plant's capacity to search the soil for nutrients and water may be hampered by reduced root development (Bosker et al. 2019). **Plant Uptake**: Research indicates that some plants are able to take up microplastics from the soil and move them to other plant tissues. The presence of microplastics in plant tissues may have an impact on the health of the plant and higher trophic levels in the food chain if the plant is consumed by animals or humans. **Oxidative Stress**: Plants that are exposed to microplastics may experience oxidative stress, which causes the production of reactive oxygen species (ROS). High ROS levels can harm cellular structures and interfere with normal metabolic processes, which can have an adverse effect on plant growth (Colzi et al. 2022). **Rhizosphere Microbiome**: The presence of microplastics in soil may affect the composition and activity of the rhizosphere microbiome, which is important for nutrient cycling and plant–microbe interactions. Changes in microbial communities can impact plant health and growth. **Release of Chemicals:** During their degradation, microplastics may emit chemicals and chemical additives. These compounds may have allelopathic consequences after being released. Microplastic ingestion and transmission cause varying degrees of morphological, physiological, biochemical, and genetic changes in plants. There is a study gap concerning the uptake of Microplastics (MPs) and Nanoplastics (NPs) by plants, associated factors that influence phytotoxicity levels, and potential remediation solutions in terms of food security and safety (Roy et al. 2022). **Photosynthesis and Respiration:** Microplastics may have an impact on the physiological functions of plants, such as respiration and photosynthesis. Reduced growth and total plant output might result from disturbances in these basic processes. **Altered Secondary Metabolites:** There is some evidence to suggest that the presence of microplastics may influence the production of secondary metabolites in plants. These compounds play roles in defense mechanisms and interactions with other organisms (Ge et al. 2021). Mitigating the impact of microplastics on plant growth requires a comprehensive understanding of these interactions and the development of sustainable practices to reduce microplastic pollution in soil.

3.4 Removal of Microplastics

Because microplastics are so minute, ubiquitous, and persistent, getting rid of them from soil is a difficult task. However, in order to solve this environmental risk, a variety of techniques and technologies are being investigated. These strategies seek to lessen the amount of microplastics in soil and lessen their negative effects on ecosystems, even if total removal may be challenging. Most often, polymers change and become embrittled as a result of environmental variables that age them, such as photodegradation or chemical degradation of additives. There are two different environments in which artificial polymer biodegradation can occur: anaerobic and aerobic. Polymers can break down into CO_2, H_2O, N_2, H_2, CH_4, salts, minerals, and biomass (mineralization) whole or in part. Transformation products that are stable or momentarily stable are the result of partial or primary degradation of the polymer chain (Klein et al. 2017). Three fundamental requirements are associated with biodegradation:

1. The presence of microorganisms capable of depolymerizing the target material and mineralizing the monomeric compounds via the right metabolic pathway's enzymes is required.
2. The environmental factors that are required for biodegradation, such as temperature, pH, moisture content, and salinity, must be met.
3. While the structure of the polymeric substrate, such as chemical bonds, degree of polymerization, degree of branching, and parameter, such as hydrophobicity or crystallinity, must not obstruct microbial actions, the morphology of polymer particles must facilitate the attachment of microorganisms and the formation of a biofilm (Sooriyakumar et al. 2022).

Here are some notable methods and technologies for the removal of microplastics from soil:

1. **Microbial Degradation**: Harnessing the natural capabilities of microorganisms to break down and degrade microplastics is a promising avenue. Some bacteria and fungi have shown the ability to metabolize certain types of plastics. Research in this area focuses on enhancing microbial degradation rates and understanding the specific conditions that facilitate this process. One possible way to lessen the negative effects of microplastics on the environment is by microbial biodegradation. Microorganisms are used in this technique to break down materials into simpler molecules (Miri et al. 2022; Qiu et al. 2022). Biodegradable plastics and spontaneous biodegradation processes provide hope for lowering the buildup of microplastics and their detrimental consequences, whereas conventional plastics can linger in the environment for hundreds to thousands of years (Arpia et al. 2021). Research has demonstrated that microplastics, which are recognized as persistent pollutants, are largely broken down by microbial activity (Mishra et al. 2022). Members of the genera *Pseudomonas*, *Bacillus*, and *Ideonella* are examples of bacteria that have been found to have the capacity to degrade plastic polymers (Atanasova et al. 2021).
2. **Phytoremediation**: Certain plant species have the capacity to absorb and accumulate microplastics from the soil. Phytoremediation involves the use of these plants to extract microplastics, which can then be harvested along with the plant biomass (Yadav et al. 2021). This method is being explored for its potential in reducing microplastic levels in contaminated soil. This is an in situ remediation measure in which plant's ability has been utilized to assimilate or detoxify contaminants from air, soil, and water (Vasavi et al. 2010). Plants used for phytoremediation absorb organic nutrients and metal ions through their roots, which then sequester, decompose, immobilize, or mobilize the materials. Some common characteristics of phytoremediation plants include their rapid growth, high biomass output, high tolerance, broad root system, flexibility, capacity to accumulate contaminants, and native status (Farming 2014). Exposure time of MPs

interacting with plant biomass seems to be an important factor and very short time was needed for MPs adhesion as the number of adhered MPs within first 7 days was the highest immediately after 24. This is consistent with the condition where *L. minor* was exposed to MPs for 12 weeks and the amount of adhered MPs, evaluated every 7 days, was similar though the whole duration of the experiment (26 ± 7% based on the initial amount of MPs in test vessels) (Rozman et al. 2023).

3. **Soil Amendments**: Biochar, a type of charcoal produced from organic materials, has been investigated for its potential to adsorb microplastics in soil. Adding biochar to contaminated soil can enhance adsorption capacity and potentially reduce the mobility of microplastics. In particular, biochar demonstrated potential as a highly effective adsorbent substance. Biochar has qualities that make it useful for removing a variety of contaminants, including microplastics, such as its abundance and high porosity. Therefore, investigating biochar-based techniques for microplastics removal can aid researchers in creating and improving new mitigation systems. One type of charcoal with a distinct and stable structure is called biochar. It is highly porous and rich in carbon. Pyrolysis, a popular thermochemical method for producing biochar, involves heating feedstock—such as biomass waste, crop leftovers, and animal litter—while reducing oxygen in the air. Volatile components are released during heating, leaving behind pure carbon in the form of biochar. Because of its special physical and chemical properties, such as its large specific surface area, flexible surface functionalization, and porous structure, biochar has drawn a lot of attention in adsorption and purification studies (Tan et al. 2015). Biochar is more cost-effective, less energy-intensive to produce, and more environmentally benign than activated carbon sorbents. Furthermore, the abundance of agricultural biomass and solid waste as fuel for biochar production renders it a renewable resource with intriguing potentials in a range of adsorption applications. In a 2020 experimental investigation, Wang et al. examined the effectiveness of two forms of biochar—corn straw and hardwood feedstock—in eliminating polystyrene microbeads with a diameter of 10 μm. The experiment increased the effectiveness of wastewater treatment plants' removal of microplastics by integrating biochar into sand filtration systems. The outcomes showed that, in contrast to an efficiency of 60–80% when utilizing the unmodified sand filtering systems, the removal efficiency attained topped 95%. Although they are not placed with the goal of eliminating microplastic, the majority of widely utilized treatment methods have some degree of this capability (Assil et al. 2021).
4. **Mechanical Removal**: In areas with high microplastic concentrations, sedimentation basins and filtration systems can be employed to physically remove microplastics from runoff water before it enters the soil. This approach is more applicable to specific point sources of contamination. In the context of plastics in the aquatic environment, mechanical degradation is a significant factor. Most of the time, environmental factors that age polymers—like photodegradation or chemical degradation of additives—cause the polymer to alter and eventually become embrittled. Next, as the material moves through various environmental

environments, friction forces cause it to shred into smaller pieces (Duwez and Nysten 2001).

5. **Electrokinetic Soil Remediation**: This technology involves the application of electric fields to the soil. In electrokinetic soil remediation, microplastics can be moved toward electrodes, facilitating their collection for subsequent removal. This method is still in the experimental stage but shows promise in certain contexts. Electrokinetic remediation method has been proposed as an effective technology for the soil remediation. In order to activate microplastics in the soil matrix and facilitate their passage toward the electrodes, it applies an electric field (Chaukura et al. 2022). The migration of charged particles in response to an electric potential gradient is the fundamental idea underlying electrokinetic remediation (Akarsu et al. 2021). Using this method, a direct current is supplied to electrodes that are positioned at opposing ends of the soil matrix. Microplastics are transported toward the electrodes with the help of electromigration and electroosmotic flow caused by the electric field created. The microplastics can be gathered and extracted from the soil once they have reached the electrodes.
6. **Chemical Degradation (Advanced Oxidation Processes)**: **Advanced oxidation processes** (AOPs) involve the use of reactive oxygen species to break down microplastics. Techniques such as ozonation, photocatalysis, and Fenton reactions are being explored for their potential in degrading microplastics in soil. However, the applicability of these methods may be limited by factors such as cost and environmental impact. Several oxidation methods have been researched to remove MP from intricate soil matrices. For example, ozonation uses ozone gas to oxidize MPs, which causes them to fragment and eventually degrade (Hu et al. 2023). Chemical oxidation eventually follows photooxidation (with UV light) or hydrolysis as the usual methods for starting the molecular deterioration of synthetic polymers in the environment (Andrady 2011). Since synthetic polymers (polyolefins, polyesters, and polyamides) are created in a wide variety of compositions, the prevailing mechanisms are highly dependent on the kind of polymer. Following the first reactions, the polymer's molecular weight decreases and the reacted groups are exposed to microbial destruction. Although photooxidation is often a quick process, the amount of additives—such as antioxidants—in a given polymer that may inhibit oxidation processes also affects how quickly it degrades. Furthermore, compared to breakdown from exposure on land, the photodegradation of plastics floating in aquatic environments occurs more slowly (Andrady et al. 1993).
7. **Nanotechnology**: Nanoparticles and nanomaterials are being investigated for their potential to capture and remove microplastics from soil. Engineered nanomaterials with high adsorption capacities could be used in remediation efforts, but their environmental impact requires careful consideration. This method depicts high reactivity as well as better effectiveness in remediation of environmental pollutants (Yadava et al. 2020). However, additional work is still needed to improve the existing microplastic removal technologies in light of the negative effects that microplastic contamination poses. Nanomaterials have become more widely used in the treatment of water and wastewater due to their structural

versatility and flexibility, which has improved treatment efficiency. In particular, high performance adsorbent and photocatalyst for the removal of microplastic from aqueous environments have been created using the special properties of nanomaterials (Goh et al. 2022). It's important to note that the effectiveness of these methods may vary depending on the soil type, microplastic characteristics, and environmental conditions. Additionally, the potential unintended consequences and ecological impacts of these removal methods need thorough assessment. A holistic approach combining multiple strategies, along with preventive measures, is likely to be the most effective in addressing the issue of microplastics in soil.

3.5 Conclusion

The presence of microplastics in solid habitats presents serious risks to wildlife, human health, and ecosystems. These tiny particles, which come from a range of sources such as plastic littering, industrial activities, and the decomposition of larger plastic trash, are found in soil, sediment, and terrestrial habitats all over the world. Microplastics have a wide range of complicated effects on solid habitats, including altered nutrient cycling, disturbance of soil structure, and ingestion by creatures that live in the soil. There is also a risk of bioaccumulation and biomagnification of microplastics as they move up the food chain. The creation and application of remediation techniques to lessen the effects of microplastics provide hope. Promising strategies for lowering microplastic pollution in solid environments include biological techniques like phytoremediation and microbial degradation combined with physical techniques like filtration, sediment dredging, and soil additives. We can work toward a sustainable future where the integrity of terrestrial ecosystems is protected and the adverse impacts of microplastics are avoided by encouraging collaboration between scientists, politicians, companies, and communities. For the sake of our planet's health and the welfare of present and future generations, we must not stop innovating, educating ourselves, and working together.

References

Akarsu C, Kumbur H, Kideys AE (2021) Removal of microplastics from wastewater through electrocoagulation-electroflotation and membrane filtration processes. Water Sci Technol 84:1648–1662

Akdogan Z, Guven B (2019) Microplastics in the environment: a critical review of current understanding and identification of future research needs. Environ Pollut 254:113011

Alimi OS, Farner Budarz J, Hernandez LM, Tufenkji N (2018) Microplastics and nanoplastics in aquatic environments: aggregation, deposition, and enhanced contaminant transport. Environ Sci Technol 52(4):1704–1724

Andrady AL (2011) Microplastics in the marine environment. Mar Pollut Bull 62(8):1596–1605

Andrady AL, Pegram JE, Song Y (1993) Studies on enhanced degradable plastics. II. Weathering of enhanced photodegradable polyethylenes under marine and freshwater floating exposure. J Environ Polym Degrad 1(2):117–126

Arpia AA, Chen WH, Ubando AT, Naqvi SR, Culaba AB (2021) Microplastic degradation as a sustainable concurrent approach for producing biofuel and obliterating hazardous environmental effects: a state-of-the-art review. J Hazard Mater 418:126381

Assil Z, Esegbue O, Mašek O, Gutierrez T, Free A (2021) Specific enrichment of hydrocarbonclastic bacteria from diesel-amended soil on biochar particles. Sci Total Environ 762:143084. https://doi.org/10.1016/J.SCITOTENV.2020.143084

Atanasova N, Stoitsova S, Paunova-Krasteva T, Kambourova M (2021) Plastic degradation by extremophilic bacteria. Int J Mol Sci 22(11):5610

Auta HS, Emenike CU, Fauziah SH (2017) Distribution and importance of microplastics in the marine environment: a review of the sources, fate, effects, and potential solutions. Environ Int 102:165–176

Avio CG, Gorbi S, Milan M, Benedetti M, Fattorini D, d'Errico G, Pauletto M, Bargelloni L, Regoli F (2015) Pollutants bioavailability and toxicological risk from microplastics to marine mussels. Environ Pollut 198:211–222

Awet TT, Kohl Y, Meier F, Straskraba S, Grün AL, Ruf T et al (2018) Effects of polystyrene nanoparticles on the microbiota and functional diversity of enzymes in soil. Environ Sci Europe 30:11. https://doi.org/10.1186/s12302-018-0140-6

Bosker T, Bouwman LJ, Brun NR, Behrens P, Vijver MG (2019) Microplastics accumulate on pores in seed capsule and delay germination and root growth of the terrestrial vascular plant Lepidium sativum. Chemosphere 226:774–781

Bostan N, Ilyas N, Akhtar N, Mehmood S, Saman RU, Sayyed RZ, Shatid AA, Alfaifi MY, Elbehairi SEI, Pandiaraj S (2023) Toxicity assessment of microplastic (MPs); a threat to the ecosystem. Environ Res 1(234):116523

Brennecke D, Duarte B, Paiva F, Caçador I, Canning-Clode J (2016) Microplastics as vector for heavy metal contamination from the marine environment. Estuar Coast Shelf Sci 178:189–195

Browne MA, Crump P, Niven SJ, Teuten E, Tonkin A, Galloway T, Thompson R (2011) Accumulation of microplastic on shorelines worldwide: sources and sinks. Environ Sci Technol 45(21):9175–9179

Butbunchu N, Pathom-Aree W (2019) Actinobacteria as promising candidate for polylactic acid type bioplastic degradation. Front Microbiol 10:2834. https://doi.org/10.3389/fmicb.2019.02834

Campanale C, Massarelli C, Savino I, Locaputo V, Uricchio VF (2020) A detailed review study on potential effects of microplastics and additives of concern on human health. Int J Environ Res Public Health 17(4):1212

Chae Y, An YJ (2018) Current research trends on plastic pollution and ecological impacts on the soil ecosystem: a review. Environ Pollut 240:387–395

Chang X, Fang Y, Wang Y, Wang F, Shang L, Zhong R (2022) Microplastic pollution in soils, plants, and animals: a review of distributions, effects and potential mechanisms. Sci Total Environ 850:157857

Chaukura N, Muzawazi ES, Katengeza G, Mahmoud AED (2022) Remediation technologies for contaminated soil systems. In: Emerging contaminants in the terrestrial-aquatic-atmosphere continuum. Elsevier, pp 353–365

Cole M, Lindeque P, Halsband C, Galloway TS (2011) Microplastics as contaminants in the marine environment: a review. Mar Pollut Bull 62(12):2588–2597

Colzi I, Renna L, Bianchi E, Castellani MB, Coppi A, Pignattelli S et al (2022) Impact of microplastics on growth, photosynthesis and essential elements in Cucurbita pepo L. J Hazard Mater 423:127238

Corradini F, Meza P, Eguiluz R, Casado F, Huerta-Lwanga E, Geissen V (2019) Evidence of microplastic accumulation in agricultural soils from sewage sludge disposal. Sci Total Environ 671:411–420

Crichton CRMH (2023) Plastic degradation in natural marine sediment and the inbuilt adaptation of Mātauranga Māori (Doctoral dissertation, ResearchSpace, Auckland)

Cverenkárová K, Valachovičová M, Mackul'ak T, Žemlička L, Bírošová L (2021) Microplastics in the food chain. Life (Basel) 11:1349. https://doi.org/10.3390/life11121349

de Souza Machado AA, Kloas W, Zarfl C, Hempel S, Rillig MC (2018) Microplastics as an emerging threat to terrestrial ecosystems. Glob Chang Biol 24:1405–1416. https://doi.org/10.1111/gcb.14020

Deng Y, Wu J, Chen J, Kang K (2023) Overview of microplastic pollution and its influence on the health of organisms. J Environ Sci Health A Tox Hazard Subst Environ Eng 58(4):412–422

Duwez AS, Nysten B (2001) Mapping aging effects on polymer surfaces: specific detection of additives by chemical force microscopy. Langmuir 17(26):8287–8292

Farming O (2014) CO and Ni: a review. Media:10–12. https://doi.org/10.1007/978-1-4020-9654-9

Fu Q, Lai JL, Ji XH, Luo ZX, Wu G, Luo XG (2022) Alterations of the rhizosphere soil microbial community composition and metabolite profiles of Zea mays by polyethylene-particles of different molecular weights. J Hazard Mater 423(Pt A):127062

Gan Q, Cui J, Jin B (2023) Environmental microplastics: classification, sources, fates, and effects on plants. Chemosphere 313:137559

Ge J, Li H, Liu P, Zhang Z, Ouyang Z, Guo X (2021) Review of the toxic effect of microplastics on terrestrial and aquatic plants. Sci Total Environ 791:148333

Glausiusz J (2014) The plastics puzzle. Nature 508:306

Goh PS, Kang HS, Ismail AF, Khor WH, Quen LK, Higgins D (2022) Nanomaterials for microplastic remediation from aquatic environment: why nano matters? Chemosphere 299:134418

Hansen E, Nilsson NH, Lithner D, Lassen C (2013) Hazardous substances in plastic materials. Klima-og forurensningsdirektoratet; Vejle, Denmark

Hu J, Lim FY, Hu J (2023) Ozonation facilitates the aging and mineralization of polyethylene microplastics from water: behavior, mechanisms, and pathways. Sci Total Environ 866:161290

Huang Z, Hu B, Wang H (2023) Analytical methods for microplastics in the environment: a review. Environ Chem Lett 21(1):383–401

Karbalaei S, Hanachi P, Walker TR, Cole M (2018) Occurrence, sources, human health impacts and mitigation of microplastic pollution. Environ Sci Pollut Res 25:36046–36063

Klein S, Dimzon IK, Eubeler J, Knepper TP (2017) Analysis, occurrence, and degradation of microplastics in the aqueous environment. Freshwater Microplastics, 58:51–67

Koelmans AA, Bakir A, Burton GA, Janssen CR (2016) Microplastic as a vector for chemicals in the aquatic environment: critical review and model-supported reinterpretation of empirical studies. Environ Sci Technol 50(7):3315–3326

Kublik S, Gschwendtner S, Magritsch T, Radl V, Rillig MC, Schloter M (2022) Microplastics in soil induce a new microbial habitat, with consequences for bulk soil microbiomes. Front Environ Sci 10:989267

Loder MGJ, Gerdts G (2015) Methodology used for the detection and identification of microplastics—a critical appraisal. In: Bergmann M, Gutow L, Klages M (eds) Marine anthropogenic litter. Springer International Publishing, pp 201–227

Mao X, Xu Y, Cheng Z, Yang Y, Guan Z, Jiang L, Tang K (2022) The impact of microplastic pollution on ecological environment: a review. Front Biosci (Landmark Ed) 27(2):46

Miller EM, Kroon FJ (2020) Bioaccumulation and biomagnification of microplastics in marine organisms: a review and meta-analysis of current data. PLoS One 15(10):e0240792

Miri S, Saini R, Davoodi SM, Pulicharla R, Brar SK, Magdouli S (2022) Biodegradation of microplastics: better late than never. Chemosphere 286:131670

Mishra S, Swain S, Sahoo M, Mishra S, Das AP (2022) Microbial colonization and degradation of microplastics in aquatic ecosystem: a review. Geomicrobiol J 39(3–5):259–269

Muhammad MH, Idris AL, Fan X, Guo Y, Yu Y, Jin X et al (2020) Beyond risk: bacterial biofilms and their regulating approaches. Front Microbiol 11:928

Othman AR, Hasan HA, Muhamad MH, Ismail NI, Abdullah SRS (2021) Microbial degradation of microplastics by enzymatic processes: a review. Environ Chem Lett 19:3057–3073

Qiu X, Qi Z, Ouyang Z, Liu P, Guo X (2022) Interactions between microplastics and microorganisms in the environment: modes of action and influencing factors. Gondwana Res 108:102–119

Roy T, Dey TK, Jamal M (2022) Microplastic/nanoplastic toxicity in plants: an imminent concern. Environ Monit Assess 195(1):27

Rozman U, Blazic A, Kalcikova G (2023) Phytoremediation: a promising approach to remove microplastics from the aquatic environment. Environ Pollut 338:122690

Saccá ML, Barra Caracciolo A, Di Lenola M, Grenni P (2017) Ecosystem services provided by soil microorganisms in soil biological communities and ecosystem resilience. Springer International Publishing, pp 9–24

Schell T, Rico A, Vighi M (2020) Occurrence, fate and fluxes of plastics and microplastics in terrestrial and freshwater ecosystems. Rev Environ Contam Toxicol 250:1–43

Schirinzi GF, Pérez-Pomeda I, Sanchís J, Rossini C, Farré M, Barceló D (2017) Cytotoxic effects of commonly used nanomaterials and MPs on cerebral and epithelial human cells. Environ Res 159:579–587

Sooriyakumar P, Bolan N, Kumar M, Singh L, Yu Y, Li Y et al (2022) Biofilm formation and its implications on the properties and fate of microplastics in aquatic environments: a review. J Hazard Mater Adv 6:100077

Surendran U, Jayakumar M, Raja P, Gopinath G, Chellam PV (2023) Microplastics in terrestrial ecosystem: sources and migration in soil environment. Chemosphere 318:137946

Tan X, Liu Y, Zeng G, Wang X, Hu X, Gu Y, Yang Z (2015) Application of biochar for the removal of pollutants from aqueous solutions. Chemosphere 125:70–85

Vasavi A, Usha R, Swamy PM (2010) Phytoremediation—An overview review. J Ind Pollut Control 26:83–88

Wang Z, Sedighi M, Lea-Langton A (2020) Filtration of microplastic spheres by biochar: removal efficiency and immobilisation mechanisms. Water Res 184:116165

Wanner P (2021) Plastic in agricultural soils—A global risk for groundwater systems and drinking water supplies?—A review. Chemosphere 264:128453

Wright SL, Kelly FJ (2017) Plastic and human health: a micro issue? Environ Sci Technol 51:6634–6647

Yadav M, Singh G, Jadeja RN (2021) Phytoremediation for heavy metal removal: technological advancements, pollutants and water management: resources, strategies and scarcity. Wiley, Hoboken, pp 128–150

Yadava M, Das M, Thakore S, Jadeja RN (2020) Surface modified magnetic nanoparticles: a new generation of nanoadsorbents for facile remediation protocols. In: Environment at crossroads challenges and green solutions. Scientific Publishers, pp 291

Yee MS, Hii LW, Looi CK, Lim WM, Wong SF, Kok YY, Tan BK, Wong CY, Leong CO (2021) Impact of microplastics and nanoplastics on human health. Nanomaterials 11:496

Zobkov MB, Esiukova EE (2018) Microplastics in a marine environment: review of methods for sampling, processing, and analyzing microplastics in water, bottom sediments, and coastal deposits. Oceanology 58:137–143

Microplastics in Aquatic Environment

4

Indrajeet Singh and Ajay Kumar

Abstract

Microplastics, which are microscopic plastic particles smaller than 5 mm in size, are becoming increasingly problematic in aquatic ecosystems. They enter the environment through ordinary plastics, industrial processes, and wastewater treatment plants. Microplastics endanger the health of living beings and the whole ecosystem by being found in a variety of aquatic systems, from freshwater to saltwater. Microplastics accumulate in aquatic habitats as a consequence of the extensive use and poor disposal of synthetic polymer substance. Aquatic species can consume these particles, potentially affecting the entire food chain. Microplastics are prevalent in aquatic ecosystems, exposing many aquatic animals to them and highlighting the broad nature of the problem. A variety of elements, including turbulent water regimes, airstream, and currents, influence the dispersion of microplastics within the water column in aquatic settings. Furthermore, biofouling can lower microplastics' surface hydrophobicity, leading them to sink and collect in aquatic sediments. The significant concentration of microplastics in the middle of the water column emphasizes the importance of addressing this issue holistically. To put it briefly, the prevalence of microplastics in aquatic habitats is a major environmental hazard that necessitates thorough understanding and efficient remedies. The existence of microplastics in both freshwater and marine habitats emphasizes the importance of tackling this issue in order to maintain aquatic life and ecosystem health. More study and coordinated efforts are required in order to lessen the influence of microplastics on aquatic habitats and the animals that live inside them.

I. Singh · A. Kumar (✉)
Department of Biotechnology, Faculty of Engineering and Technology, Rama University, Kanpur, Uttar Pradesh, India

A. Kumar, V. Singh (eds.), *Microplastics Pollution and its Remediation*,
https://doi.org/10.1007/978-981-97-4068-0_4

Keywords

Microplastic · Aquatic · Environmental concern

4.1 Introduction

"Plastics" are man-made materials that are composed of polymers, which are high-molecular-weight substances that can be either naturally occurring or synthesized. Due to their affordability, adaptability, and widespread use, plastics have seen heavy application across numerous industries since the 1950s portability, robustness, and longevity. About 9% of the roughly 6.3 billion tons of plastic that were formed worldwide in the last 60 years was recycled addicted to derived raw materials, whereas as regards 12% was burned (Alabi et al. 2019). Plastic trash decomposes at a very slow rate; the biodegradation of a plate with a thickness of 1 millimeter might take anywhere from a few decades to one hundred years, according to the chemical makeup of the garbage and its surroundings (Chamas et al. 2020). When plastic items break down, they break into little bits. Microplastics, or particles of plastic that are less than 5 mm in size, are a kind of litter. Research indicates that plastic particles lesser than 100 nm in diameter should be referred to as "nanoplastics" (NP). Microplastics (MP) can be categorized according to their source. On one hand, there is primary MP, which is intentionally created for various buyers and manufacturing uses such as onslaught abrasives, cosmetics, drug delivery polymer carriers, sandblasting agents, plastic-coated fertilizers, etc. On the other hand, there is secondary MP, which is created naturally when bulk plastic waste breaks down (Hirt and Body-Malapel 2020; Yee et al. 2021). The microscopic particles of MP allow them to be carried far and wide by the wind and water. Therefore, you may find these particles in everything from the air and soil to the water and arctic ice to the deepest parts of the ocean and even in living things (Fackelmann and Sommer 2019).

The chemical and physical properties of MP determine its effects on living things. The chemical composition of MP is a reflection of both the polymer and the additives. Polyethylene (PE) is the most abundant polluting MP; its particles are made of polystyrene and polypropylene. Plasticizers, antioxidants, UV stabilizers, lubricants, colors, and flame retardants are all examples of additives that give plastic its unique qualities, either as inert or reinforcing fillers. Particle size and shape, elasticity, shear strength, and surface charge are all physical characteristics of MP. There are several shapes into which MP can be classified: cellulose, grains, granules, garbage, films, and foams. Microplastics are excellent adsorbents of contaminants and microbes due to their high surface-to-volume ratio (Campanale et al. 2020; Wright et al. 2020; Li et al. 2022).

In order to solve the issue of pollution caused by microplastics, there needs to be creative research and systematic management of synthetic waste materials, the elimination of microplastics in sewage treatment systems, or increased usage of either bioplastic or plastic that is readily decomposed. Nevertheless, in order to decrease contamination by microplastics, it is crucial to comprehend the gravity and consequences of microplastic. Researchers have begun to focus on the consequences of microplastics, both directly and indirectly, due to the growing body of literature on the topic of microplastics' prevalence and pervasiveness around the world. Consequently, the potential variables that affected microplastic's ability to sorb or bioaccumulate pollutants, the dangers that microplastic poses to marine life and human health, and the prevalence and distribution of microplastic were the primary foci of this chapter.

4.2 Types of Microplastics (MP_S)

Microplastics (MPs) come in various types and shapes, and they are a significant environmental concern, particularly in aquatic habitats. The different types of microplastics include polyethylene terephthalate (PET), polyethylene (PE), polystyrene (PS), polypropylene (PP), polyamide, polyester, polyvinylchloride (PVC), and epoxy resin. Microplastic might be a wide variety of different sources for their origin, such as plastic bags, polystyrene foam, disposable plastic items, synthetic textiles, and beauty products with microbeads. They can also be the result of the degradation of larger plastic debris over time. The size of microplastics varies, with microfibers, the most common type, being between 0.1 and 0.8 mm in size. Microplastics can enter the environment through human use and have been detected in numerous places that have freshwater, particularly drinking water (Khalid et al. 2021).

4.3 Sources of MP Contamination in Aquatic Environments

Microplastic contamination is a growing concern in aquatic environments. The environmental issues of concern in both scientific and societal aspects are shown in Fig. 4.1. Microplastics (MPs) are small plastic particles that can be found in various forms, such as microbeads, microplastic particles, and plastic debris. They are widely used in various industries, including personal care products, textiles, and agriculture, which have led due to their ubiquitous occurrence in marine ecosystems.

The degradation of larger plastic items and materials and abrasion and fragmentation of larger plastic items can lead to the formation of microplastics (Duis and Coors 2016). Microplastics can be emitted into the surroundings from personal care products. It is possible for microplastics to be emitted into the surroundings that originate from everyday plastic items, such as exfoliants and cosmetics (Vivekanand et al. 2021). Industrial activities, various industries, including textiles, construction, and automotive sectors, release microplastics into the environment (Ricciardi et al. 2021). Microplastics can be introduced into aquatic environments by wastewater treatment plants. Wastewater treatment plants introduce microplastics into aquatic environments by disposing of sewage sludge into the oceans (Vivekanand et al. 2021). Surface runoff, wind, and soil erosion can disperse microplastics in aquatic

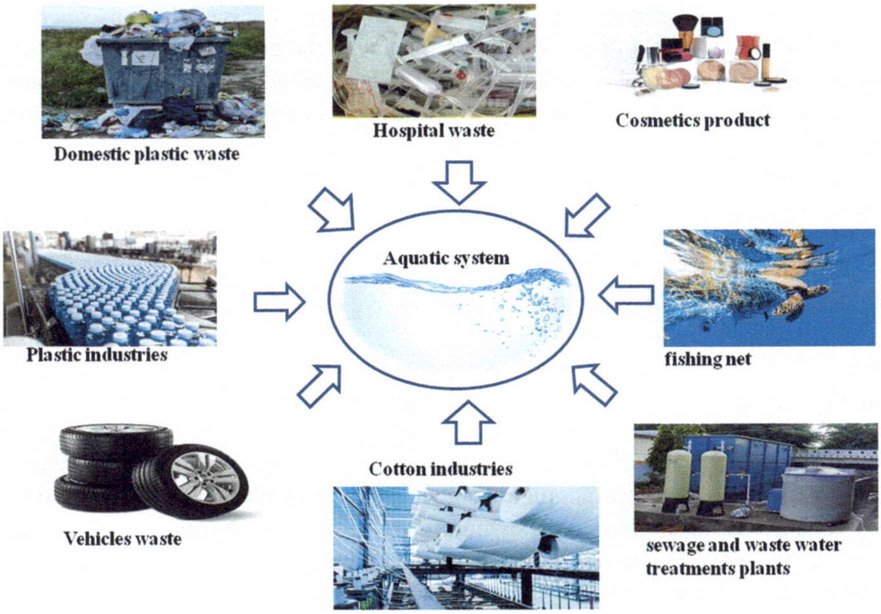

Fig. 4.1 Various source of microplastic in aquatic system

ecosystems (Ashrafy et al. 2023). Beach litter introduces fragments of plastic and other degraded materials into aquatic habitats (Du et al. 2021).

Many aquatic animals are capable of ingesting and excreting microplastics in the aquatic environment; however, there is at present insufficient proof of bioaccumulation or biomagnification. Microplastics can also be found in water bodies. Fortunately, the extensive occurrence of microplastics in marine ecosystems poses a concern for marine life and human health (Du et al. 2021).

4.4 Negative Implications of Microplastic Pollution on Marine Organisms

A decrease in the amount of food consumed, damage to the tissues, stress due to oxidation, alterations in the expression of genes associated to the immune system, a decrease in the amount of time spent feeding, genotoxicity (Li et al. 2021; Raju et al. 2022; Bhuyan 2022). Putting at risk multiple hierarchical levels of biological organization, Interference with the food webs of the deep sea and with the natural carbon cycles of the ocean, Species that create marine habitats that are at danger (Corinaldesi et al. 2021). Additionally, microplastics have the ability to absorb a wide variety of hazardous hydrophobic pollutants from the environment around them, which results in the indirect transfer of these contaminants into the food chain. Because of their minute size, marine organisms are more likely to accidentally consume these plastic particles, which can have a negative impact on their overall health. New scientific research is required in order to clarify the many elements that determine the presence of microplastics in the marine ecosystem and its biological implications on marine biota. The potential harm that microplastics could bring to marine organisms is a big issue (Chatterjee and Sharma 2019).

4.5 Impacts of Microplastics in Aquatic Ecosystems

It is possible to take in microplastics by aquatic organisms such as plankton, fish, and benthic organisms, which can result in trophic transfer (Wright et al. 2013). The effects of microplastics on aquatic life are summarized in Table 4.1.

4.6 Microplastics Affect the Food Chain in the Marine Environment

Microplastics may have an enormous effect throughout the ocean's food web. Microplastics, when consumed by marine organisms, can alter gene expression, cause tissue inflammation, reduce reproductive success, and physically harm animal organs. In addition, they can impair immune function, limit growth, and prevent reproduction in marine animals. Microplastics can enter the food chain through direct consumption by marine animals or by adhering to the surface of microorganisms that serve as prey for fish and other higher-ranking predators. Furthermore, the presence of adsorbed contaminants on microplastics can potentially endanger marine creatures due to the considerably higher concentration of chemical pollutants on these particles compared to the surrounding water. Hence, the examination

Table 4.1 Problems caused by tiny plastic particles in the ocean

Marine organisms	Impacts	References
Copepods	Lessened reproductive success, mortality, eating ability, and overall fat content	Cole et al. (2014)
Crustaceans	Changes in behavior and impairments in fertility, offspring development, food intake, enzyme activity, and metabolism	Devriese et al. (2015)
Benthos	Microplastics that sink or settle in the environment are consumed by benthic dispersion and deposits feeders	Egbeocha et al. (2018)
Coral reefs	Calcification slowed down the coral reef's skeletal development and mineralization	Reichert et al. (2018)
Molluscs	Caused alterations in consumption and fertility instability in oysters; decreased filtering/feeding capacity in mussels; microplastics may also enter the human food chain through infected molluscan species	Sussarellu et al. (2016)
Marine plants	Several mangrove sediments included MPs, most notably polyethylene, polypropylene, and polystyrene	Li et al. (2019)
Humans	More than 90% of micro- and nanoplastics consumed by humans are eliminated by the excretory system	da Costa et al. (2016)

of microplastics' influence on the oceanic food chain emerges as a crucial domain of apprehension, necessitating additional investigation and focus to uphold the well-being of marine populations and the overall marine ecosystem (Cverenkárová et al. 2021). Polystyrene MPs of a size of 10 m can be ingested by aquatic organisms such shrimp, paropa, polychaetes, and ciliates, and polychaetes and pinworms can pass absorbed MP particles to shrimp at higher levels of the food chain. Evidence of MPs being passed up the food chain includes tiny pieces of plastic found in otoliths (scallop bones used by fish for direction and balance), lantern fish, and sea lion feces. The retention length of MPs in the internal organs of the predator component is greater than the period between intakes to removal, which means that the possibility of MP intake by predator elements in a food chain is higher. This is due to the fact that the duration of MP preservation in the organs of the predator element is higher. This matter highlights the occurrence of MPs transmission in the food chain. This means that MPs are amplified in aquatic habitats (Carbery et al. 2018).

4.7 Long-Term Effects of Microplastic Exposure in Marine Life

The extended duration effects of microplastic contamination in marine environments are harmful to aquatic creatures and may even pose a risk to public health if contaminated seafood is consumed. When marine organisms are exposed to microplastics, a few of the long-term consequences include the following:

Growth delay: Microplastics can cause growth delays in marine organisms, affecting their development and overall health (Raju et al. 2022). Exposure to microplastics can lead to increased oxidative stress in aquatic organisms, which may have a harmful effect on their health and survival (Raju et al. 2022). Reduction of feeding activity: Microplastics can negatively affect the feeding behavior of marine organisms, leading to reduced food intake and potential nutritional deficiencies (Raju et al. 2022). Genotoxicity: Microplastics can cause genetic damage in marine organisms, which may lead to genetic disorders and reduced fitness (Raju et al. 2022). Bioaccumulation and biodegradation: Microplastics can bioaccumulate in marine organisms and degrade into smaller particles or fragments, increasing their bioavailability and potential impact on aquatic ecosystems (Issac and Kandasubramanian 2021). Impaired physiological functions: It is possible for fish to sustain structural damage to their intestines, livers, gills, and brains as a result of exposure to microplastics, which ultimately impacts their metabolic rate, habits, and reproduction (Zolotova et al. 2022). Population decline and ecological imbalances: Microplastics can negatively impact the sustenance, growth, reproduction, and continued existence of organisms in aquatic ecosystems, which might result in population decreases as well as ecological disparities or instabilities (Chatterjee and Sharma 2019). The long-term effects of microplastic exposure in marine life highlight the need for further research and action to tackle the growing matter of synthetic contamination in aquatic ecosystems. This includes developing effective strategies for waste management, reducing plastic production, and monitoring how microplastics are affecting the marine environment and their implications (Smith et al. 2018).

4.8 Health Risks Associated with Microplastics in Marine Environment

Microplastics (MPs) are becoming a significant concern in aquatic environments, posing a variety of health dangers to both fish and humans. The following are some of the potential health hazards connected with microplastics in aquatic environments: Microplastics are able to affect a variety of physical situations in fish, as they can spread poisonous substances and dangerous germs to fish (Bhuyan 2022). The presence of microplastics has the potential to be harmful to human beings through both chemical and physical pathways. They are linked to chemicals that are used in manufacturing and have the ability to be absorbed by the environment, which raises concern about the toxicity of these substances (Chatterjee and Sharma 2019). Microplastics have the potential to cause structural damage to the gastrointestinal tract, liver, gills, and brain of fish, in addition to having an impact on the fish's behavior, fertility, and physiological stability (Smith et al. 2018). Nanoplastics, which are a form of microplastic, have the potential to cause long-term harm, including neurological damage, liver damage, and coronary artery disease, in addition to genotoxicity and developmental harm (Yuan et al. 2022). Microplastics can negatively impact the well-being of aquatic organisms, including a decline in feeding performance and richness, a slowdown of larval growth along with development, increased oxygen consumption, and stimulation of reactive oxygen species production (Smith et al. 2018). There is a growing body of data indicating that microplastic contamination is occurring in aquatic environments; however, the influence of particular plastic particle variables, including the ability to absorb pathogens and the stage of breakdown, is yet unknown. In order to bring clarity to these issues and gain a better understanding of the probable dangers that are connected with microplastics in marine habitats, further investigation efforts are required (Smith et al. 2018).

4.9 Detection and Removal of Microplastic Contaminated Water

Contamination of the natural ecosystem can be caused by microplastics that float in the water or are buried in the sediment (Yu et al. 2018). Microplastic contamination samples are collected from various locations and examined in a controlled laboratory setting. Consequently, separation of the tiny particles from the big collections for a good recovery becomes mandatory (Yu et al. 2018). Therefore, it is crucial to collect, remove, and analyze microplastics from a variety of sources (Cole et al. 2011; Zhang et al. 2021a). Surface trawls are commonly used to collect water samples for scientific research (Harrison et al. 2011). a set of bongo nets (Sankhla et al.

2020). Benthal uses (Cole et al. 2011; Zhang et al. 2021a) trawlers for his fishing. Isolating microplastic from water has been shown to be simpler than isolating bioremediation from sediments. Beach, estuary, and seafloor samples can be used to get insight into microplastic concentrations (Harrison et al. 2011; Sankhla et al. 2020; Vianello et al. 2013). There is less likelihood of collecting soil or sediments on land using stainless steel cutlery or a box device (Vianello et al. 2013), but in the oceans, sand and shells may contain microplastics (Mohan et al. 2021; Van Cauwenberghe et al. 2013; Kiendrebeogo et al. 2021; Eriksen et al. 2013). Purification/extraction and then quantification/identification are the two stages of microplastic analysis (Yu et al. 2018; Ramage et al. 2022). Before proceeding with further quantification and/or identification of the microplastics, it is necessary to separate them from the primary medium (Yu et al. 2018; Ramage et al. 2022). Density distinction is the foremost step in initially separating that entails incorporating the segments with a recognized fluid-like saturated solution of salt (Yu et al. 2018). Using this technique, low-density particles can be transformed into high-density ones (particles with higher densities sink, while those with lower densities float). This makes it simpler to recover pollutants (Ramage et al. 2022). Microplastics were previously detected and identified using the naked eye or microscopes.

Microplastics' characteristics were studied (Ramage et al. 2022), including their clearness, rigidity, profile, and color. The amount of observational error linked with these procedures is amplified when the observer's own subjectivity is taken into account. Meanwhile, hue and clarity might change depending on the surrounding conditions (Yu et al. 2018). Samples from clams (gills and organic activity glands; mostly digestive gland) were extracted in hydrated oxide and alcohol before being microwave-treated to identify microplastic pollution (Rebelein et al. 2021; Avio et al. 2015; Makhdoumi et al. 2021). Methanolic solutions are refined using high-performance liquid chromatography [HPLC] (Yu et al. 2018). Several studies (Fries et al. 2015; Xia et al. 2021; Eriksen et al. 2013; Pastorino et al. 2021; Napper et al. 2021) have demonstrated that microplastic absorbs pyrene. The samples are thermally treated in controlled conditions, the discharged vaporish complex is unbound, and then the whole thing is put into a gigahertz column for analysis (Ramage et al. 2022). The sample size must be bigger than 500 mm (Ramage et al. 2022) for this technique to be practical. Natural plastic additives might be used to evaluate the area just once. The absence of a solvent in this procedure makes it robust against contamination from the environment (Ramage et al. 2022). A database of polymer spectra is available for research. Each experiment will employ a single particle with a well-known mass. Some chemicals, including pyroxene, have access to a pyrolysis database (Ramage et al. 2022). Using optical microscopes like stereomicroscopes and also by ocular examination (Schnepf et al. 2023), many study studies have distinguished microplastics by analyzing their morphological properties of persecution, color, and shape. Surface morphology can be calculated with certain methods (Desforges et al. 2014; Huang et al. 2023; Reisser et al. 2013; Zhang et al. 2021a, b; Ziajahro Mi et al. 2021). ESEM-EDS and SEMEDS are used to determine the building blocks of plastics and the surface morphological properties. Infrared spectroscopy remodeled by the Fourier transform can be used to determine the microplastic area's polymeric composition (Campanale et al. 2023). Unevenly shaped microplastics (Zhang et al. 2021a; Campanale et al. 2023) were found using spectrometry, where infrared spectra were matched with those of well-known compounds derived from infrared spectral libraries. The back-scattered beam differs from the irradiating laser because the section's molecules and atoms have interacted with the laser light (Yu et al. 2018; Ramage et al. 2022). Particles with a diameter of more than one millimeter can be studied with Raman spectrometry (RS) techniques, which have been validated via extensive scientific study. It's the only proven method for analyzing particles connecting 1 and 20 mm in size. Due of its inadequate water responsiveness and excellent abstraction resolution, RS-connected study (Yu et al. 2018; Ramage et al. 2022) enables the examination of microscopic particles between 1 and 20 mm in size.

Nontransparent and dark particles can be analyzed using RS technique; the RS method is commonly used to do rapid chemical mapping, allowing for rapid and automatic data collection/processing (Korda et al. 1997; Pan et al. 2021; National Research Council 1993; Azubuike et al. 2016; Sathicq et al. 2021). It's wonderful to see disruptions caused by light from natural, organic, and inorganic contaminants, hindering the recognition of microplastics (Ramage et al. 2022). Using this procedure, a high-resolution image of the sample can be created (Cole et al. 2013; Wang et al. 2016; Alfaro-Núñez et al. 2021; Hartman Jr. 1975). Since there is currently no market-available database of identification details, samples must be coated in a high vacuum (Yu et al. 2018; Ramage et al. 2022). Other methods involve using a hydrophobic dye that is absorbed by the microplastic surface and fluoresces under blue light (Ramage et al. 2022) to serve as a tag. It is common practice to count and misrepresent primary microplastics, which have a sample size below the microscale. This technique is simple and fast for low-value microplastics (Yu et al. 2018; Ramage et al. 2022). It is important to keep track of the number of fluorescent particles. The dye could potentially discolor nonconventional particles, such as organic waste. Because of this, the prevalence of microplastics may be overestimated (Ramage et al. 2022). Figure 4.2 depicts the chemical characteristics of microplastics and the effects they have. Bioremediation refers to the employment of microbes (often fungi and bacteria) to cleanse an area (Hartman Jr. 1975; Cooper and Corcoran 2010; Golwala et al. 2021). Debris may have piled up due to the lack of microbes, and life-sustaining nutrients may have been depleted be laid to rest in a cemetery. Therefore, bacteria are a blessing for the environment (Sathicq et al. 2021; Shahnawaz 2016; Shah et al. 2008; Nguyen et al. 2019). With the help of bacteria that are good for the environment, bioremediation can restore polluted areas for less money and with less environmental impact (Cole et al. 2013). These microbes seek out nutrients essential to sustain life (Alfaro-Núñez et al. 2021). They are after anything made of carbon, be it plastic or anything else. Primary microplastics, which range in size from 1 to 5 mm (Alfaro-Núñez et al. 2021; Hartman Jr. 1975) are manufactured from the most important resources. The deterioration or littering of macroplastics are two examples of the human-caused activities that contribute to the production of secondary microplastics (Shahnawaz 2016). The seven main sources of microplastics are commonly referred to as "town mud," "marine coatings,"

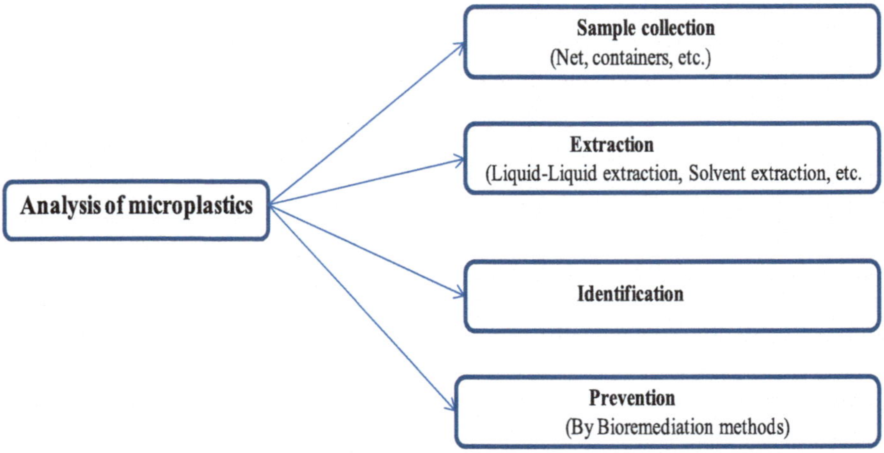

Fig. 4.2 Techniques for analysis of microplastic particle in aquatic sample

"tending product," "plastic pellets," "road markings," "artificial textiles," and "tires" (Shahnawaz 2016). Millions of marine animals (about 180 species) are killed by microplastics every year, according to global data (Alfaro-Núñez et al. 2021; Hartman Jr. 1975; Cooper and Corcoran 2010). There have been a number of approaches used to control the spread of microplastics in the past. The following methods were employed to address the issue of synthetic waste: fire (which releases various harmful gases into the air when burned), exercise (which turns waste into usable form), construction (roads and rail networks were built using plastic waste), gas production (which is entirely hypothetical), as well as low landing (which is totally unrealistic). Bioremediation is considered to be the most lucrative and ecological approach (Golwala et al. 2021; Shah et al. 2008). Bioremediation can either take place away from the pollutant's source (ex situ bioremediation) or at the pollutant's source (in situ bioremediation). When it comes to in situ bioremediation, everything is taken care of right where the contamination is (Nguyen et al. 2019; Kawasaki and Player 2015). Ex situ bioremediation involves moving toxins away from the original contaminated site and degrading them in a laboratory setting, where variables like pH and temperature may be controlled. (Out-of-place bioremediation is favored over its in-place counterpart) When compared to other bioremediation methods, composting is widely regarded as superior (Shahnawaz 2016).

4.10 Conclusions and Future Perspective of Microplastics

The present state and prospective developments regarding microplastic contamination are of growing concern appropriate to their impact on the surroundings and health of human beings. Microplastics have been found in various remote and pristine areas, including the snow and ice of cryospheric regions such as the Arctic and the Tibetan Plateau. They are measured an promising and constant contaminant with the potential to threaten worldwide environmental systems and human health. There is a lot of plastic being made, and it has become an essential part of our everyday lives. The proliferation of plastics in our society has resulted in significant environmental damage. These are key contributors to this form of pollution. Plastics don't biodegrade for thousands of years. One of the primary causes of this destruction is the widespread use of plastics, including microplastics. It increases the danger of ecotoxicity since it pollutes and contaminates the environment. Microplastics are a major problem because they destroy aquatic life by contaminating both freshwater and saltwater. Scientists have developed many techniques to sense and recognize such dangers, enabling the prevention of further damage. To encourage brief text and prospect study connecting experts on this topic, this overview highlights several crucial areas connected to microplastics. To mitigate the potential for further damage, experts must identify this new danger and establish appropriate responses. Large-scale public education campaigns are required to reduce or eliminate the use of plastic. We need a more viable and effective alternative to plastic in order to reduce or eliminate its widespread use. Future-focused research is essential for designing effective approaches for managing and disposing of plastic garbage. To better manage plastic, society should support and encourage the use of bioremediation techniques.

References

Alabi OA, Ologbonjaye KI, Awosolu O, Alalade OE (2019) Public and environmental health effects of plastic wastes disposal: a review. J Toxicol Risk Assess 5(21):1–3

Alfaro-Núñez A, Astorga D, Cáceres-Farías L, Bastidas L, Villegas CS, Macay K, Christensen JH (2021) Microplastic pollution in seawater and marine organisms across the tropical eastern Pacific and Galápagos. Sci Rep 11:1–8

Ashrafy A, Liza AA, Islam MN, Billah MM, Arafat ST, Rahman MM, Rahman SM (2023) Microplastics pollution: a brief review of its source and abundance in different aquatic ecosystems. J Hazard Mater Adv 9:100215

Avio CG, Gorbi S, Milan M, Benedetti M, Fattorini D, d'Errico G, Pauletto M, Bargelloni L, Regoli F (2015) Pollutants bioavailability and toxicological risk from microplastics to marine mussels. Environ Pollut 198:211–222

Azubuike CC, Chikere CB, Okpokwasili GC (2016) Bioremediation techniques-classification based on site of application: principles, advantages, limitations and prospects. World J Microbiol Biotechnol 32:180

Bhuyan MS (2022) Effects of microplastics on fish and in human health. Front Environ Sci 10:250

Campanale C, Massarelli C, Savino I, Locaputo V, Uricchio VF (2020) A detailed review study on potential effects of microplastics and additives of concern on human health. Int J Environ Res Public Health 17(4):1212

Campanale C, Savino I, Massarelli C, Uricchio VF (2023) Fourier transform infrared spectroscopy to assess the degree of alteration of artificially aged and environmentally weathered microplastics. Polymers 15(4):911

Carbery M, O'Connor W, Palanisami T (2018) Trophic transfer of microplastics and mixed contaminants in the marine food web and implications for human health. Environ Int 115:400–409

Chamas A, Moon H, Zheng J, Qiu Y, Tabassum T, Jang JH, Abu-Omar M, Scott SL, Suh S (2020) Degradation rates of plastics in the environment. ACS Sustain Chem Eng 8(9):3494–3511

Chatterjee S, Sharma S (2019) Microplastics in our oceans and marine health. Field actions science reports. J Field Actions 1(Special Issue 19):54–61
Cole M, Lindeque P, Halsband C, Galloway TS (2011) Microplastics as contaminants in the marine environment: a review. Mar Pollut Bull 62:2588–2597
Cole M, Lindeque P, Fileman E, Halsband C, Goodhead R, Moger J, Galloway TS (2013) Microplastic ingestion by zooplankton. Environ Sci Technol 47:6646–6655
Cole M, Webb H, Lindeque PK, Fileman ES, Halsband C, Galloway TS (2014) Isolation of microplastics in biota-rich seawater samples and marine organisms. Sci Rep 4(1):4528
Cooper DA, Corcoran PL (2010) Effects of mechanical and chemical processes on the degradation of plastic beach debris on the Island of Kauai, Hawaii. Mar Pollut Bull 60:650–654
Corinaldesi C, Canensi S, Dell'Anno A, Tangherlini M, Di Capua I, Varrella S, Willis TJ, Cerrano C, Danovaro R (2021) Multiple impacts of microplastics can threaten marine habitat-forming species. Commun Biol 4(1):431
Cverenkárová K, Valachovičová M, Mackuľak T, Žemlička L, Bírošová L (2021) Microplastics in the food chain. Life (Basel) 11(12):1349. https://doi.org/10.3390/life11121349
da Costa JP, Santos PS, Duarte AC, Rocha-Santos T (2016) (Nano) plastics in the environment—sources, fates and effects. Sci Total Environ 566:15–26
Desforges J-PW, Galbraith M, Dangerfield N, Ross PS (2014) Widespread distribution of microplastics in subsurface seawater in the NE Pacific Ocean. Mar Pollut Bull 79:94–99. https://doi.org/10.1016/j.marpolbul.2013.12.035
Devriese LI, Van der Meulen MD, Maes T, Bekaert K, Paul-Pont I, Frère L, Robbens J, Vethaak AD (2015) Microplastic contamination in brown shrimp (Crangon crangon, Linnaeus 1758) from coastal waters of the southern North Sea and channel area. Mar Pollut Bull 98(1–2):179–187
Du S, Zhu R, Cai Y, Xu N, Yap PS, Zhang Y, He Y, Zhang Y (2021) Environmental fate and impacts of microplastics in aquatic ecosystems: a review. RSC Adv 11(26):15762–15784
Duis K, Coors A (2016) Microplastics in the aquatic and terrestrial environment: sources (with a specific focus on personal care products), fate and effects. Environ Sci Eur 28:2
Egbeocha CO, Malek S, Emenike CU, Milow P (2018) Feasting on microplastics: ingestion by and effects on marine organisms. Aquat Biol 27:93–106
Eriksen M, Mason S, Wilson S (2013) Microplastic pollution in the surface waters of the Laurentian Great Lakes. Mar Pollut Bull 77:177–182
Fackelmann G, Sommer S (2019) Microplastics and the gut microbiome: how chronically exposed species may suffer from gut dysbiosis. Mar Pollut Bull 143:193–203
Fries E, Dekiff JH, Willmeyer J, Nuelle MT, Ebert M, Remy D (2015) Identification of polymer types and additives in marine microplastic particles using pyrolysis-GC/MS and scanning electron microscopy. Environ Sci Process Impacts 15:1949–1956
Golwala H, Zhang X, Iskander SM, Smith AL (2021) Solid waste: an overlooked source of microplastics to the environment. Sci Total Environ 9:144581
Harrison J, Ojeda J, Romero-González M (2011) The applicability of reflectance micro-Fourier-transform infrared spectroscopy for the detection of synthetic microplastics in marine sediments. Sci Total Environ 416:455–463
Hartman WJ Jr (1975) An evaluation of land treatment of municipal wastewater and physical siting of facility installations. Office of the Chief of Engineers (Army), Washington, DC
Hirt N, Body-Malapel M (2020) Immunotoxicity and intestinal effects of nano- and microplastics: a review of the literature. Part Fibre Toxicol 17:1–22
Huang Z, Hu B, Wang H (2023) Analytical methods for microplastics in the environment: a review. Environ Chem Lett 21(1):383–401
Issac MN, Kandasubramanian B (2021) Effect of microplastics in water and aquatic systems. Environ Sci Pollut Res 28:19544–19562
Kawasaki ES, Player A (2015) Nanotechnology, nanomedicine, and the development of new, effective therapies for cancer. Nanomedicine 2:101–109
Khalid N, Aqeel M, Noman A, Khan SM, Akhter N (2021) Interactions and effects of microplastics with heavy metals in aquatic and terrestrial environments. Environ Pollut 290:118104

Kiendrebeogo M, Estahbanati MK, Mostafazadeh AK, Drogui P, Tyagi RD (2021) Treatment of microplastics in water by anodic oxidation: a case study for polystyrene. Environ Pollut 15:116–168

Korda A, Santas P, Tenente A, Santas R (1997) Petroleum hydrocarbon bioremediation: sampling and analytical techniques, in situ treatments and commercial microorganisms currently used. Appl Microbiol Biotechnol 48:677–686

Li R, Zhang L, Xue B, Wang Y (2019) Abundance and characteristics of microplastics in the mangrove sediment of the semi-enclosed Maowei Sea of the South China Sea: new implications for location, rhizosphere, and sediment compositions. Environ Pollut 244:685–692

Li Y, Sun Y, Li J, Tang R, Miu Y, Ma X (2021) Research on the influence of microplastics on marine life. IOP Conf Ser Earth Environ Sci 631(1):012006. IOP Publishing

Li C, Jiang B, Guo J, Sun C, Shi C, Huang S, Liu W, Wu C, Zhang Y (2022) Aging process of microplastics in the aquatic environments: aging pathway, characteristic change, compound effect, and environmentally persistent free radicals formation. Water 14(21):3515

Makhdoumi P, Amin AA, Karimi H, Pirsaheb M, Kim H, Hossini H (2021) Occurrence of microplastic particles in the most popular Iranian bottled mineral water brands and an assessment of human exposure. J Water Process Eng 1:101708

Mohan H, Rajput SS, Jadhav EB, Sankhla MS, Sonone SS, Jadhav S, Kumar R (2021) Ecotoxicity, occurrence, and removal of pharmaceuticals and illicit drugs from aquatic systems. Biointerface Res Appl Chem 11:12530–12546

Napper IE, Baroth A, Barrett AC, Bhola S, Chowdhury GW, Davies BF, Duncan EM, Kumar S, Nelms SE, Niloy MN, Nishat B (2021) The abundance and characteristics of microplastics in surface water in the transboundary Ganges River. Environ Pollut 274:116348

National Research Council (1993) In situ bioremediation: when does it work? National Academies Press, Atlanta

Nguyen B, Claveau-Mallet D, Hernandez LM, Xu EG, Farner JM, Tufenkji N (2019) Separation and analysis of microplastics and nanoplastics in complex environmental samples. Acc Chem Res 52:858–866

Pan Z, Liu Q, Jiang R, Li W, Sun X, Lin H, Jiang S, Huang H (2021) Microplastic pollution and ecological risk assessment in an estuarine environment: the Dongshan Bay of China. Chemosphere 262:127876

Pastorino P, Pizzul E, Bertoli M, Anselmi S, Kušće M, Menconi V, Prearo M, Renzi M (2021) First insights into plastic and microplastic occurrence in biotic and abiotic compartments, and snow from a high-mountain lake (Carnic Alps). Chemosphere 265:129121. https://doi.org/10.1016/j.chemosphere.2020.129121

Raju P, Santhanam P, Perumal P (2022) Impacts of microplastics on marine organisms: present perspectives and the way forward. Egypt J Aquat Res 48(3):205–209

Ramage SJ, Pagaling E, Haghi RK, Dawson LA, Yates K, Prabhu R, Hillier S, Devalla S (2022) Rapid extraction of high-and low-density microplastics from soil using high-gradient magnetic separation. Sci Total Environ 831:154912

Rebelein A, Int-Veen I, Kammann U, Scharsack JP (2021) Microplastic fibers—underestimated threat to aquatic organisms? Sci Total Environ 1:146045

Reichert J, Schellenberg J, Schubert P, Wilke T (2018) Responses of reef building corals to microplastic exposure. Environ Pollut 237:955–960

Reisser J, Shaw J, Wilcox C (2013) Marine plastic pollution in waters around Australia: characteristics, concentrations, and pathways. PLoS One 8:e80466. https://doi.org/10.1371/journal.pone.0080466

Ricciardi M, Pironti C, Motta O, Miele Y, Proto A, Montano L (2021) Microplastics in the aquatic environment: occurrence, persistence, analysis, and human exposure. Water 13:973

Sankhla MS, Kumar R, Shefali (2020) New and advanced technologies in aquaculture to support environmentally sustainable development. Microbial biotechnology: basic research and applications, vol 1. Springer, pp 249–263

Sathicq MB, Sabatino R, Corno G, Di Cesare A (2021) Are microplastic particles a hotspot for the spread and the persistence of antibiotic resistance in aquatic systems? Environ Pollut 10:116896

Schnepf U, von Moers-Meßmer MA, Brümmer F (2023) A practical primer for image-based particle measurements in microplastic research. Microplastics Nanoplastics 3(1):16

Shah AA, Hasan F, Hameed A, Ahmed S (2008) Biological degradation of plastics: a comprehensive review. Biotechnol Adv 26:246–265. https://doi.org/10.1016/j.biotechadv.2007.12.005

Shahnawaz M (2016) Studies on collection, isolation, screening and molecular characterization of polythene degrading bacteria from the rhizosphere of Avicennia marina (Forsk.) Vierh. collected from West Coast of India. Ph. D. thesis. Department of Botany, Savitribai Phule Pune University, Pune. https://doi.org/10.1007/s11356-016-6542-3

Smith M, Love DC, Rochman CM, Neff RA (2018) Microplastics in seafood and the implications for human health. Curr Environ Health Rep 5(3):375–386

Sussarellu R, Suquet M, Thomas Y, Lambert C, Fabioux C, Pernet ME, Le Goïc N, Quillien V, Mingant C, Epelboin Y, Corporeau C (2016) Oyster reproduction is affected by exposure to polystyrene microplastics. Proc Natl Acad Sci U S A 113(9):2430–2435

Van Cauwenberghe L, Vanreusel A, Mees J, Janssen CR (2013) Microplastic pollution in deep-sea sediments. Environ Pollut 182:495–499

Vianello A, Boldrin A, Guerriero P, Moschino V, Rella R, Sturaro A, Da Ros L (2013) Microplastic particles in sediments of lagoon of Venice, Italy: first observations on occurrence, spatial patterns and identification. Estuar Coast Shelf Sci 130:54–61

Vivekanand AC, Mohapatra S, Tyagi VK (2021) Microplastics in aquatic environment: challenges and perspectives. Chemosphere 282:131151

Wang J, Tan Z, Peng J, Qiu Q, Li M (2016) The behaviors of microplastics in the marine environment. Mar Environ Res 113:7–17

Wright SL, Thompson RC, Galloway TS (2013) The physical impacts of microplastics on marine organisms: a review. Environ Pollut 178:483–492

Wright SL, Ulke J, Font A, Chan KL, Kelly FJ (2020) Atmospheric microplastic deposition in an urban environment and an evaluation of transport. Environ Int 136:105411

Xia B, Sui Q, Sun X, Zhu L, Wang R, Cai M, Chen B, Qu K (2021) Microplastic pollution in surface seawater of Sanggou Bay, China: occurrence, source and inventory. Mar Pollut Bull 1:111899

Yee MS, Hii LW, Looi CK, Lim WM, Wong SF, Kok YY, Tan BK, Wong CY, Leong CO (2021) Impact of microplastics and nanoplastics on human health. Nanomaterials 11(2):496

Yu Y, Zhou D, Li Z, Zhu C (2018) Advancement and challenges of microplastic pollution in the aquatic environment: a review. Water Air Soil Pollut 229:1–18

Yuan Z, Nag R, Cummins E (2022) Human health concerns regarding microplastics in the aquatic environment-from marine to food systems. Sci Total Environ 823:153730

Zhang D, Fraser MA, Huang W, Ge C, Wang Y, Zhang C, Guo P (2021a) Microplastic pollution in water, sediment, and specific tissues of crayfish (Procambarus clarkii) within two different breeding modes in Jianli, Hubei Province, China. Environ Pollut 272:15939

Zhang K, Hamidian AH, Tubić A, Zhang Y, Fang JK, Wu C, Lam PK (2021b) Understanding plastic degradation and microplastic formation in the environment: a review. Environ Pollut 23:116554

Ziajahro Mi S, Neale PA, Silveira IT, Chua A, Leusch FD (2021) An audit of microplastic abundance throughout three Australian wastewater treatment plants. Chemosphere 263:128294

Zolotova N, Kosyreva A, Dzhalilova D, Fokichev N, Makarova O (2022) Harmful effects of the microplastic pollution on animal health: a literature review. PeerJ 10:e13503

Microplastics in Terrestrial Ecosystems

5

Mohammad Toha, Md Rayhanul Islam Rayhan, and Md Mostafizur Rahman

Abstract

Globally, the generation of plastics has increased dramatically since 1950 due to population growth, urbanization, and industrialization, making our environment a menace of plastic pollution. Despite the exponential growth of microplastics (MPs), the adoption of effective MP remediation technologies and strategies remains constrained. As a result, a massive number of plastics are converted into a plethora of MPs. These various types of MPs have already been ubiquitously dispersed in our environmental compartments, making our environment a sacred place for these emerging pollutants. Consequently, MPs in terrestrial ecosystems have been widely reported in recent years due to their potential hazards to eco-

M. Toha
Department of Environmental Science, Bangladesh University of Professionals, Dhaka, Bangladesh

Laboratory of Environmental Health and Ecotoxicology, Department of Environmental Sciences, Jahangirnagar University, Dhaka, Bangladesh

M. R. I. Rayhan
Laboratory of Environmental Health and Ecotoxicology, Department of Environmental Sciences, Jahangirnagar University, Dhaka, Bangladesh

Centre for Environmental Research and Development (cERD), Dhaka, Bangladesh

M. M. Rahman (✉)
Department of Environmental Science, Bangladesh University of Professionals, Dhaka, Bangladesh

Laboratory of Environmental Health and Ecotoxicology, Department of Environmental Sciences, Jahangirnagar University, Dhaka, Bangladesh

Centre for Environmental Research and Development (cERD), Dhaka, Bangladesh

Department of Environmental Sciences, Jahangirnagar University, Dhaka, Bangladesh
e-mail: rahmanmm@juniv.edu

A. Kumar, V. Singh (eds.), *Microplastics Pollution and its Remediation*,
https://doi.org/10.1007/978-981-97-4068-0_5

systems, crops, and human health. Moreover, this diverse assortment of MPs served as a multidimensional reservoir of other emerging contaminants, including heavy metals, antibiotics, and persistent organic pollutants, which pose significant ecological hazards. Although various types of MPs with horrible levels have been found in terrestrial ecosystems, comprehensive knowledge regarding the source, fate, transport, effects, and remediation techniques of MPs in terrestrial ecosystems is still limited in developing nations. In these circumstances, conducting a comprehensive study on the source, fate, transport, effects, and mitigation of MP pollution is crucial. Hence, this chapter aims to add increasing knowledge regarding the source, fate, transport, effects, and remediation techniques of MP pollution in terrestrial ecosystems.

Keywords

Source · Fate · Transport · Effects · Remediation techniques · Mitigation strategies

5.1 Introduction

Since the commercialization of plastics in 1950, various plastics in massive amounts have been used globally due to their low cost, tiny weight, flexibility, and durability (Mai et al. 2024). The global production of plastics was only 1.5 MT in 1950, which increased 215 times higher in 2015 with 322 MT (Surendran et al. 2023). Moreover, it has been predicted that the global production of plastic waste will be 670 MT in 2040 (Delangiz et al. 2022). Although a massive amount of plastic is being generated exponentially, the recycling, reuse, and recovery rates of these plastics are inadequate in our modern world (Rahman et al. 2024). According to the research studies, plastics' recycling, incineration, and remediation rates were 9%, 12%, and 79%, respectively (Plastics Europe 2017). Consequently, global mismanaged plastic waste was 60–99 MT in 2015, which is expected to reach 155–265 MT in 2060 (Wong et al. 2020). Moreover, most of this mismanaged plastic waste is generated in Africa and Asia. This horrible state indicates that if the plastic waste generation rate and poor waste management continue, the impact of plastic waste will be devastating for our entire ecosystems. However, after the use, most plastics are discarded into the surrounding environment. Then, these larger plastics go through weathering, UV radiation, biodegradation, and physical-mechanical processes, giving birth to the issue of microplastics (MPs) (<5 mm) (Jarosz et al. 2024). After that, these MPs can rapidly spread the surrounding environment, especially terrestrial and aquatic ecosystems, through surface runoff and other environmental factors (Thacharodi et al. 2024). Hence, MPs have been ubiquitously found in all the spheres of the Earth, from aquatic to atmospheric. In terrestrial soil compartments, MPs are transported through various processes such as advection, dispersion, settling, adsorption, degradation, and aggregation (Xu et al. 2024). Moreover, MPs could go through deeper parts of the soil by air, water, and wind (Wong et al. 2020). MPs in terrestrial ecosystems can generally be created from primary and secondary sources (Mai et al. 2024). Primary MPs are found in various products, including toothpaste, body wash, scrubs, and cosmetics that enter the terrestrial environment (Wu et al. 2024). Secondary MPs are produced when plastic products break down due to physical, chemical, and biological processes. For instance, fragments of large plastic bottles (Surendran et al. 2023). Moreover, MP contamination in soil compartments occurs due to various anthropogenic and natural sources such as road dust, fertilizers, synthetic grass, highways, landfills, textiles, tourism, plastic mulching films, irrigation tools, biosolids, wastewater treatment plants, and greenhouse materials (Dissanayake et al. 2022; Kataria et al. 2024). Research studies revealed that sewage sludge generated (430000–63,000) tons and (300000–44,000) tons of MPs in the farmland of Europe and North America, respectively (Wong et al. 2020). Moreover, (4.83×10^2–1.25×10^5) MPs/m^3 were generated from the irrigation water of China (Yu et al. 2021). Consequently, it has been discovered that the average amount of MPs in Chinese farmland was 1076 items/kg in 2020 (Büks and Kaupenjohann 2020). This number rose to 1430–3410 items/kg in 2023 (Wang et al. 2024). Similar to China, many MPs have been found in the soils of other Asian countries, Europe, North America, Australia, and Africa (Sa'adu and Farsang 2023). In the European countries of Germany, Greece, Spain, and the Netherlands, the following MP concentrations were documented: 14.60 items/g, 301 ± 140 items/kg, 2242 ± 984 items/kg, and 888 ± 500 items/kg (Sa'adu and Farsang 2023). MP levels in Tunisian farming soils were between 0.21 and 1.50 items/g, while the concentration levels in Sydney, Australia's industrial soil, were 67,500 mg/kg (Boughattas et al. 2021; Wong et al. 2020). A high level of MPs has also been found in Chili, with a concentration of 2000–400 MPs/kg (Boughattas et al. 2021). Although the concentration and composition of MPs vary from place to place, some plastic polymers such as PET, PP, polyvinyl chloride (PVC), PS, LDPE, and high-density polyethylene (HDPE) are found in all types of terrestrial compartments (Hossain et al. 2023). This significant concentration of MPs around the globe poses a serious threat to human and ecological health, as these emerging contaminants can also interact with antibiotics, heavy metals, pesticides, and persistent organic pollutants (Wan et al. 2023; Sikder et al. 2024). MPs readily adsorb persistent organic pollutants (POPs) like PAHs, PCBs, pesticides, and herbicides and create a complex bond, influencing soil organism (earthworm) distribution and movement, which is considered to be a synergistic threat to ecosystems (Zhang et al. 2021). Moreover, these MPs create a strong bond with heavy metals (HMs), triggering the geochemical processes of soil that can alter the pore formation and physical and chemical properties of soil (Baho et al. 2021). For example, Boots et al. (2019) found that polyethylene polymers can drastically decrease soil pH. The combined interaction of MPs and antibiotics may lead to antibiotic resistance genes (ARGs) that seriously impact whole ecosystems (Wei et al. 2024). Moreover, this combination can reduce the exchange of gases and compounds that could lead to ecological health deterioration (Wei et al. 2024). MPs can also change the biomass of plant communities through nutrient immobilization, affecting soil microbial communities and plant root systems (Baho et al. 2021). Qi et al. (2020) found that the upper and lower parts of wheat trees and the *Lolium perenne* seeds were greatly affected by the MP polymers of PE. Moreover, these polymers decreased these plants' reproductive growth and germination (Yu et al. 2021; Saud et al. 2023). MPs can enhance the growth of invasive species such as *Calamagrostis* and promote allelopathy that can severely affect plant communities (Yu et al. 2021). Lozano and Rillig (2020) found that MPs can act as a growth factor of *Hieracium*, while allelopathic effects inhibit the *Festuca* plant. Biota species of *Eisenia fetida, Lumbricus terrestris, Isopods, Folsomia candida, Achatina fulica,* Nematodes or roundworm, and Snails (*Achatina fulica*) also faced several effects of MPs such as increasing oxidative stress, mortality rate, neurotoxicity, changing feeding behavior, and damage DNA (Yu et al. 2021; Wong et al. 2020). Moreover, MPs' presence and impact have also been observed in the terrestrial fauna (Sika deer, camel, goat, sheep, elephant, anecic earthworm, and terrestrial birds) (Wong et al. 2020). Notably, 27%, 50%, and 94% of goats, sheep, and terrestrial birds' digestive systems contained MPs (Wong et al. 2020). MPs can infiltrate the human body via the food chain and pose significant health risks (Bhatia et al. 2024; Stapleton et al. 2023). MPs harm gut health, weaken the immune system, and raise the chance of getting Type 2 diabetes, heart disease, and several inflammatory diseases (Pattanayak 2022). Hence, the menace of MPs is present everywhere in terrestrial ecosystems. However, terrestrial ecosystems contain 4–23 times more MPs than marine ecosystems (Wong et al. 2020). Moreover, most MPs in marine ecosystems come from land-based sources, showing the MP contamination interlinkage between terrestrial and marine ecosystems (Wong et al. 2020). Therefore, it is crucial to understand the source, fate, and transport of terrestrial MPs to reduce the contamination of other environmental compartments. Although various physical, chemical, and biological remediation techniques are implemented in developed nations to control MP pollution, developing nations are still far behind from this technological advancement. Thus, understanding the probable cost-effectiveness of MPs' remediation techniques can enhance awareness regarding MPs' remediation techniques in developing nations. Furthermore, most existing research regarding MP pollution in terrestrial compartments only focuses on the potential toxicity of MPs to plants. The presence of MPs in terrestrial ecosystems is limited to comprehensive knowledge in developing nations compared to developed nations. In these circumstances, a comprehensive study on the source, fate, transport, effects, and mitigation of MP pollution is required to protect terrestrial ecosystems and public health. Hence, this chapter aims to discuss the source, fate, transport, effects, and remediation and mitigation techniques of MP pollution in terrestrial ecosystems. This chapter is important because it adds to the increasing amount of knowledge regarding MP pollution in terrestrial ecosystems, which is important, especially in developing nations where anthropogenic activities heavily pollute terrestrial resources.

5.2 Sources of Terrestrial Microplastics (MPs)

Due to anthropogenic activities, MP contamination in terrestrial compartments occurs from various sources (Rodríguez-Seijo et al. 2018b). However, soil acts as a primary medium for MPs entering plant communities. MPs contamination in the soil environment can occur in two ways. They are:

Primary MPs: In this process, MPs are intentionally produced in tiny sizes of <5 mm for commercial purposes. These tiny MPs enter the terrestrial compartments through drainage systems and wastewater effluent plants. For example, microbeads in beauty products.

Secondary MPs: In this process, mesoplastics are broken down through the degradation process to create small-sized plastics known as MPs. The sources of MPs are illustrated in Fig. 5.1.

However, some of the familiar sources of MPs are detailed below:

1. Sewage sludge

   Sewage treatment plants are widely recognized as important sources of terrestrial MPs. In raw sewage, MP fibers and beads from clothing and cosmetic products are significantly concentrated (Wong et al. 2020). Various polymers

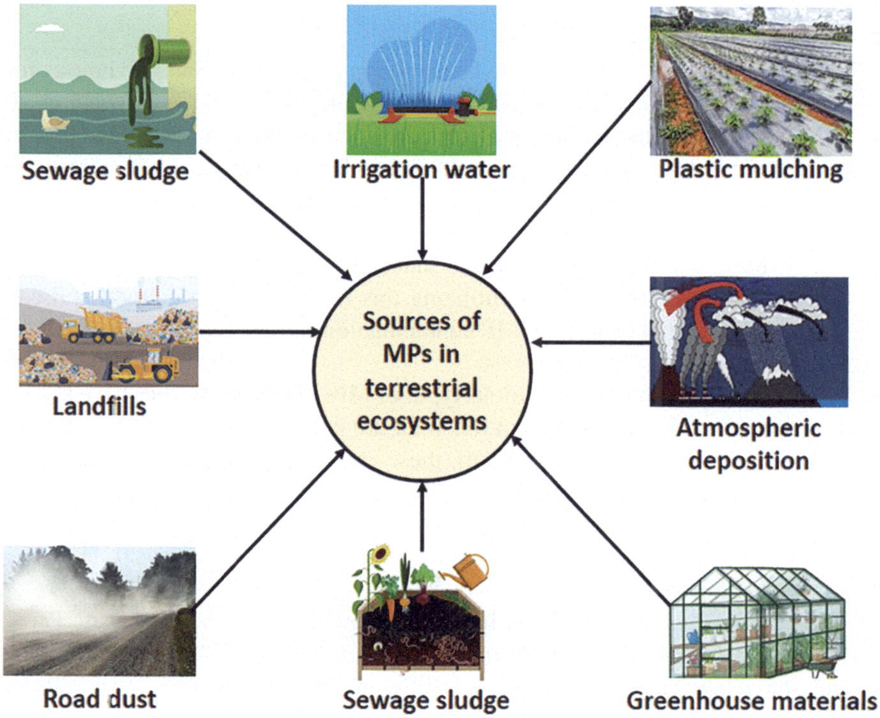

Fig. 5.1 Sources of microplastics in terrestrial ecosystems

such as LDPE, HDPE, PP, PE, and PS with 170,900 MPs/kg have been found in sewage sludge (Gatidou et al. 2019; Magni et al. 2019). The amount and type of MPs depend on the removal efficiency of the sewage treatment plant. In Europe, 1.25×10^2 t–8.50×10^2 t of MPs are discharged into the farmland soil annually (Surendran et al. 2023). Notably, if sludge compost is used directly as fertilizer, it could be the primary source of MPs in the soil (Yu et al. 2021). Recently, Canada, the Netherlands, Ireland, and the UK have banned microbeads in beauty products due to their harmful effects (Surendran et al. 2023).

2. Plastic mulching

   Plastic mulching is generally used to obstruct weed activities, retaining and maintaining soil moisture and temperatures (Yu et al. 2021). In plastic mulching, LDPE films and PU foams may contaminate MPs in agricultural soil (Wong et al. 2020). 33–56% of MPs resulting from plastic mulching were present in the depth range of (0–100) cm (Surendran et al. 2023). Moreover, it has been found that 8885 MPs/kg and 2899 MPs/kg are present in the topsoil and deep subsoil, respectively. Hence, most MPs are present in the topsoil of 0–10 cm (Li et al. 2022).

3. Irrigation water

   Various types of irrigation water, such as surface water, purified sewage, and groundwater, contain many MPs. Gatidou et al. (2019) found PET, PE, PP, and PVA polymers with a concentration of 4.83×10^2–1.25×10^5 MPs/m^3 in irrigation water.

4. Agro-sector greenhouse materials

   Recently, plastic materials have been used in greenhouses and polyhouses for cultivation protection. This plastic material contaminates the soil. Wang et al. (2022) found that the hierarchy of concentration of MPs in abandoned greenhouses, normal greenhouses, and simple greenhouses were 2215.56 ± 1549.86 items/kg > 891.11 ± 316.71 items/kg, and 632.50 ± 566.93 items/kg, respectively.

5. Composting

   Fertilizers are made by household and industrial waste recycling products (Surendran et al. 2023). Hence, utilizing this fertilizer in agricultural soil can lead to the potential sources of MP contamination.

6. Surface run-off

   Road dust is composed of various types of MPs. Research studies have shown that almost 146 k tons of MPs in tire wear are emitted per year (O'Brien et al. 2023). Due to the stormwater run-off, these various types of road dust MPs can enter the terrestrial environment. Despite wastewater treatment, MPs persist in effluents at (0.40–1) items/L, potentially entering the soil through increased use of untreated or partially treated wastewater for irrigation (Yu et al. 2021).

7. Atmospheric deposition

   Airborne microplastic fibers have been found in different parts of Paris, France, and China. About 17% were made of pure synthetic fibers (PET and PE), and the other 12% were a mix of different materials (Zhao et al. 2023). Significantly, 3–10 tons of these fibers accumulated in the Paris agglomeration due to atmospheric fallout (Akdogan and Guven 2019). Zhao et al. (2023) have predicted that atmospheric MP fiber will be produced at a rate ranging from 0 to 36.64 g/km^2 in the coming years. Furthermore, in indoor environments, polyesters were the most prevalent synthetic polymers (Ageel et al. 2022). Hence, atmospheric fallout can also contaminate the terrestrial environment.

8. Landfills

   Untreated leachate from landfills can serve as an MP source within soil environments (Rahman et al. 2024). MPs will be emitted into the environment with the rise in plastic manufacture and usage. On the other hand, Rahman et al. (2023) discovered that the concentration of MPs in landfill leachate can vary between 0.420 and 24.60 items/L. Since traditional treatment techniques cannot efficiently extract MPs from leachate, managing landfills should be a top priority for developing countries.

Hence, it can be concluded that terrestrial environments are contaminated through many sources.

5.3 Occurrence, Fate, and Transport of Microplastics in Terrestrial Ecosystems

Although a significant number of studies have been conducted on MP contamination in terrestrial ecosystems, the knowledge regarding the fate and transport of MPs is still limited. However, existing research revealed that various anthropogenic and environmental factors distribute MPs in terrestrial ecosystems. Terrestrial ecosystems receive MPs from both primary and secondary sources. In this system, lightweight MPs are transported from the soil layer to another area with the help of wind (Kumar et al. 2023). On the other hand, if heavy-weight MPs are not moved away through the run-off water, these MPs settle down in the top soil layer (Surendran et al. 2023). These highly dense MPs undergo several degradation processes that tremendously influence environmental factors. These MPs are degraded due to the interaction with soil colloids, UV radiation, thermal oxidation, physical abrasion, and microbial action (Dissanayake et al. 2022). The fate and transport of MPs in terrestrial soil compartments are demonstrated in Fig. 5.2.

However, it has been noteworthy that the influence of these factors depends on the seasons, regions, and conditions (Surendran et al. 2023). For example, sunlight plays a vital role on MP degradation in the morning but does not impact the night. Furthermore, in desert regions, the influence of temperature on MP degradation is very high in summer, whereas polar regions face it in winter (Liu et al. 2021). During the degradation process, the physiochemical properties of MPs are modified and release various chemicals such as pigments, phthalates, retardants, stabilizers, and oxygenated products (Surendran et al. 2023). However, it has been found that the degradation process is slow and takes even two or three decades (Wong et al. 2020). For example, Raddadi and Fava (2019) found that plastic polymers such as PVC remained the same after 35 years of soil incubation, whereas PP lost 0.40% of its total weight after 1 year. After the topsoil degradation process, MPs move toward subsoil due to the influence of soil biota, soil cracks and aggregates, tillage, and

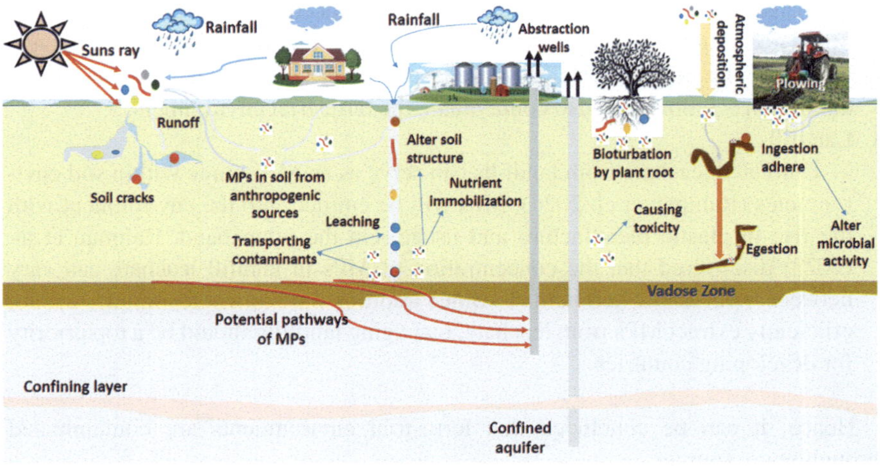

Fig. 5.2 Fate and transport of microplastics in terrestrial ecosystems

plowing (Rillig et al. 2017a). Various chemical fertilizers, organic manures, and plant root activity influence MP movement (Surendran et al. 2023). In this subsoil layer, the level of temperatures, oxygen, and UV radiation is comparatively shallow to the top soil layer, reducing the degradation rate of MPs and creating severe photo-oxidative fermentation (Dissanayake et al. 2022). However, several research studies found that anecic earthworms, hoppers, mites, mycelia (indigenous fungal species), mosquito larvae, collembola, small soil arthropods, and digging mammals have a great influence on the vertical distribution of MPs in the soil environment (Kumar et al. 2023). Anecic earthworms ingest the MPs and move toward the deeper soil layer (Rillig et al. 2017a). Rillig et al. (2017b) found that earthworms transported PE beads with the size of 710–2800 μm in the depth of 10 cm of soil within 21 days. Moreover, mites and digging mammals can enhance the transportation of MPs toward deeper soil by adhesion, grasping, and pushing (Surendran et al. 2023). Collembola and digging mammals chew the MPs and redistribute this into the new place of soil (Rillig et al. 2018). However, it has been found that Zn, dissolved organic matter, carbon, nitrogen, organic phosphorous, nitrogen, ions (NH_4^+, PO_4^{3-}), and enzyme activities have a significant influence on the movement of MPs in the soil environment (Hodson et al. 2017; Liu et al. 2022). The movement of MPs in deeper soil also depends on the physical characteristics (size, shape, density) of MPs (Mai et al. 2024). Surface roughness and oxygen-containing groups of MPs can increase its mobility and help to reach the groundwater water environment from a deeper soil layer environment (Ren et al. 2021). Table 5.1 shows the global status of MP contamination in terrestrial soil environments.

However, the MPs contamination varies from soil to soil. It has been found that agricultural soil, farmland soil, industrial soil, beach soil, floodplain soil, and tree soil contained 0.54 mg/kg, 78 ± 12.91 items/kg, 67,500 mg/kg, 220 items/kg, 593 items/kg, and 42,960 items/kg of MPs, respectively (Wong et al. 2020). Mai et al.

Table 5.1 Global status of microplastics (MPs) contamination in terrestrial soil environment

City	Country	Concentration (items/kg)	Composition	References
Tamil Nadu	India	1500	PP, PET, PE	Karthika et al. (2024)
Dhaka	Bangladesh	2.13×10^4	EVA, PVC, LDPE, HDPE	Hossain et al. (2023)
Narayanganj	Bangladesh	151	PA, PE, PU	Tajwar et al. (2023)
Sagian	Pakistan	876 ± 194	PET, PP, PE	Luqman et al. (2023)
Pingquan	China	2261.60	PET, PES	Zhang et al. (2021)
Yunnan	China	9780.60	PP, PE, PS, PET	Huang et al. (2021)
Xishuangbanna	China	6043.80	PES, PP, PE	Xu et al. (2022)
Shandong	China	4292.4	PS, PES, PVC	Wang et al. (2021)
Hebei	China	4143.90	PVC, PS	Wang et al. (2021)
Yeoju	Korea	664	PE, PP	Choi et al. (2020)
Arusha	Tanzania	Up to 1500	PET, PS, HDPE	Kundu et al. (2022)
–	Mauritius	231	PP, PA	Ragoobur et al. (2021)
–	Tunisia	852.24 ± 124.2	PEVA, PP	Boughattas et al. (2021)
Llia, Western Greece	Greece	301 ± 140	PE	Isari et al. (2021)
Murcia region	Spain & Netherland	2242 ± 984	–	Van Schothorst et al. 2021
–	Switzerland	<593	PE, PS, PVC, PP, SBR	Scheurer and Bigalke (2018)
Melipilla	Chile	360	PP, PE	Corradini et al. (2019)
Ontario	Canada	4144	PP, PE, PS	Crossman et al. (2020)
Sydney	Australia	(0.03–6.70) wt%	PVC, PE, PS	Fuller and Gautam (2016)

(2018) stated that (0.45 ± 0.25) w/w MPs were found in the earthworms, whereas 364 particles were found in 16 birds with a size range of 150 μm–8.5 mm.

Lastly, the hydrophobicity of MPs and their adsorption capacity can attract other hydrophobic pollutants and carry them from the soil layer to our food chain. Hence, MPs are potential carriers of harmful pollutants such as PAHs, PCBs, pesticides, herbicides, heavy metals, and antibiotics. Therefore, it is essential to comprehend the parameters controlling this adsorption process to evaluate the actual risks associated with MP contamination and create effective mitigation plans. Furthermore, ongoing investigation is required to navigate the intricate realm of MPs and their influence on the health of our soil.

5.4 Effects of Microplastics on the Terrestrial Ecosystems

Since a massive amount of MPs have emerged in terrestrial compartments, the significant impacts of MPs on soil physiochemical properties, soil biota, terrestrial plants, and human health are being observed around the globe. The brief discussions of the effects of MPs on the terrestrial environment are given below.

5.4.1 Effects of Microplastics on the Physical Properties of Terrestrial Soil

Soil is an essential environmental matrix for ecosystem functions and provides habitat. Due to the massive loading of plastics, global soil systems have already shown the significant impact of MPs on soils' physical and chemical properties (Bodor et al. 2024). After entering the terrestrial compartments, MPs can make a complex system with soil compartments through various external forces such as anthropogenic activities and bioturbation (Dissanayake et al. 2022). After that, these complex systems change the soil's physical and chemical characteristics, including soil pH, water-stable aggregates, organic carbon, bulk density, water retention capacity, nutrient cycling, pore characteristics, hydraulic properties, and soil conductivity (de Miranda et al. 2024). Notably, the changes in the physicochemical properties of soil are highly dependent on MP physical properties (shape, size, polymer, and concentration) (Kumar et al. 2023). For example, fibers (>30 μm) can trigger the formation of tiny pores in the soil, soil bulk density, and the distribution of soil aggregates (de Souza Machado et al. 2019). Moreover, linear-shaped polymers can increase soil clumping through blending more homogenously than nonlinear polymers. Qi et al. (2020) found that volatile compounds of dodecanal decreased and increased the EC and pH of the soil, respectively. Adding a higher concentration of MPs in rough-textured soils can drastically reduce infiltration rates and soil organic carbon (Rillig et al. 2017b). Furthermore, it can harm the soil's capacity to retain water and lead to a lack of oxygen (de Souza Machado et al. 2018a). Moreover, the presence of these contaminants hinders the evaporation process, potentially resulting in elevated soil moisture levels (Kumar et al. 2023). It has been observed that MPs-contaminated soil showed fewer macro-aggregates than the soil without MPs (Dissanayake et al. 2022). de Souza Machado et al. (2018a) examined the impacts of polyacrylic fibers, polyamide beads, polyester fibers, and polyethylene fragments on loamy sand soil. This investigation found that high contamination of these polymers in soil decreased the bulk densities and water-stable aggregates within 5 weeks. On the other hand, Ren et al. (2021) found that PS and PP at concentrations of up to 6% and varying sizes (20, 200, and 500 μm) substantially reduced the saturated hydraulic conductivity (Ks) values in loam, clay, and sandy soils, whereas, clay soils observed most. The impact of MPs also resulted in a significant decrease in the saturated and residual water contents (θs and θr) of clay soils and an increase in the slope of the water characteristic curve (SWCC). It has been observed that smaller MPs result in a significant reduction in soil porosity and aeration when compared to more prominent MPs (Ma et al. 2023; Dissanayake et al. 2022). Furthermore, the incorporation of MPs into the soil resulted in a change in the distribution of pore sizes, leading to a drop in the hydraulic conductivity of saturated soils.

5.4.2 Effects of Microplastics on the Chemical Properties of Terrestrial Soil

Although plastic is resistant to degradation in soil, degraded plastic polymers can show significant chemical reactions with soil particles (Dissanayake et al. 2022). Moreover, these reactions may lead to changes in the soil's physiochemical properties within a short period. For example, MP polymers HDPE were found to change the pH value to 0.61 from the base value within one month (Boots et al. 2019). Moreover, Li et al. (2021) observed that MP polymers PE reduced the acidic soil pH and increased the pH of alkaline soil. This study's findings indicate that the fluctuation of soil pH depends on the soil type. Moreover, the fluctuation of soil pH can affect nutrient availability, plant growth, and crop productivity. Liu et al. (2017) observed that high concentrations of MPs increased the level of dissolved organic carbon, nitrogen, phosphorus, nitrate, phosphate, humic substances, and fulvic acid. Moreover, Palansooriya et al. (2023) reported that LDPE (3–7)% dramatically reduced the total exchangeable cations (Na^+, K^+, Mg^{2+}, and Ca^{2+}) and fluctuated the EC level. On the other hand, Qi et al. (2020) found that soil pH and EC decreased and increased, respectively, after 2 months of incubation. Hence, the effects of MPs on soil chemical properties depend on the polymer type, concentration, incubation time, soil type, and other environmental and anthropogenic factors.

5.4.3 Effects of Microplastics on Soil Nutrient Cycling

Over the past decades, nutrient cycling has been significantly affected due to the massive concentration of MPs in terrestrial environments (Ma et al. 2024). However, nutrient cycling facilitates the movement and transformation of vital elements such as carbon, nitrogen, and phosphorus throughout an ecological system (Aralappanavar et al. 2024). In ecologically sound environments, these quantities remain in equilibrium (Rillig et al. 2017a). Generally, MPs are heavily composed of carbon and release significant carbon into the environment (Kumar et al. 2023). This significant release of carbon in terrestrial environments can enhance the growth of microbes. Interestingly, microbes rapidly take both excess carbon and other nutrients from the environment, which could lead to a nutrient-free environment. As a result, this nutrient immobilization situation creates a devastating impact on plant growth and root development. Qi et al. (2020) observed that MPs quickly altered the carbon, nitrogen, and phosphorous cycle. Moreover, excessive MPs could directly alter the nitrogen cycling process of nitrogen fixation, mineralization, immobilization, nitrification, and denitrification. On the other hand, MPs bind with soil minerals, creating soil aggregates that modify the soil carbon cycle (Kumar et al. 2023). Moreover, MPs can inhibit native Soil Organic Matter (SOM) storage while promoting MP-derived Dissolved Organic Carbon (DOC). However, MPs affect above and below-ground carbon flow by impacting CO_2 assimilation and emission (Kumar et al. 2023). For example, PE and PVC increase root carbon allocation while decreasing soil allocation (Kumar et al. 2023). Low-dose PVC powder enhances rhizodeposition and root growth, increasing soil carbon (Zhang et al. 2024a). However, these emerging contaminants also greatly affect the carbon mineralization process. MPs can have diverse effects on the total phosphorus content in soil, depending on their type and composition (Kumar et al. 2023). Research studies revealed that PS can significantly decrease the overall phosphorus content in soil (Sun et al. 2022; Gao et al. 2022). On the other hand, PVC can increase phosphorus availability in acidic red soils (Aralappanavar et al. 2024). This might have occurred due to their interaction with soil properties or the release of plasticizer additives, potentially affecting phosphorus cycling in these specific soil types. Zhu et al. (2018) found that MPs reduced the microbial families of Isosphaeraceae, Xanthobacteraceae, and Rhizobiaceae by altering the organic matter decomposition rate and nutrient cycle. Hence, as the nutrient cycle is the sole factor for an ecosystem, it is crucial to conduct more studies regarding this issue.

5.4.4 Combined Effects of MPs and Other Pollutants on Soil Physiochemical Properties and Nutrient Cycling

Although the knowledge regarding the combined effects of MPs and other pollutants on soil physiochemical properties and nutrient cycling is limited, few existing research found some interconnection (Ni et al. 2024; Ding et al. 2021; Wang et al. 2019b). It has been found that MPs, along with other organic and inorganic soil pollutants, can create a complex bond that can lead to more eco-toxicological impacts on terrestrial ecosystems (Baho et al. 2021). However, the combined effects of MPs and other pollutants are a new challenge for the world community. Hence, it is crucial to conduct more research on this issue to understand better the combined effects of these pollutants on the soil environment.

5.4.5 Effects of Microplastics on Soil Biota

Soil organisms greatly influence terrestrial ecosystems' function and soil quality preservation (Morris and Blackwood 2024). Unfortunately, over the years, it has been observed that MPs can seriously threaten the various types of soil biota (Chukwuemeka et al. 2024). The impact level depends on the physical characteristics of MPs, chemical composition, environmental and anthropogenic factors, and behavior of soil organisms (Dissanayake et al. 2022). For example, Wu et al. (2021) observed that more minor size MPs showed more toxicity rather than more prominent size MPs, as small size MPs reduce soil porosity and aeration and quickly penetrate the organism's cell wall and cell membrane. Moreover, small size MPs can rapidly decrease the soil fertility rate. MPs are bound by weak van der Waals forces that can help them quickly leach into the environment (Khalid et al. 2020). However, it has been found that sometimes soil biota takes MPs as food, which can create severe problems in their body, such as reduced growth rate, energy depletion, and even mortality (Rillig et al. 2017a; Ge et al. 2021; Rodríguez-Seijo et al. 2018a). Low density polyethylene (LDPE) and Polyethylene terephthalate (PET) polymers at the concentration of 0.20–3% altered the alpha diversity of bacterial communities (de Souza Machado et al. 2018b). Moreover, this 3% LDPE increased the respiration eight times than the base value and created a nutrient immobilization situation (Chen et al. 2020).

Boughattas et al. (2021) have found the impact of MPs on soil biota such as *L. terrestris, E. fetida, and Eisenia Andrei.* This study stated that MPs can decrease these soil biotas' reproduction and growth rates. Rodriguez-Seijo et al. (2017) observed the gut damage and histopathological damage of earthworms due to the PE pellets at 125 mg/kg and 62.50 mg/kg, respectively. On the other hand, Wang et al. (2019a) have observed the oxidative stress of *E. fetida* and trigger the antioxidant upregulation of various soil biota. Moreover, it has been found that 2% of PS polymers showed a massive mortality rate of 40%. According to Fei et al. (2020), the presence of 1% PE and PVC in acidic soil decreased the diversity and abundance of soil bacteria and restricted the activity of certain hydrolysis enzymes. MPs with a wide surface area give soil bacteria and fungi a place to live, which alters the distribution and makeup of soil microbial communities and the migration of soil microorganisms (Prata et al. 2020). However, MP can create neurological damage in invertebrate organisms and changes in membrane permeability, antioxidative capacity, and histology (Dissanayake et al. 2022). According to Gaylor et al. (2013), 15% (v/v) LDPE can reduce the growth rate, form a reproduction tunnel, and increase the mortality rate of *L. terrestris.* On the other hand, *L. terrestris* showed no change in the presence of HDPE with concentrations of 236, 1261, and 4505 mg/kg (Lwanga et al. 2017a). However, at 60% (w/w) LDPE concentration, *L. terrestris* faced oxidative stress (Lwanga et al. 2017b). Lei et al. (2018) observed that *Caenorhabditis elegans*, snails, and springtails faced problems such as intestinal functioning, energy metabolism, reproduction, and growth rate. The exposure of PET to *A. fulica* can reduce food intake and excretion, glutathione peroxidase, and total antioxidant capacity (Rai et al. 2023). Moreover, PS with a concentration of 1000 μg/kg can damage the intestinal cells and DNA of soil biota (Rai et al. 2023). PE and PS can alter the gut microbiome and movement of *Lobella sokamensis* (Li et al. 2023). Table 5.2 shows the effects of MPs on soil biota.

Thus, MPs significantly impact soil biota. Moreover, as trophic transfer of these MPs has already been observed from the soil to chicken feces, it is crucial to focus on more scientific studies for better understanding.

Table 5.2 Effects of microplastics (MPs) on soil biota

Soil biota	Type of MPs	Size	Effects	References
Eisenia fetida	PE and PS	≤300 μm and ≤250 μm	Increased peroxidase, catalase, and lipid peroxidation	Wang et al. (2019b)
Eisenia fetida	LDPE	5 mm and (0.25 μm–1 mm)	Absence of MPs uptake, inhibition of TBARs and AChE	Rodríguez-Seijo et al. (2018b)
Lumbricus terrestries	PE	<150 μm	Growth rate significantly reduced, mortality increased	Lwanga et al. (2017a)
L. terrestries	Microfibers	–	Burrowing and feeding behavior modulated	Prendergast-Miller et al. (2019)
Isopods	PE	183–193 mm	No effect	Kokalj et al. (2018)
Eisenia andrei	PE	250–1000 μm	Damage the expression of immune system gene	Rodríguez-Seijo et al. (2018a)
Folsomia candida	PE	<150 μm	Inhibit reproduction abilities	Ju et al. (2019)
Nematodes or roundworm	PS	(1–5) μm	Damage to GABAergic and cholinergic neurons	Lei et al. (2018)
Terrestrial worm	PA	(13–18) μm	Reduced reproduction capacity	Lahive et al. (2019)
Lobella sokamensis	PE and PS	(0.47–1155) μm	Disrupt the movement	Kim and An (2019)
Mice	PS	5 μm	Create metabolic disorder	Jin et al. (2019)
Bacterivorous nematode	LDPE	(57 ± 40) μm	Affect reproduction	Schöpfer et al. (2020)
Soil enzymes	PP and PP	–	Alter microbial community structure	Yi et al. (2021)

5.4.6 Effects of Microplastics on Soil Flora and Fauna

Although many MP's research have been conducted, the effects of MPs on soil flora and fauna are yet to be comprehensively discovered. Consequently, information regarding the effects of MPs on soil flora and fauna is highly scarce. The first study regarding the effects of MPs on soil flora and fauna was conducted in 2018. In this investigation, Qi et al. (2018) found the negative impact of MPs on wheat growth. Moreover, de Souza Machado et al. (2019) found the negative impacts of MPs on spring onions. This study discovered that plants' effects due to MPs depend on the MP's size and concentration. Since the presence of MPs can alter the soil structure, water holding availability, nutrition content, and bulk density, these changes might directly affect the plant growth, root system, and nutrient uptake process (Qi et al.

2018; de Souza Machado et al. 2019). Moreover, adding MPs could affect the soil microbial communities and bioactivity, leading to decreased plant performance (Li et al. 2023). However, the effect level depends on the soil type and plant species (Dissanayake et al. 2022). It has been found that MP polymer PS and LDPE decreased the plant height, weight, and fruit number (Bosker et al. 2019). Moreover, paddy plants exposed to 20 nm PS were found to have devastating impacts on plant health (Zhou et al. 2021a). However, 50 nm PS was found in the primary root systems of onion plants, indicating the penetration ability of MPs. 0.20 μm PS was also found in the lettuce, altering their cellular membranes and biogeochemistry (Zhou et al. 2021b). Germination is a stage where seed plants' metabolic reactions occur (Kumari et al. 2022; Li et al. 2020). The impact of MPs has been observed in the germination process of various plant species. It has been found that various concentrations of PS (0, 10, 50, 100, and 500 mg/L) can reduce the seed germination rate of paddy, *Trifolium repens, and Orychophragmus violaceus* (Giorgetti et al. 2020). Jiang et al. (2019) stated that the interaction between MPs and plants could cause ecotoxicity and genotoxicity. For example, 100 mm PS can tremendously reduce the growth of fava beans. Polystyrene MPs can induce cytotoxicity to plants (Maity et al. 2020). However, during the degradation process, plastics may produce lactic acid and harmful chemicals that can reduce the growth of tomatoes (Ullah et al. 2021). Plastic polymer PE can decrease leaves' soluble protein and sugar content (Baho et al. 2021). Although various studies have found the negative impacts of MPs on terrestrial flora, many studies have not found any negative impacts. For example, Boots et al. (2019) found that acrylic polymers have a significant impact on the germination of *perennial ryegrass*, whereas Judy et al. (2019) found no impact of PS, PE, and PVC polymers on the germination of wheat. Moreover, scanning electron microscopy (SEM) revealed that PE did not penetrate the lettuce roots; instead, they adhered to the root's surface (Kumari et al. 2022). Hence, it can be concluded that the exact effects of MPs on terrestrial plants are still unknown. Moreover, continuous research studies should be conducted to develop practical and real situations. The effects of MPs on terrestrial plants are given in Table 5.3.

However, although various types of MPs have been detected in terrestrial fauna, knowledge of the effects of MPs on terrestrial fauna still needs to be discovered. Some existing research focused on the presence of MPs in terrestrial fauna and stated potential effects. Table 5.4 shows the occurrence and effects of MPs on terrestrial fauna.

The impact of MPs on terrestrial fauna-related study is largely scarce, so it is crucial to focus on this issue.

5.5 Mitigation Strategies of Microplastics Pollution in Terrestrial Ecosystems

The prevalence of MPs is increasing in terrestrial environments on a global scale (Mai et al. 2024). After being released into terrestrial environments, plastics can find their way into wetlands, lakes, and rivers (Mingming et al. 2024). As these MPs

Table 5.3 The effects of microplastics (MPs) on terrestrial plants

Plant species	Types of MPs	MPs amount	Incubation time	Experiment type	Effects	References
Cucumber (*Cucumis sativus* L.)	Polystyrene (PS)	5 and 0.1 μm	Greenhouse	60 days	Increased plant height, root length, dry weight of the aboveground and underground parts	Liu et al. (2024)
Lettuce (*Lactuca sativa* L.)	Polystyrene, polyethylene, and polypropylene	0.01% and 0.1% w/w	Pot experiment	48 days	Above ground length and fresh weight drastically reduced	Lian et al. (2024)
Cotton plants	Carboxyl-modified polystyrene microplastics (PS-COOH)	100, 300, and 500 mg/L	Hydroponic culture system	15 days	Dose-dependent phenotypic change occurred; oxidative stress observed	Li et al. (2024)
Vicia faba	PS	(10–100) mg/L	–	2	CAT activity decreased and induced genotoxicity and oxidative stress	Jiang et al. (2019)
Cucurbita pepo	PP PE, PVC and PET	0.02%, 0.1%, and 0.2% w/w	–	28	Reduced chlorophyll content and photosynthetic efficiency	Colzi et al. (2022)
Brassica rapa	PS	10 mg/kg	–	21	Affects plant photosynthes is and growth parameters	Ren et al. (2021)
Rice (*Oryza sativa* L.)	PS	Low, medium, high	Suspension and greenhouse + field trials	21 days	Higher doses of PS caused ≈40% decrease in shoot biomass, compared with that of 25.9% in field trials	Wu et al. (2020)
Perennial ryegrass (*Lolium perenne* L.)	Biodegradable polylactic acid and HDPE	1 g kg^{-1}	Soil	30 days	Reduced germination of seeds when exposed to PLA. Additionally, compared to PE, the root biomass was lower with PLA. The amount of chlorophyll did not vary significantly	Boots et al. (2019)

Zea mays L. var. Jubilee)	PE	0.0125, 100 mg L^{-1}	Hydroponic	10–15 days	The tested MP did not go to the shoots; instead, it collected on the surface of the roots. PE hampered the maize plant's ability to absorb nutrients and water	Urbina et al. (2020)
Rice seedlings	PS, PTFE	40–200 mg/L	–	7	Biomass of rice decreased, net photosynthetic rate and chlorophyll content of rice decreased	Dong et al. (2020)
Spring onion (*Allium fistulosum*)	Polyester (PES) high-density polyethylene (HDPE), polypropylene (PP), polystyrene (PS), and polyethylene terephthalate (PET)	PES 0.2% HDPE, PS, PP, PET 2%	1.5 months	Pot experiment	1. PES and PS increased root biomass 2. PA decreased the ratio between root and leaf dry biomass 3. All tested microplastics significantly increased total root length and decreased root average diameter	de Souza Machado et al. (2019)
Wheat (*Triticum aestivum*)	PHBV	1%	25 days	Greenhouse	The plants died after 25 days due to the addition of PHBV. It raised the dissolved organic C content and biomass of the soil	Zhou et al. (2021a)

Table 5.4 The occurrence and effects of MPs on terrestrial fauna

Country	Species	Occurrence	Effects	References
Iran	Goat (Capra aegagrus hircus)	27% digestive tract	Inflammatory, cytotoxicity, genotoxicity	Omidi et al. (2012); Urli et al. (2023)
United Arab Emirates	Camel (Camelus dromedaries)	Stomach 55%	Oxidative stress, cells damage, and cell viability	Plastic Soup Foundation (2018); Urli et al. (2023)
Zimbabwe	African Elephant (Loxodonta)	Stomach	Cytotoxicity and genotoxicity	Gogo (2016)
Japan	Sika Deer (Cervus nippon)	Digestive tract (64%)	Oxidative stress, reduce growth rate, genotoxicity	Coffey (2019); Provencher et al. (2022)
China	Terrestrial Birds	Gastrointestinal tracts (94.1%)	Inflammation, damage gastrointestinal tract, and mortality	Zhao et al. (2016); Lusher et al. (2022)
Mexico	Chicken (Gallus gallus domesticus)	Feces (98%)	Inflammation, damage gastrointestinal tract, and mortality	Lwanga et al. (2017a)

are toxic and nonbiodegradable, the accumulation of these poisons within the human body can lead to the development of chronic ailments such as lung cancer, cerebral impairment, respiratory complications, and renal dysfunction (Lin et al. 2024). Moreover, agricultural soil is unsuitable for production if the levels of MPs are higher than allowed (Sun et al. 2024). Hence, the mitigation of MP contamination in terrestrial ecosystems has significant importance. Removing existing plastics from the environment and reducing future pollution are the two main goals of the current mitigation strategies. These strategies address changes in behavior, technology, and regulations. However, integrated waste management systems are crucial at the beginning of the mitigation stage. Primary and secondary MP pollution is growing and becoming a big problem globally. Therefore, it is necessary to implement comprehensive measures to regulate the production of MP. A possible way to address this worldwide concern is to decrease the use of plastic packaging and enhance the recycling and reutilization of certain plastics. The global MP contamination problem can be solved by incorporating responsible and sustainable methods into the manufacturing, using, and disposing of diverse plastic materials. The subsequent strategy involves implementing regulatory measures, encompassing many laws, rules, and voluntary pledges (Chen 2015). Enforcing taxation and levies is essential for raising public consciousness about plastic consumption and reducing plastic waste generation. Together, these initiatives aim to mitigate the harmful consequences of plastic pollution by lowering the quantity of plastic in the environment.

According to Thacharodi et al. (2024), technological innovation is crucial in eliminating existing plastic particles. Researchers worldwide have utilized various methodologies that have exhibited efficacy in eradicating MPs. For example, reverse osmosis, membrane bioreactor, biofilter, coagulation/flocculation, and other advanced techniques are commonly employed in wastewater treatment systems to eliminate various pollutants, including MPs (Rahman et al. 2024). Moreover, conventional primary and secondary treatment procedures can eliminate approximately 99% of MPs (Thacharodi et al. 2023). Over the years, biological remediation techniques have been considered safe, sustainable, cheap, ecologically friendly, and alternatives to conventional treatment procedures (Talukdar et al. 2024). Bioremediation uses naturally occurring microorganisms to transform hazardous substances into less or nontoxic ones. According to Venkatesh et al. (2015), indigenous microorganisms such as bacteria, fungi, plants, and earthworms are employed to remove or decompose detrimental pollutants and waste materials effectively.

Moreover, polluted regions can be cleaned up by utilizing native microorganisms. Various factors influence bioremediation, including pH, temperature, soil texture and composition, nutrient inputs, and oxygen availability (Zhou et al. 2024). Hence, it is a procedure that mitigates the toxicity of MPs by the adsorption, absorption, or accumulation of microorganisms. Pollutants can be eliminated by in situ and ex situ bioremediation methods. In situ bioremediation uses microorganisms to interact directly with substrates dissolved and sorbed in contaminants. MPs can induce toxicity in microorganisms by interfering with oxidative phosphorylation, osmotic equilibrium, polypeptides, and nucleic acids at higher concentrations (Thacharodi et al. 2023). MP-tolerant microorganisms can be employed for the bioremediation of MPs. Lastly, microorganisms can eliminate these MPs under MP stress (Urra et al. 2019). Microorganisms ought to transform toxic substances into a less dangerous and bioavailable state. These processes include MP accumulation in the cell wall, redox reactions, extracellular or intracellular trapping, and cell membrane transfer (Jasmin et al. 2020). It has been found that plasmids primarily induce resistance in MP. Sun et al. (2023) stated that chromosomal components may be involved. However, Das et al. (2012) reported that several bacteria, such as *Achromobacter, Bacillus, and Pseudomonas*, can degrade PVC. Balasubramanian et al. (2010) observed that *Pseudomonas* can destroy 20% of HDPE within 120 days. Additionally, it has been found that various Fungi, such as *Aspergillus niger, A versicolor, Penicillium pinophilum, P. chrysogenum, P. frequentan, and P. oxalicum,* exhibit the ability to breakdown polyethylene (PE). *Verticillium lecanii, Fusarium redolens, Acremonium kiliense*, and *Glioclodium virens* are additional organisms that have the potential to degrade polyethylene (Thacharodi et al. 2024).

In modern days, nanoparticles have been employed to eliminate MPs from terrestrial environments. Generally, nanomaterials adsorb MPs. MPs are eliminated by zeolite, metal-oxide, and other nanoparticles. Metal-oxide nanoparticles have been widely used in recent years. The stability of metallic nanoparticles renders them infrequently employed as adsorbents. With certain exceptions, Tang and Hadibarata (2021) discovered that multistage treatment technologies enable water treatment facilities to eliminate MP more effectively. Primary clarifiers employ flocculants and coagulants to facilitate sedimentation. Activated sludge is employed in secondary treatment to eliminate soluble organic pollutants. Membrane bioreactors generate effluent of superior quality as compared to secondary clarifiers. Phosphate and nitrogen are eliminated via A2O systems in specialized tanks (Tang and Hadibarata 2021). Ultrafiltration, reverse osmosis, and activated carbon filtration are examples of membrane techniques used in tertiary treatment.

Before being released, materials can be disinfected using ozone, chlorination-dechlorination, or UV light. The primary, secondary, and tertiary wastewater treatment plants can eliminate 16.5%–98.4%, 78.1%–100%, and 87.3%–99.9% of MPs, correspondingly (Tang and Hadibarata 2021). Using granule-activated carbon (GAC) in conjunction with heat regeneration and electrocoagulation enhances the efficiency of MP removal. According to Kim and Park (2021), conventional STPs cannot remove MPs smaller than 200 μm (Kim and Park 2021). However, utilizing GAC treatment with thermal regeneration proves to be more efficient in eliminating MPs within 20–50 μm. Electric coagulation can enhance the efficacy of removal for 30 min. Electrocoagulation enhances the size and density of MPs, rendering it a very efficient technique for removing MPs. Electrocoagulation, standard biological treatment, and GAC with thermal regeneration are all viable approaches to mitigate the environmental risks of microhazardous contaminants such as MPs. The remediation technique of MP pollution in terrestrial ecosystems is shown in Table 5.5.

Hence, it can be concluded that integrated waste management systems, awareness, standard law, policies, global collaborative initiatives, and cost-effective remediation techniques will aid in building a solid framework for controlling MP contamination, encouraging more environmentally friendly behaviors, and minimizing environmental impact.

5.6 Future Directions

- Limited research has been conducted on the environmental and anthropogenic factors that influence the fate and transport of MPs in terrestrial ecosystems. Since MPs are transboundary pollutants, greater emphasis must be placed on the influence of meteorological factors on their transport through terrestrial compartments. Hence, investigating how climatic conditions influence the movement and settlement of MPs in terrestrial environments is critical for creating effective methods to reduce the environmental dissemination of these emerging contaminants.
- Although the quantity and morphological features of MP contamination in terrestrial compartments have been the subject of numerous studies, a more comprehensive examination of MPs geographical and temporal distribution is still required. Therefore, it is imperative to conduct spatiotemporal distribution research on MPs to evaluate their life cycle and the historical occurrence of MP pollution in a specific terrestrial environment.
- Few studies have been conducted on how MPs affect soil physiochemical structures. Hence, future research should be focused on the interaction between MPs and soil aggregates and their properties.

Table 5.5 Remediation techniques of microplastic pollution

Method	Number of MPs eliminated/ efficiency	Advantages	Disadvantages	References
Biofilter	79%	Large MP particles have been retained effectively by the biofilter, which was able to dramatically reduce the MP abundance in the treated effluent	Deeper layers did not guarantee the full retention of microplastics and only produced a small increase in treatment effectiveness. The final effluent from the filter did not contain any particles larger than 100 μm	Liu et al. (2020)
Coagulation/ flocculation	99.4%	Can remove smaller MPs; cost-effective; high efficiency	Chemicals are required	Rajala et al. (2020); Thacharodi et al. (2024)
Photocatalytic degradation	97% average mass loss	No chemicals required; environment friendly	Produced a modest level of by-products (less than 10 mg/L COD)	Lee et al. (2020)
Membrane bioreactor	98.3%	Using combined advanced treatment methods with porous membranes	Shape dependency of the removal percentage, membrane fouling	Lares et al. (2018)
Constructed wetland	>89%	Operational cost affordable; green water treatment facility	Large area required; maintenance costly; risk of contamination	Zhou et al. (2022)
Reverse osmosis	86%	Most economical and efficient	MPs having irregular shapes can abrade surfaces and decrease filtration efficiency	Tadsuwan and Babel (2021)
Biodegradation	Depends on microbes	Versatility, low running costs, practically applicable in many contexts, ease and reliability for large-scale application	Difficult-to-control environmental factors, challenges in conducting large-scale analysis, and difficulty in identifying the right microbial community	Arossa et al. (2019); Hall et al. (2015)

- Although some of the studies have been conducted focusing on the impact of MPs on soil biota, flora, and fauna, knowledge regarding this issue is still limited. Hence, understanding the exact impacts of MPs on soil biota, flora, and fauna is crucial to reducing the effect of MPs on it.

- It is essential to comprehend the combined impacts of MPs and numerous harmful pollutants, such as POPs, chemical additives, and heavy metals, as MPs are adsorbed on these pollutants. Therefore, future research should be conducted to understand the long-term presence of these contaminants in different environmental compartments and their possible eco-toxicological consequences.
- Hybrid remediation techniques that contribute to achieving sustainable development goals (SDGs) and improved efficacy should be the primary focus of future research.
- Despite the enactment of various legal strategies by multiple nations to address MP contamination, there is still a lack of standardized legislative approaches. Consequently, implementing efficient policy and management techniques can reduce the impact of MPs on environmental components.

5.7 Conclusions

Our current world lives in the Age of Plastic, where the menace of microplastics (MPs) is potentially discovered everywhere in our environments. The significant accumulation of MPs in terrestrial soils poses a considerable risk due to their capability to disrupt soil composition, impede nutrient and water absorption, and possibly hinder plant growth. Furthermore, these substances have the potential to disturb microbial communities and release potentially harmful compounds, posing risks to both human health and soil quality. MPs can be readily consumed by marine and terrestrial organisms, enabling their entry into the human food chain. The three principal routes by which MPs can penetrate the body's interior are ingestion, inhalation, and dermal contact. Consequently, MPs have been detected in several human organs and biological fluids, indicating their ability to penetrate the body via multiple entry points. Developing countries face significant repercussions due to insufficient treatment infrastructure and a lack of understanding regarding MPs' origin, fate, and transit. Additionally, no specific laws, policies, or rules govern MPs in developing countries. Consequently, it is critical to resolve their presence and prevent further pollution. This study highlights the importance of identifying the source, fate, and transport as a first step toward tackling the issue of MP pollution. Furthermore, it is imperative to establish standardized laws and regulations regarding this issue in developing nations. Acquiring information regarding the sources of these pollutants enables stakeholders to develop more targeted strategies for reducing the environmental impact of MPs. Furthermore, biological remediation systems should be explored and deployed in developing countries due to their efficacy and cost-effectiveness in eliminating MPs. Overall, mitigating MP requires a multipronged approach to ensure a sustainable terrestrial environment.

References

Ageel HK, Harrad S, Abdallah MAE (2022) Occurrence, human exposure, and risk of microplastics in the indoor environment. Environ Sci Processes Impacts 24(1):17–31

Akdogan Z, Guven B (2019) Microplastics in the environment: a critical review of current understanding and identification of future research needs. Environ Pollut 254:113011

Aralappanavar VK, Mukhopadhyay R, Yu Y, Liu J, Bhatnagar A, Praveena SM, Sarkar B (2024) Effects of microplastics on soil microorganisms and microbial functions in nutrients and carbon cycling—A review. Sci Total Environ 924:171435

Arossa S, Martin C, Rossbach S, Duarte CM (2019) Microplastic removal by red sea giant clam (*Tridacna maxima*). Environ Pollut 252:1257–1266. https://doi.org/10.1016/j.envpol.2019.05.149

Baho DL, Bundschuh M, Futter MN (2021) Microplastics in terrestrial ecosystems: moving beyond the state of the art to minimize the risk of ecological surprise. Glob Chang Biol 27(17):3969–3986

Balasubramanian V, Natarajan K, Hemambika B, Ramesh N, Sumathi CS, Kottaimuthu R, Rajesh Kannan V (2010) High-density polyethylene (HDPE)-degrading potential bacteria from marine ecosystem of Gulf of Mannar, India. Lett Appl Microbiol 51(2):205–211. https://doi.org/10.1111/j.1472-765X.2010.02883.x

Bhatia SK, Kumar G, Yang YH (2024) Understanding microplastic pollution: tracing the footprints and eco-friendly solutions. Sci Total Environ 914:169926. https://doi.org/10.1016/j.scitotenv.2024.169926

Bodor A, Feigl G, Kolossa B, Mészáros E, Laczi K, Kovács E et al (2024) Soils in distress: the impacts and ecological risks of (micro) plastic pollution in the terrestrial environment. Ecotoxicol Environ Saf 269:115807

Boots B, Russell CW, Green DS (2019) Effects of microplastics in soil ecosystems: above and below ground. Environ Sci Technol 53(19):11496–11506. https://doi.org/10.1021/acs.est.9b03304

Bosker T, Bouwman LJ, Brun NR, Behrens P, Vijver MG (2019) Microplastics accumulate on pores in seed capsule and delay germination and root growth of the terrestrial vascular plant Lepidium sativum. Chemosphere 226:774–781

Boughattas I, Hattab S, Zitouni N, Mkhinini M, Missawi O, Bousserrhine N, Banni M (2021) Assessing the presence of microplastic particles in Tunisian agriculture soils and their potential toxicity effects using Eisenia andrei as bioindicator. Sci Total Environ 796:148959

Büks F, Kaupenjohann M (2020) Global concentrations of microplastics in soils—a review. Soil 6(2):649–662

Chen CL (2015) Regulation and management of marine litter. In: Marine anthropogenic litter, pp 395–428

Chen H, Wang Y, Sun X, Peng Y, Xiao L (2020) Mixing effect of polylactic acid microplastic and straw residue on soil property and ecological function. Chemosphere 243:125271

Choi YR, Kim Y, Yoon J, Dickson N, Kim K (2020) Plastic contamination of forest, urban, and agricultural soils: a case study of Yeoju in the Republic of Korea. J Soils Sediments 21:1962–1973

Chukwuemeka IS, Li G, Mo Y, Jacques KJ (2024) Impacts of microplastics and urbanization on soil health: an urgent concern for sustainable development. Green Anal Chem 8:100095

Coffey H (2019) Japan's famous Nara Deers killed by tourists' plastic waste. Independent

Colzi I, Renna L, Bianchi E, Castellani MB, Coppi A, Pignattelli S et al (2022) Impact of microplastics on growth, photosynthesis and essential elements in Cucurbita pepo L. J Hazard Mater 423:127238

Corradini F, Meza P, Eguiluz R, Casado F, Huerta-Lwanga E, Geissen V (2019) Evidence of microplastic accumulation in agricultural soils from sewage sludge disposal. Sci Total Environ 671:411–420

Crossman J, Hurley RR, Futter M, Nizzetto L (2020) Transfer and transport of microplastics from biosolids to agricultural soils and the wider environment. Sci Total Environ 724:138334. https://doi.org/10.1016/j.scitotenv.2020.138334. PMID: 32408466.

Das G, Bordoloi NK, Rai SK, Mukherjee AK, Karak N (2012) Biodegradable and biocompatible epoxidized vegetable oil modified thermostable poly (vinyl chloride): thermal and performance characteristics post biodegradation with Pseudomonas aeruginosa and Achromobacter sp. J Hazard Mater 209:434–442. https://doi.org/10.1016/j.jhazmat.2012.01.043

de Miranda CO, de Souza JJLL, Schaefer CEGR, Lwanga EH, Villela FNJ (2024) Short-term impacts of polyethylene and polyacrylonitrile microplastics on soil physicochemical properties and microbial activity of a marine terrace environment in maritime Antarctica. Environ Pollut 347:123791

de Souza Machado AA, Lau CW, Till J, Kloas W, Lehmann A, Becker R, Rillig MC (2018a) Impacts of microplastics on the soil biophysical environment. Environ Sci Technol 52(17):9656–9665

de Souza Machado AA, Kloas W, Zarfl C, Hempel S, Rillig MC (2018b) Microplastics as an emerging threat to terrestrial ecosystems. Glob Chang Biol 24(4):1405–1416

de Souza Machado AA, Lau CW, Kloas W, Bergmann J, Bachelier JB, Faltin E, Becker R, Görlich AS, Rillig MC (2019) Microplastics can change soil properties and affect plant performance. Environ Sci Technol 53(10):6044–6052. https://doi.org/10.1021/acs.est.9b01339

Delangiz N, Aliyar S, Pashapoor N, Nobaharan K, Asgari Lajayer B, RodríguezCouto S (2022) Can polymer-degrading microorganisms solve the bottleneck of plastics' environmental challenges? Chemosphere 294:133709. https://doi.org/10.1016/j.chemosphere.2022.133709

Ding J, Zhu D, Wang Y, Wang H, Liang A, Sun H, Chen L (2021) Exposure to heavy metal and antibiotic enriches antibiotic resistant genes on the tire particles in soil. Sci Total Environ 792:148417

Dissanayake PD, Kim S, Sarkar B, Oleszczuk P, Sang MK, Haque MN, Ok YS (2022) Effects of microplastics on the terrestrial environment: a critical review. Environ Res 209:112734

Dong Y, Gao M, Song Z, Qiu W (2020) Microplastic particles increase arsenic toxicity to rice seedlings. Environ Pollut 259:113892

Fei Y, Huang S, Zhang H, Tong Y, Wen D, Xia X, Barceló D (2020) Response of soil enzyme activities and bacterial communities to the accumulation of microplastics in an acid cropped soil. Sci Total Environ 707:135634. https://doi.org/10.1016/j.scitotenv.2019.135634

Fuller S, Gautam A (2016) A procedure for measuring microplastics using pressurized fluid extraction. Environ Sci Technol 50(11):5774–5780

Gao B, Li Y, Zheng N, Liu C, Ren H, Yao H (2022) Interactive effects of microplastics, biochar, and earthworms on CO2 and N2O emissions and microbial functional genes in vegetable-growing soil. Environ Res 213:113728

Gatidou G, Arvaniti OS, Stasinakis AS (2019) Review on the occurrence and fate of microplastics in sewage treatment plants. J Hazard Mater 367:504–512

Gaylor MO, Harvey E, Hale RC (2013) Polybrominated diphenyl ether (PBDE) accumulation by earthworms (Eisenia fetida) exposed to biosolids-, polyurethane foam microparticle-, and penta-BDE-amended soils. Environ Sci Technol 47(23):13831–13839

Ge J, Li H, Liu P, Zhang Z, Ouyang Z, Guo X (2021) Review of the toxic effect of microplastics on terrestrial and aquatic plants. Sci Total Environ 791:148333

Giorgetti L, Spanò C, Muccifora S, Bottega S, Barbieri F, Bellani L, Castiglione MR (2020) Exploring the interaction between polystyrene nanoplastics and Allium cepa during germination: internalization in root cells, induction of toxicity and oxidative stress. Plant Physiol Biochem 149:170–177

Gogo J (2016) Electric fence for Vic Falls dumpsite. … 8 elephants dead from consuming plastics. Herald

Hall N, Berry K, Rintoul L, Hoogenboom M (2015) Microplastic ingestion by scleractinian corals. Mar Biol 162(3):725–732. https://doi.org/10.1007/s00227-015-2619-7

Hodson ME, Duffus-Hodson CA, Clark A, Prendergast-Miller MT, Thorpe KL (2017) Plastic bag derived-microplastics as a vector for metal exposure in terrestrial invertebrates. Environ Sci Technol 51(8):4714–4721

Hossain MN, Rahman MM, Afrin S, Akbor MA, Siddique MAB, Malafaia G (2023) Identification and quantification of microplastics in agricultural farmland soil and textile sludge in Bangladesh. Sci Total Environ 858:160118

Huang J, Chen H, Zheng Y, Yang Y, Zhang Y, Gao B (2021) Microplastic pollution in soils and groundwater: characteristics, analytical methods and impacts. Chem Eng J 425:131870

Isari EA, Papaioannou D, Kalavrouziotis IK, Karapanagioti HK (2021) Microplastics in agricultural soils: a case study in cultivation of watermelons and canning tomatoes. Water 13(16):2168

Jarosz K, Borek-Dorosz A, Drozdek M, Rokicińska A, Kiełbasa A, Janus R, Michalik M (2024) Abiotic weathering of plastic: experimental contributions towards understanding the formation of microplastics and other plastic related particulate pollutants. Sci Total Environ 917:170533

Jasmin C, Anas A, Singh D, Purohit HJ, Gireeshkumar TR, Nair S (2020) Aberrations in the microbiome of cyanobacteria from a tropical estuary polluted by heavy metals. Mar Pollut Bull 160:111575

Jiang X, Chen H, Liao Y, Ye Z, Li M, Klobučar G (2019) Ecotoxicity and genotoxicity of polystyrene microplastics on higher plant Vicia faba. Environ Pollut 250:831–838. https://doi.org/10.1016/j.envpol.2019.04.055

Jin Y, Lu L, Tu W, Luo T, Fu Z (2019) Impacts of polystyrene microplastic on the gut barrier, microbiota and metabolism of mice. Sci Total Environ 649:308–317

Ju H, Zhu D, Qiao M (2019) Effects of polyethylene microplastics on the gut microbial community, reproduction and avoidance behaviors of the soil springtail, Folsomia candida. Environ Pollut 247:890–897

Judy JD, Williams M, Gregg A, Oliver D, Kumar A, Kookana R, Kirby JK (2019) Microplastics in municipal mixed-waste organic outputs induce minimal short to long-term toxicity in key terrestrial biota. Environ Pollut 252:522–531. https://doi.org/10.1016/j.envpol.2019.05.027

Karthika S, Sivasubramanian K, Dhevagi P, Thavamani P, Rajkishore SK, Amrutha VM, Lakshmanan A (2024) Distribution pattern and risk assessment of microplastics contamination in different agricultural systems. Int J Environ Anal Chem, 1–20

Kataria N, Yadav S, Garg VK, Rene ER, Jiang JJ, Rose PK et al (2024) Occurrence, transport, and toxicity of microplastics in tropical food chains: perspectives view and way forward. Environ Geochem Health 46(3):1–36

Khalid N, Aqeel M, Noman A (2020) Microplastics could be a threat to plants in terrestrial systems directly or indirectly. Environ Pollut 267:115653

Kim SW, An YJ (2019) Soil microplastics inhibit the movement of springtail species. Environ Int 126:699–706

Kim KT, Park S (2021) Enhancing microplastics removal from wastewater using electro-coagulation and granule-activated carbon with thermal regeneration. Processes 9(4):617

Kokalj AJ, Horvat P, Skalar T, Kržan A (2018) Plastic bag and facial cleanser derived microplastic do not affect feeding behaviour and energy reserves of terrestrial isopods. Sci Total Environ 615:761–766

Kumar A, Mishra S, Pandey R, Yu ZG, Kumar M, Khoo KS, Show PL (2023) Microplastics in terrestrial ecosystems: un-ignorable impacts on soil characterises, nutrient storage and its cycling. TrAC Trends Anal Chem 158:116869

Kumari A, Rajput VD, Mandzhieva SS, Rajput S, Minkina T, Kaur R, Glinushkin AP (2022) Microplastic pollution: an emerging threat to terrestrial plants and insights into its remediation strategies. Plants 11(3):340

Kundu MN, Komakech HC, Lugomela G (2022) Analysis of macro-and microplastics in riverine, riverbanks, and irrigated farms in Arusha, Tanzania. Arch Environ Contam Toxicol 82(1):142–157

Lahive E, Walton A, Horton AA, Spurgeon DJ, Svendsen C (2019) Microplastic particles reduce reproduction in the terrestrial worm Enchytraeus crypticus in a soil exposure. Environ Pollut 255:113174

Lares M, Ncibi MC, Sillanpää M, Sillanpää M (2018) Occurrence, identification and removal of microplastic particles and fibers in conventional activated sludge process and advanced MBR technology. Water Res 133:236–246. https://doi.org/10.1016/j.watres.2018.01.049

Lee JM, Busquets R, Choi IC, Lee SH, Kim JK, Campos LC (2020) Photocatalytic degradation of polyamide 66; evaluating the feasibility of photocatalysis as a microfibre-targeting technology. Water 12(12):3551. https://doi.org/10.3390/w12123551

Lei L, Wu S, Lu S, Liu M, Song Y, Fu Z, He D (2018) Microplastic particles cause intestinal damage and other adverse effects in zebrafish Danio rerio and nematode Caenorhabditis elegans. Sci Total Environ 619:1–8

Li H, Zhang L, Lu H, Ma J, Zhou X, Wang Z, Yi C (2020) Macro-/nanoporous Al-doped ZnO/cellulose composites based on tunable cellulose fiber sizes for enhancing photocatalytic properties. Carbohydr Polym 250:116873

Li HZ, Zhu D, Lindhardt JH, Lin SM, Ke X, Cui L (2021) Long-term fertilization history alters effects of microplastics on soil properties, microbial communities, and functions in diverse farmland ecosystem. Environ Sci Technol 55(8):4658–4668

Li S, Ding F, Flury M, Wang Z, Xu L, Li S et al (2022) Macro-and microplastic accumulation in soil after 32 years of plastic film mulching. Environ Pollut 300:118945

Li Z, Yang Y, Chen X, He Y, Bolan N, Rinklebe J, Sonne C (2023) A discussion of microplastics in soil and risks for ecosystems and food chains. Chemosphere 313:137637

Li W, Zhao J, Zhang Z, Ren Z, Li X, Zhang R, Ma X (2024) Uptake and effect of carboxyl-modified polystyrene microplastics on cotton plants. J Hazard Mater 466:133581. https://doi.org/10.1016/j.jhazmat.2024.133581

Lian Y, Shi R, Liu J, Zeb A, Wang Q, Wang J, Yu M, Li J, Zheng Z, Ali N, Liu W (2024) Effects of polystyrene, polyethylene, and polypropylene microplastics on the soil-rhizosphere-plant system: phytotoxicity, enzyme activity, and microbial community. J Hazard Mater 465:133417. https://doi.org/10.1016/j.jhazmat.2023.133417

Lin B, Wang L, Chen Q, Liu Z, Liu B, Wen S, Wei C (2024) Health assessment based on exposure to microplastics in tropical agricultural soil. J Hazard Mater 465:133372

Liu H, Yang X, Liu G, Liang C, Xue S, Chen H, Geissen V (2017) Response of soil dissolved organic matter to microplastic addition in Chinese loess soil. Chemosphere 185:907–917

Liu F, Nord NB, Bester K, Vollertsen J (2020) Microplastics removal from treated wastewater by a biofilter. Water 12(4):1085. https://doi.org/10.3390/w12041085

Liu P, Shi Y, Wu X, Wang H, Huang H, Guo X, Gao S (2021) Review of the artificially-accelerated aging technology and ecological risk of microplastics. Sci Total Environ 768:144969

Liu H, Wang X, Shi Q, Liu Y, Lei H, Chen Y (2022) Microplastics in arid soils: impact of different cropping systems (Altay, Xinjiang). Environ Pollut 303:119162

Liu B, Li R, Zhuang H, Lin Z, Li Z (2024) Effects of polystyrene microplastics on the phenylpropane metabolic pathway in cucumber plants. Environ Exp Bot 220:105671. https://doi.org/10.1016/j.envexpbot.2024.105671

Lozano YM, Rillig MC (2020) Effects of microplastic fibers and drought on plant communities. Environ Sci Technol 54(10):6166–6173

Luqman M, Shahid T, Awan MUF, Kashif SUR, Arooj F, Awan AR (2023) Quantification and characterization of microplastics (MPs) pollution in peri-uburban agricultural lands of Lahore, Pakistan. PLoS One 18(10):e0291760

Lusher AL, Provencher JF, Baak JE, Hamilton BM, Vorkamp K, Hallanger IG, Gabrielsen GW (2022) Monitoring litter and microplastics in Arctic mammals and birds. Arctic Sci 8(4):1217–1235

Lwanga EH, Gertsen H, Gooren H, Peters P, Salánki T, van der Ploeg M, Besseling E, Koelmans AA, Geissen V (2017a) Incorporation of microplastics from litter into burrows of Lumbricus terrestris. Environ Pollut 220:523–531. https://doi.org/10.1016/j.envpol.2016.09.096

Lwanga EH, Vega JM, Quej VK, de los Angeles Chi J, del Cid LS, Chi C, Segura GE, Gertsen H, Salánki T, van der Ploeg M, Koelmans AA (2017b) Field evidence for transfer of plastic debris along a terrestrial food chain. Sci Rep 7(1):14071. https://doi.org/10.1038/s41598-017-14588-2

Ma J, Xu M, Wu J, Yang G, Zhang X, Song C, Long L, Chen C, Xu C, Wang Y (2023) Effects of variable-sized polyethylene microplastics on soil chemical properties and functions and microbial communities in purple soil. Sci Total Environ 868:161642. https://doi.org/10.1016/j.scitotenv.2023.161642

Ma Y, Yang K, Yu H, Tan W, Gao Y, Lv B (2024) Effects and mechanism of microplastics on organic carbon and nitrogen cycling in agricultural soil: a review. Soil Use Manag 40(1):e12971

Magni S, Binelli A, Pittura L, Avio CG, Della Torre C, Parenti CC, Regoli F (2019) The fate of microplastics in an Italian wastewater treatment plant. Sci Total Environ 652:602–610

Mai L, Bao LJ, Shi L, Wong CS, Zeng EY (2018). A review of methods for measuring microplastics in aquatic environments. Environmental Science and Pollution Research, 25:11319-332.

Mai L, Bao LJ, Wong CS, Zeng EY (2024) Microplastics in the terrestrial environment. In: Microplastic contamination in aquatic environments. Elsevier, pp 229–247

Maity S, Chatterjee A, Guchhait R, De S, Pramanick K (2020) Cytogenotoxic potential of a hazardous material, polystyrene microparticles on Allium cepa L. J Hazard Mater 385:121560. https://doi.org/10.1016/j.jhazmat.2019.121560

Mingming H, Huaqiang L, Zhu T, Chenxi Z, Zihan Z, Jiang Q, Shengkai T (2024) Toxic effects of micro (nano)-plastics on terrestrial ecosystems and human health. TrAC Trends Anal Chem 172:117517

Morris SJ, Blackwood CB (2024) The ecology of soil biota and their function. In: Soil microbiology, ecology and biochemistry, pp 275–302

Ni B, Zhang TL, Cai TG, Xiang Q, Zhu D (2024) Effects of heavy metal and disinfectant on antibiotic resistance genes and virulence factor genes in the plastisphere from diverse soil ecosystems. J Hazard Mater 465:133335

O'Brien S, Rauert C, Ribeiro F, Okoffo ED, Burrows SD, O'Brien JW, Thomas KV (2023) There's something in the air: a review of sources, prevalence and behaviour of microplastics in the atmosphere. Sci Total Environ 874:162193

Omidi A, Naeemipoor H, Hosseini M (2012) Plastic debris in the digestive tract of sheep and goats: an increasing environmental contamination in Birjand, Iran. Bull Environ Contamin Toxicol 88:691–694

Palansooriya KN, Sang MK, El-Naggar A, Shi L, Chang SX, Sung J, Ok YS (2023) Low-density polyethylene microplastics alter chemical properties and microbial communities in agricultural soil. Sci Rep 13(1):16276

Pattanayak S (2022) Plastics and their additives reached the blood and tissue spaces: what are the possible consequences?. https://doi.org/10.52635/eamr/12.2.128-133

Plastic Soup Foundation (2018) Camels continue to die of plastic in the desert [WWW Document]. https://www.plasticsoupfoundation.org/en/2018/01/camels-continue-to-die-of-plastic-in-the-dessert/. Accessed 23 Nov 2019

Plastics Europe (2017) Recommendations towards the EU plastics strategy discussion paper from the interest group plastics of the European network

Prata JC, da Costa JP, Lopes I, Duarte AC, Rocha-Santos T (2020) Environmental exposure to microplastics: an overview on possible human health effects. Sci Total Environ 702:134455. https://doi.org/10.1016/j.scitotenv.2019.134455

Prendergast-Miller MT, Katsiamides A, Abbass M, Sturzenbaum SR, Thorpe KL, Hodson ME (2019) Polyester-derived microfibre impacts on the soil-dwelling earthworm Lumbricus terrestris. Environ Pollut 251:453–459

Provencher JF, Au SY, Horn D, Mallory ML, Walker TR, Kurek J, Lusher A (2022) Animals and microplastics: ingestion, transport, breakdown, and trophic transfer. In: Polluting textiles. Routledge, pp 33–62

Qi Y, Yang X, Pelaez AM, Lwanga EH, Beriot N, Gertsen H, Geissen V (2018) Macro-and microplastics in soil-plant system: effects of plastic mulch film residues on wheat (Triticum aestivum) growth. Sci Total Environ 645:1048–1056

Qi Y, Ossowicki A, Yang X, Lwanga EH, Dini-Andreote F, Geissen V, Garbeva P (2020) Effects of plastic mulch film residues on wheat rhizosphere and soil properties. J Hazard Mater 387:121711

Raddadi N, Fava F (2019) Biodegradation of oil-based plastics in the environment: existing knowledge and needs of research and innovation. Sci Total Environ 679:148–158

Ragoobur D, Huerta-Lwanga E, Somaroo GD (2021) Microplastics in agricultural soils, wastewater effluents and sewage sludge in Mauritius. Sci Total Environ 798:149326

Rahman MM, Sultan MB, Alam M (2023) Microplastics and adsorbed micropollutants as emerging contaminants in landfill: a mini review. Curr Opin Environ Sci Health 31:100420

Rahman MM, Toha M, Sikder S (2024) Micro/Nanoplastics pollution from landfill sites: a comprehensive review on the formation, distribution, effects and potential mitigation. In: Technical landfills and waste management, vol 1: Landfill impacts, characterization and valorisation, pp 247–265

Rai M, Pant G, Pant K, Aloo BN, Kumar G, Singh HB, Tripathi V (2023) Microplastic pollution in terrestrial ecosystems and its interaction with other soil pollutants: a potential threat to soil ecosystem sustainability. Resources 12(6):67

Rajala K, Grönfors O, Hesampour M, Mikola A (2020) Removal of microplastics from secondary wastewater treatment plant effluent by coagulation/flocculation with iron, aluminum and polyamine-based chemicals. Water Res 183:116045. https://doi.org/10.1016/j.watres.2020.116045

Ren Z, Gui X, Xu X, Zhao L, Qiu H, Cao X (2021) Microplastics in the soil-groundwater environment: aging, migration, and co-transport of contaminants—a critical review. J Hazard Mater 419:126455

Rillig MC, Ziersch L, Hempel S (2017a) Microplastic transport in soil by earthworms. Sci Rep 7(1):1362. https://doi.org/10.1038/s41598-017-01594-7

Rillig MC, Muller LA, Lehmann A (2017b) Soil aggregates as massively concurrent evolutionary incubators. ISME J 11(9):1943–1948

Rillig MC, de Souza Machado AA, Lehmann A, Klümper U (2018) Evolutionary implications of microplastics for soil biota. Environ Chem 16(1):3–7

Rodriguez-Seijo A, Lourenço J, Rocha-Santos TAP, Da Costa J, Duarte AC, Vala H, Pereira R (2017) Histopathological and molecular effects of microplastics in Eisenia andrei Bouché. Environ Pollut 220:495–503

Rodríguez-Seijo A, da Costa JP, Rocha-Santos T, Duarte AC, Pereira R (2018a) Oxidative stress, energy metabolism and molecular responses of earthworms (Eisenia fetida) exposed to low-density polyethylene microplastics. Environ Sci Pollut Res 25:33599–33610

Rodríguez-Seijo A, Santos B, da Silva EF, Cachada A, Pereira R (2018b) Low-density polyethylene microplastics as a source and carriers of agrochemicals to soil and earthworms. Environ Chem 16(1):8–17

Sa'adu I, Farsang A (2023) Plastic contamination in agricultural soils: a review. Environ Sci Eur 35(1):13

Saud S, Yang A, Jiang Z, Ning D, Fahad S (2023) New insights in to the environmental behavior and ecological toxicity of microplastics. J Hazard Mater Adv 10:100298

Scheurer M, Bigalke M (2018) Microplastics in Swiss floodplain soils. Environ Sci Technol 52(6):3591–3598

Schöpfer L, Menzel R, Schnepf U, Ruess L, Marhan S, Brümmer F, Kandeler E (2020) Microplastics effects on reproduction and body length of the soil-dwelling nematode Caenorhabditis elegans. Front Environ Sci 8:41

Sikder S, Toha M, Mostafizur Rahman M (2024) An overview on municipal solid waste characteristics and its impacts on environment and human health. In: Technical landfills and waste management, vol 1: Landfill impacts, characterization and valorisation, pp 135–155

Stapleton MJ, Ansari AJ, Hai FI (2023) Antibiotic sorption onto microplastics in water: a critical review of the factors, mechanisms and implications. Water Res 233:119790. https://doi.org/10.1016/j.watres.2023.119790

Sun X, Zhang X, Xia Y, Tao R, Zhang M, Mei Y, Qu M (2022) Simulation of the effects of microplastics on the microbial community structure and nitrogen cycle of paddy soil. Sci Total Environ 818:151768

Sun Z, Zhao M, Chen L, Gong Z, Hu J, Ma D (2023) Electrokinetic remediation for the removal of heavy metals in soil: limitations, solutions and prospection. Sci Total Environ 903:165970. https://doi.org/10.1016/j.scitotenv.2023.165970

Sun X, Withana PA, Palansooriya KN, Vithanage M, Yang X, Lee SR, Ok YS (2024) Impacts of microplastics on terrestrial plants: a critical review. Land Degrad Dev 35(5):1629–1643

Surendran U, Jayakumar M, Raja P, Gopinath G, Chellam PV (2023) Microplastics in terrestrial ecosystem: sources and migration in soil environment. Chemosphere 318:137946
Tadsuwan K, Babel S (2021) Microplastic contamination in a conventional wastewater treatment plant in Thailand. Waste Manag Res 39(5):754–761. https://doi.org/10.1177/0734242X20982055
Tajwar M, Hasan M, Shreya SS, Rahman M, Sakib N, Gazi MY (2023) Risk assessment of microplastic pollution in an industrial region of Bangladesh. Heliyon 9(7):e17949
Talukdar A, Kundu P, Bhattacharya S, Dutta N (2024) Microplastic contamination in wastewater: sources, distribution, detection and remediation through physical and chemical-biological methods. Sci Total Environ 916:170254
Tang KHD, Hadibarata T (2021) Microplastics removal through water treatment plants: its feasibility, efficiency, future prospects and enhancement by proper waste management. Environ Challenges 5:100264
Thacharodi A, Hassan S, Singh T, Mandal R, Khan HA, Hussain MA, Pugazhendhi A (2023) Bioremediation of polycyclic aromatic hydrocarbons: an updated microbiological review. Chemosphere 328:138498. https://doi.org/10.1016/j.chemosphere.2023.138498
Thacharodi A, Meenatchi R, Hassan S, Hussain N, Bhat MA, Arockiaraj J, Pugazhendhi A (2024) Microplastics in the environment: a critical overview on its fate, toxicity, implications, management, and bioremediation strategies. J Environ Manag 349:119433
Ullah R, Tsui MTK, Chen H, Chow A, Williams C, Ligaba-Osena A (2021) Microplastics interaction with terrestrial plants and their impacts on agriculture. J Environ Qual 50(5):1024–1041
Urbina MA, Correa F, Aburto F, Ferrio JP (2020) Adsorption of polyethylene microbeads and physiological effects on hydroponic maize. Sci Total Environ 741:140216. https://doi.org/10.1016/j.scitotenv.2020.140216
Urli S, Corte Pause F, Crociati M, Baufeld A, Monaci M, Stradaioli G (2023) Impact of microplastics and nanoplastics on livestock health: an emerging risk for reproductive efficiency. Animals 13(7):1132
Urra J, Alkorta I, Mijangos I, Epelde L, Garbisu C (2019) Application of sewage sludge to agricultural soil increases the abundance of antibiotic resistance genes without altering the composition of prokaryotic communities. Sci Total Environ 647:1410–1420
Van Schothorst B, Beriot N, Huerta Lwanga E, Geissen V (2021) Sources of light density microplastic related to two agricultural practices: the use of compost and plastic mulch. Environments 8(4):36
Venkatesh KR, More N, Kanoujia S (2015) Chromium accumulation in Eisenia fetida in modified vermi compost supplemented with tannery sludge. J Agroecol Nat Resour Manag 1(4):135–141
Wan L, Cheng H, Liu Y, Shen Y, Liu G, Su X (2023) Global meta-analysis reveals differential effects of microplastics on soil ecosystem. Sci Total Environ 867:161403
Wang J, Coffin S, Sun C, Schlenk D, Gan J (2019a) Negligible effects of microplastics on animal fitness and HOC bioaccumulation in earthworm Eisenia fetida in soil. Environ Pollut 249:776–784
Wang J, Liu X, Li Y, Powell T, Wang X, Wang G, Zhang P (2019b) Microplastics as contaminants in the soil environment: a mini-review. Sci Total Environ 691:848–857
Wang Z, Fu D, Gao L, Qi H, Su Y, Peng L (2021) Aged microplastics decrease the bioavailability of coexisting heavy metals to microalga Chlorella vulgaris. Ecotoxicol Environ Saf 217:112–199
Wang K, Chen W, Tian J, Niu F, Xing Y, Wu Y, Xu L (2022) Accumulation of microplastics in greenhouse soil after long-term plastic film mulching in Beijing, China. Sci Total Environ 828:154544
Wang Y, Wang Y, Shao T, Wang R, Dong Z, Xing B (2024) Antibiotics and microplastics in manure and surrounding soil of farms in the Loess Plateau: occurrence and correlation. J Hazard Mater 465:133434
Wei L, Su Z, Yue Q, Huang X, Wei M, Wang J (2024) Microplastics, heavy metals, antibiotics, and antibiotic resistance genes in recirculating aquaculture systems. TrAC Trends Anal Chem 172:117564

Wong JKH, Lee KK, Tang KHD, Yap PS (2020) Microplastics in the freshwater and terrestrial environments: prevalence, fates, impacts and sustainable solutions. Sci Total Environ 719:137512

Wu X, Liu Y, Yin S, Xiao K, Xiong Q, Bian S, Yang J (2020) Metabolomics revealing the response of rice (Oryza sativa L.) exposed to polystyrene microplastics. Environ Pollut 266:115159

Wu X, Lu J, Du M, Xu X, Beiyuan J, Sarkar B, Wang H (2021) Particulate plastics-plant interaction in soil and its implications: a review. Sci Total Environ 792:148337

Wu F, Ha X, Wang S, Li J, Gao Y (2024) Microplastic occurrences, transport, and quantification and associated effects on primary productivity and carbon cycling processes in freshwater ecosystems. TrAC Trends Anal Chem 172:117611

Xu G, Yang L, Xu L, Yang J (2022) Soil microplastic pollution under different land uses in tropics, southwestern China. Chemosphere 289:133176

Xu L, Wang Y, Wei F, Dai Z, Zhang M (2024) Transport behavior of microplastics in soil–water environments and its dependence on soil components. Environ Pollut 346:123542

Yi M, Zhou S, Zhang L, Ding S (2021) The effects of three different microplastics on enzyme activities and microbial communities in soil. Water Environ Res 93(1):24–32

Yu L, Zhang J, Liu Y, Chen L, Tao S, Liu W (2021). Distribution characteristics of microplastics inagricultural soils from the largest vegetable production base in China. Science of the Total Environment, 756, 143860.

Zhang J, Zou G, Wang X, Ding W, Xu L, Liu B, Chen Y (2021) Exploring the occurrence characteristics of microplastics in typical maize farmland soils with long-term plastic film mulching in northern China. Front Mar Sci 8:1–13

Zhang K, Wang M, Li Y, Zhang X, Xiao K, Ma C, Chen Y (2024a) Wheat (Triticum aestivum L.) seedlings performance mainly affected by soil nitrate nitrogen under the stress of polyvinyl chloride microplastics. Sci Rep 14(1):4962

Zhao X, Zhou Y, Liang C, Song J, Yu S, Liao G, Wu C (2023) Airborne microplastics: occurrence, sources, fate, risks and mitigation. Sci Total Environ 858:159943

Zhao S, Zhu L, Li D (2016). Microscopic anthropogenic litter in terrestrial birds from Shanghai, China: Not only plastics but also natural fibers. Science of the Total Environment, 550, 1110-1115.

Zhou CQ, Lu CH, Mai L, Bao LJ, Liu LY, Zeng EY (2021a) Response of rice (Oryza sativa L.) roots to nanoplastic treatment at seedling stage. J Hazard Mater 401:123412

Zhou J, Wen Y, Marshall MR, Zhao J, Gui H, Yang Y, Zang H (2021b) Microplastics as an emerging threat to plant and soil health in agroecosystems. Sci Total Environ 787:147444

Zhou X, Zhao Y, Pang G, Jia X, Song Y, Guo A, Ji M (2022) Microplastic abundance, characteristics and removal in large-scale multi-stage constructed wetlands for effluent polishing in northern China. Chem Eng J 430:132752. https://doi.org/10.1016/j.cej.2021.132752

Zhou S, Wang L, Liu J, Zhang C, Liu X (2024) Microplastics' toxic effects and influencing factors on microorganisms in biological wastewater treatment units, vol 89. Water Sci Technol, p wst2024040

Zhu BK et al (2018) Exposure to nanoplastics disturbs the gut microbiome in the soil oligochaete Enchytraeus crypticus. Environ Pollut 239:408–415

Microplastic in Glacier Region: Presence, Distribution, and Possible Sources

6

Hardik Giri Gosai and Monika Yadav

Abstract

The role cf human behavior in the global spread of marine debris cannot be understated. A growing awareness of environmental issues is evident in remote regions like glacier areas. Although researchers have discovered glacier ice as a temporary deposit for microplastics, the processes that release and transport these pollutants through the ocean ecosystem remain unclear. Moreover, there is an ongoing debate about the actual significance of microplastics from nearby sources in the glacier environment. This chapter focuses on providing an understanding of the dispersal patterns of microplastics in various habitats, such as biota, snow, ocean, and sediment. By shedding light on their origins and routes of travel in this critical region, one can determine the impact of microplastics on the environment.

Keywords

Environmental contamination · Glacier · Microplastic · Marine litter · Remote place

H. G. Gosai
Department of Environmental Studies, Faculty of Science, The Maharaja Sayajirao University of Baroda, Vadodara, Gujarat, India
e-mail: hardik.g.g-envphd@msubaroda.ac.in

M. Yadav (✉)
Department of Applied Sciences and Humanities (Environmental Science), Faculty of Engineering and Technology, Rama University, Kanpur, India

A. Kumar, V. Singh (eds.), *Microplastics Pollution and its Remediation*,
https://doi.org/10.1007/978-981-97-4068-0_6

6.1 Introduction

Plastics are long-chain polymeric materials created by humans, which are widely employed in many facets of life because of their exceptional qualities that include low cost, great plasticity and flexibility, corrosion resistance, thermal and electrical insulation, and lightweight. Numerous channels allow a notable quantity of plastic garbage to reach the environment, leading to major pollution issues as a result of poor management and disposal techniques. Once discharged into the environment, plastic trash can degrade into smaller bits of rubbish by using a range of biological, physical, and chemical processes. The particles are referred to as "microplastics," and microplastic contamination is now a major worry on a worldwide scale. The greater diversity and complexity of kinds, shapes, sizes, and compositions of microplastics seen in real environments is considered to be associated with their toxicity (Zhang et al. 2021).

Since most plastic materials are extremely robust and decompose slowly in the environment, there is an enormous buildup of plastic garbage. The waste products of various plastic materials can be categorized as nanoplastics (5 mm) in terms of size. As microplastic trash is more difficult to gather for recycling and quickly distributes in natural environments, microplastic contamination is a serious issue for the environment that affects both health of human and the food chain. Microplastic fragments generally fall into two categories: primary fragments, which are designed for a specific purpose, and secondary fragments, which are residue left behind when bigger plastic pieces break down. The impact of plastic particles on nature has still not been completely comprehended, even though it has been shown that they affect plant and animal life and are present within the food chain. Moreover, airborne pieces of plastic can build up in the lungs of human beings and have several negative health impacts (Stefánsson et al. 2021).

Toxic substances could be introduced into marine life via microplastics. Heavy metals, hydrocarbons, herbicides, pesticides, and POPs (persistent organic pollutants) are a few examples of the toxins that the marine environment has stuck to their surface. The marine biota is negatively impacted by these pollutants in many ways, such as decreased carcinoma, oxidative stress, cellular or subcellular damage, immunological response, and inflammation. Chemical additives are added to plastic items during production to enhance their final characteristics (Faggio et al. 2018; Aliko et al. 2018; Gosai et al. 2024; Gosai and Mankodi 2024a, b; Sureda et al. 2018; Sharma et al. 2024).

The most prevalent plastic debris found floating in marine habitats worldwide includes plastic bags, bottles, fishing lines, and packaging straps. Food wrappers, cigarette butts, straws/stirrers, and drink bottles are the most often discovered objects on coastal beaches. In contrast, the primary sources of freshwater plastic contamination were found to be lids and bottles, food wrappers, cigarette butts, and bags. Despite the prevalence of plastic debris in urban and agricultural soils, there is currently limited research on microplastic components in terrestrial ecosystems (Parolini et al. 2021). For example, it is predicted that more than five trillion microplastics float on the ocean's surface, and that number will likely rise as a result of macroplastic degradation. On the other hand, it is believed that contamination on land is more than 20 times than that in aquatic ecosystems. These studies have all shown that aquatic and terrestrial ecosystems worldwide are contaminated by both macroplastics and microplastics. Furthermore, it has been shown by recent monitoring surveys that plastic contamination affects even the so-called distant locations. Numerous plastic objects, primarily microplastics, have been discovered in the Southern Ocean, the Arctic, Antarctica, and the deep water (Parolini et al. 2021).

In a similar vein, microplastics have also been found in high-mountain ecosystems, including glacier waters (Antacli et al. 2024), soils in high-altitude areas (Feng et al. 2023), seasonal snow (Wen et al. 2024), Plateau lakes (Zhang et al. 2024), and surface waters from plain and mountainous lakes (Malygina et al. 2021). According to these findings, microplastics may be able to travel through the atmosphere to these far-off places. Increased anthropogenic activity in the Southern Ocean and Antarctica has the potential to negatively impact environmental, scientific, and historic assets found in the region's marine and terrestrial ecosystems as well as the cryosphere. The scientific community is paying close attention to the growing problem of plastic contamination in Southern ocean and Antarctica. Given the role that the polar regions play in maintaining the balance of the global climate, maintaining the environmental sustainability of Antarctic marine ecosystems and preserving it are two of the core objectives leading to the sustainable development of our planet (Caruso et al. 2022).

It has recently been discovered that polar ice contains both nanoplastics and microplastics (Hamilton et al. 2024; Rosso et al. 2024; Riboni et al. 2024). Scientists are increasingly concerned and becoming more aware of the issue of plastic pollution in Antarctica and the Southern Ocean. These regions have traditionally been considered pristine environments because of their extreme climate and isolated location. Nonetheless, there is a problem with plastic waste because of fishing, tourism, and research (Anfuso et al. 2020; Antacli et al. 2024; Ambrosini et al. 2019; Cózar et al. 2017; Golubeva and Gradova 2024). Plastics, both micro and nano, can get swirled in growing sea ice in polar locations, alter further, and then resurface as food for animals that eat sea ice during the ice's cyclical melting. These mechanisms, which have been well reported, indicate that the quantities of microplastics in the Arctic Sea ice can be multiple orders of magnitude more than those in the adjacent seas (Mountford and Morales Maqueda 2021).

6.2 Fate and Dispersion of Microplastics in Glacier

One of the largest biomes on the Earth is sea ice, which covers more than 35 million km^2 of the planet's surface every year, all year long (or roughly 8% of the world's oceans) (McQuaid 2023). During the austral winter and spring, sea ice covers an area of around 17–20 million km^3 as it spreads away from the continent into the Southern Ocean. There are two types of sea ice found in the Southern Ocean: pack ice, which travels naturally in accordance with winds and currents, and land-fast ice, which is sea ice that is fixed to a shore or ice shelf. Unlike the multiyear ice that is typical in the Arctic, the majority of the ice in the Antarctic is first-year ice, with more than 80% of the sea ice thawing each season (Thomas and Dieckmann 2003). In 2005, microplastics were first inadvertently found in ice cores. The key sink for microplastic litter is the Arctic Sea ice since mean concentrations of 6.3 × 103 particles/L in fast-melting ice and 2.3 × 104 particles/L in pack ice have been observed (Peeken et al. 2018). Sea ice concentrations of microplastic in Baltic Sea varied from 8 to 41 particles/L of melting ice, according to a follow-up field research (Geilfus et al. 2019).

Figure 6.1 shows multiple sources of microplastics in the glacier region. Most of the fibers of microplastic discovered in sea ice, biota, sediments, marine surface water of Antarctic are likely from adjacent sources, including research boats, tourism, sewage, recreational and scientific research stations, and fishing activities. Microplastics may spread quickly once they get into the water, so the Arctic's marine ecosystem is at risk (Materić et al. 2022; Waluda et al. 2020; Kanhai et al. 2018; Cózar et al. 2017; Mu et al. 2019). Plastics may be carried to the Arctic Ocean by ocean currents; the annual flow of plastics to this area is estimated to be 62,000–105,000 tons (Zarfl and Matthies 2010). Furthermore, more maritime activity is likely to result from the Arctic's melting sea ice, which could lead to the immediate microplastic discharge into Arctic Ocean. The origins, modes of transportation, and final destinations of microplastics in the Arctic remain largely unknown. It may be significant because sea ice has been suggested as a major seasonal absorber and means of transportation for these pollutants. Additionally, local anthropogenic activities like shipping, many long-distance transportation

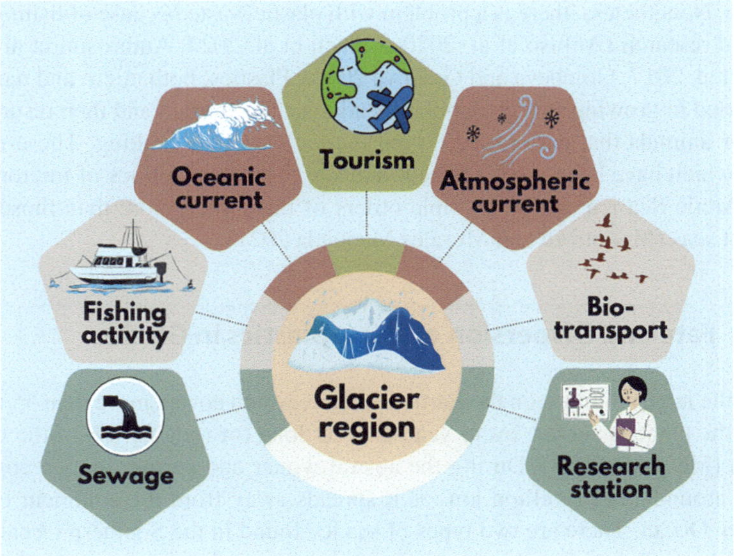

Fig. 6.1 Source of microplastics in glacier

mechanisms, such as riverine input, biotransport, and oceanic currents, allow plastic pollution to reach the Arctic Ocean (Kanhai et al. 2018).

Given that lower-density microplastics are constantly moving through the layer of water above ice and then forth again, model-based results indicate that they are transported both inside and outside the shelf. The results imply that the Barents Sea may be contaminated by lighter microplastic fragments that are trapped in the ice layer and may originate from sources in the Arctic Region. Microplastics embedded in the ice move faster than those in the water's top layers. If there isn't any freezing, fewer lower-density microplastics flow over the Fram Strait (Golubeva and Gradova 2024). Bio-transport is an additional long-distance route that plastics might use to reach the polar regions. Plastic consumption was seen in the Northern Fulmar (Fulmaris glacialis) and Arctic Thick-billed Murre (Uria lomvia). Further study demonstrates that, as the seabirds shifted from their wintering habitats in the North Atlantic Ocean to the Arctic, they took the toxins with them (Kanhai et al. 2018). The most effective technique to control microplastics will most likely be determined by the source of the ice. Microplastics were found in abundance on the experiment's open sea ice tank's surface, which indicates that wet or dry atmospheric deposition might be a mechanism for microplastics to permeate Arctic sea ice. Nonetheless, they were unable to validate these experimental findings by analyzing Baltic Sea ice cores collected in situ (Geilfus et al. 2019).

6.3 Microplastic in the Glacier's Snow and Ice

Glaciers are the biggest freshwater reservoirs on Earth; they make up a sizable portion of the land surface, and are crucial to the overall water cycle despite the paucity of studies on microplastic pollution. Microplastic fragments were found in snow cores taken from a secluded portion of the Vatnajökull ice cap's center. The samples contained a variety of microplastic particle types with diverse morphologies. Two plastic fibers measured 1300 μm and three plastic fragments had equal-area diameters (EAD) ranging from around 30 μm to 3000 μm. Polyamide, polyvinyl chloride, acrylonitrile butadiene styrene, and polyurethane were among the plastic compounds that were found (Stefánsson et al. 2021). An order of magnitude less microplastic is projected to be present in snow generally on the Canadian Archipelago and Greenland's northern coast (870 items m^3) than in ice cores of sea during the similar assessment (Kim et al. 2021). Microplastics smaller than 11 μm were found in sea ice cores at five places in the Transpolar Drift, which is two to three times greater than what was revealed in Central Arctic studies. Peeken et al. (2018) reported that the majority of microplastics in ice cores of sea were blue in hue. Microplastics can enter snow from nearby sources, travel short distances, or even travel far distances, as evidenced by a recent study of isolated alpine catchments (Allen et al. 2019).

All 19 field samples of Antarctic snow taken in the Ross Island region included microplastics; 109 of these particles were identified as such using μFTIR spectroscopy. The microplastic's average concentration in melted snow was 29.4 particles/ L^{-1}, and 13 distinct types of polymer were observed in all snow samples from all sites, with polyethylene terephthalate (PET) being the most frequent polymer and fibers being the most prevalent morphotype (Aves et al. 2022). In Kanhai et al.'s (Kanhai et al. 2020) investigation, microplastic contents in sea ice cores are orders many times greater than in surface waters under similar floats of ice (0–18 particles/ m^3). The degree of solar energy absorbed as a result of the existence of sea ice microplastics may influence its melting as well as albedo (Geilfus et al. 2019). The study of European and Arctic snow samples had a microplastic concentration of 0–14.4 × 10^3 N liter^{-1} in theArctic snow, which was significantly lower than the European snow microplastics concentration, i.e., 0.19–154 × 10^3 N L^{-1} (Bergmann et al. 2019).

6.4 Microplastic in the Glacier's Biota

There is mounting evidence that practically all marine ecosystems, especially distant ones like the Arctic and Antarctic, are already experiencing microplastic contamination (<5 mm in size). It has been determined that Antarctic saltwater and sediments contain microplastics; however, there is not much information about how they reach Antarctic marine food webs and whether higher predators swallow them. Numerous studies conducted all around the world have demonstrated that biota interacts with microplastics mainly through ingestion. Since the 1970s, numerous investigations conducted in the Arctic and its surrounding regions have reported cases of plastic ingestion (Tirelli et al. 2022). Plastics would soon become accessible for cryopelagic populations during sea ice melt, including zooplankton and primary producers that are exclusively connected to sea ice. The possible ecotoxicological effects of contact with micro- and mesoplastics are more likely to affect animals in the food web of sea ice that are exposed to larger amounts of nano- and microplastics dispensed from sea ice (Caruso et al. 2022).

Recent research on Antarctic collembolan Cryptopygus antarticus gathered from King George Island (South Shetland Islands) revealed evidence of microplastics in its gut. The composition of the microplastic fragments matched that of a sizable piece of microplastic foam that had washed up on the island's coasts. This work provides the first and only proof that microarthropods like Cryptopygus antarcticus can eat microplastic, which can then enter the Antarctic soil food webs (Bergami et al. 2020a). Marine microplastics can be of various forms, such as spheres, rods, irregular bits, and fibers. There is evidence of their existence in coastal regions and the oceans. The 60 samples produced 603 microfibers with a mean prevalence of 2.40/100 m^{-3}. Microfibers come in various colors (black, red, blue, and transparent), range of lengths, and with a diameter of 10–22 μm. The majority of these microfibers were found to be polymers made primarily of polyethylene glycols, polyurethanes, polyethylene terephthalates, and polyamides. They were found to be entangled in a variety of different zooplankton species. The Antarctic continent's biodiversity may be declining as a result of these microfibers (Absher et al. 2019).

Based on the investigation of Mancuso et al. (2023), black particles made up 42% of all the particles detected in specimens of Antarctic fish, specifically

Trematomus bernacchii in the Ross Sea, Antarctica. Blue (28%), translucent (7%), red (9%), yellow (4%), pale blue (6%), and white (4%) particles were next in order. Further concern arises regarding the dissemination of nano- and microplastics throughout terrestrial and marine ecosystems, including their potential impact on Antarctic food chains. This apprehension stems from the natural breakdown of polystyrene (PS) particles into nanoplastics by Antarctic krill, as well as the ingestion of small microplastic fragments by Antarctic collembolan from discarded foam along the coastline (Bergami et al. 2020b). In penguin scats, seven distinct polymer kinds were found. The majority (60%) of the particles were microfibers made of polyester, with semisynthetic cellulose polymer fiber Rayon being second (15%); these fibers were only found on the subantarctic island of Signy in the South Orkney Islands of Antarctica. According to Bessa et al. (2019), the single polymers polyacrylonitrile, polypropylene, polyethylene, and polyacrylate were identified as the residual particles, which were fragments and films. There have been reports of Adélie penguins (*Pygoscelis adeliae*) becoming entangled in fishing lines, instances of macroplastics being discovered in the pellets from adult emperor penguin (Apenodytes forsteri) and south polar skuas (Catharacta maccormicki), and cases of inadvertently eating plastic weaved rope and giving it to its chick in the form of its food lump (Golubev 2020). Overall, it confirms that microplastics are present in different levels of biota in glacier region.

6.5 Microplastic in the Surrounding Seawater of the Glacier

The clean surroundings of Antarctica and its value as an ideal region in research on global dynamics are both being threatened by the growing human population there. The majority of freshwater on the Earth is found in continental ice, which is also a major source of information about the Earth's past, including meteorite impacts, paleoclimate, and the presence of extremophilic species in lakes beneath glaciers. However, despite mounting worries about microplastic contamination in the Southern Ocean region surrounding Antarctica, there are still a lot of unexplored sites in this large region. Antacli et al. (2024) found microplastic in each sample in their current study. The microfilter samples had a 0.44 mean concentration 0.44 ± 0.44 MPs/L and varied from 0.02 to 2.14 MPs/L. The average quantity of MPs/L in the Niskin bottle samples was 19.03 ± 18.21, with a range of 0.40–55.67 MPs/L. Anthropogenic black, transparent, and pinkish microfibers were the most common forms of microplastics. They varied in dimension from 0.06 to 7.96 mm (Niskin bottle) and 0.11–3.6 mm (Microfilter). Zhang et al. (2022) conducted research in Antarctica's Southern Ocean, namely, in the Ross Sea. According to the research, there may be two microplastic contamination sources in Antarctic water: the continent itself and the ocean's southerly flow at the middle and lower latitudes. The amount of microplastic present in surface water was 0.10 ± 0.14 items/m^3, with the Ross Sea's water having the greatest concentration. The average amount of microplastic in the groundwater was 1.66 ± 1.20 items/m^3. The Dumont d'Urville Sea in the Southern Ocean, which is located north of the coast of Adélie Land in

East Antarctica, was found to have the highest documented richness. Polyester contributed for 87.3% of surface water microplastics, whereas polyethylene (22.8%), polypropylene (33.1%), and polyester (28.7%) were the most frequent microplastics in subsurface water.

Microplastic was detected in 65% and 11.4% in surface and subsurface samples from the Antarctic Weddell Sea, respectively, with average levels of 0.01 MP m^{-3} and 0.04 MP m^{-3}. The findings are consistent with values that have been reported for South of the Polar Front. One hundred two particles (13.2%) encompassed synthetic polymers whereas 56 particles (7.3%) contained natural polymers (Leistenschneider et al. 2021). Microplastics are widely distributed throughout the Arctic water (Cózar et al. 2017). The study conducted in the Arctic Central Basin, which examined several water masses, found that the average value was lower, around 0.7 particles/m^3. These results supported the theory that microplastics are being transported from surface waters to the seafloor by confirming that they are dispersed throughout the column of water (Kanhai et al. 2018). In the vicinity of Nyælesund, Svalbard, Norway, there have been some of the highest recorded concentrations of microplastics (61.2 particles per m^3) in Arctic surface waters (Maria 2019). The Canadian Arctic Archipelago's subsurface waters contained microplastics from 0 to 282 items m^3 range, along with an average of 31 ± 17 items m^3 and a size larger than 50 μm. Seventy-one percent of the microplastics were polyester and acrylic fibers, making up the bulk of the microplastics (Jones-Williams et al. 2021). The research also notes that the amount of microplastics detected in the marginal ice zone's seawater (158 ± 155 items L^{-1}) is significantly higher than that of the Arctic Sea ice, indicating that most of the microplastics were likely released during sea ice melting (von Friesen et al. 2020). Morgana et al. (2018) found a typical level of concentration 2.4 microplastics/m^3 of subsurface water in Northeast Greenland, which was bigger than 700 μm size. In the meantime, the Greenland Sea saw a rise in marine water microplastics between 2005 and 2014, which may be related to the region's reduced sea ice extent in 2014 or rising plastic production and use (Amélineau et al. 2016).

6.6 Microplastic in the Glacier's Nearby Sediments

The sub-Antarctic island of South Georgia's intertidal sediments include microplastic detritus and small microplastics (>20 μm). In Barnes et al. 2009, Barnes and colleagues released the initial report confirming the presence of microplastics. There has been little research into painting litter and microplastics in the Southern Ocean's coastal sediments. An investigation conducted on underwater sediment cores of the Antarctic Peninsula, South Georgia, as well as the South Sandwich Islands—that is, the Antarctic marine region impacted most by research, tourist attractions, and fishing—discovered that the majority of samples contained microplastics that were spread fairly evenly across the three marine areas (Cunningham et al. 2020). As stated by Anfuso et al. (2020), two most prominent activities on the Antarctic Peninsula, going back to the late twentieth century, are academic studies at 75 scientific stations and tourism. Utilizing Raman spectroscopy, the plastic debris of Admiralty Bay was detected (King George Island, Antarctica) at 17beaches; it was also used to assess the quantity and substance of beach litter. There was a typical value of 0.16 items m^{-1}. Wood items made up 47.27% of the total and were composed of processed wood fragments. The following groups accounted for less than 2% of the total: hard pieces of plastic (9.68%, high-density polyethylene or polyvinyl chloride), foam (21%), rubber particles (2.81%), metal (3.37%), foamed plastic bits (2.66%, polystyrene), and miscellaneous objects. Rubber, hard plastic, paint pieces, paper/cardboard, and metal debris are essentially artifacts left over from historical whaling and research excursions. Materials like polyurethane and expanded polystyrene have distinct origins.

The first investigation of microplastic particles (>125 μm) in sea sediments from 14 Canadian Arctic locations looked at both microplastics and microfibers, or semi-synthetic, synthetic, as well as cellulose modified by human. Between 2014 and 2017, samples from all around the Canadian Arctic were gathered. The verified microplastic concentrations in the sediments varied from 0.6 to 4.7 particles g^{-1} dry weight (dw). Eighty-two percent of the APs were microfibers, and 15% were fragments (Adams et al. 2021). The microplastics in the coastal sediments of western Svalbard, Norway, ranged from 2 to 26 items kg^{-1} (Carlsson et al. 2021). According to reports, microplastic concentrations in coastal sediments are comparatively higher in areas where human activity is present (von Friesen et al. 2020). Ambrosini et al. (2019) discovered 2–7 microplastic fragments per sample, which equates to an average of 74.4 ± 28.3 SE items kg^{-1} of sediment (dry weight). There was no variation detected between the fine and sparse supraglacial debris (78.3 ± 30.2 items kg^{-1}) from Italian glaciers and cryoconite (70.5 ± 32.9 items kg^{-1}). In all pooled samples, fibers made up 65.2% of the items and fragments, 34.8%. The color of the fibers and microplastic fragments varied, exhibiting blue, light blue, red, black, violet, and translucent hues. Polyester was used to make the majority of plastic products overall, with polyamide, polyethylene, and polypropylene coming in second. Seven distinct polymer types were found by μFTIR research from the Southern Ocean and Antarctica: polyesters (PEst, including alkyd), polyvinylchloride (PVC), polystyrene (PS), polyurethane (PU), polypropylene (PP), acrylic polymers (AP), and rubber (TPE) (Cunningham et al. 2020).

6.7 Conclusions

The glacier regions are linked to global occurrences and act as recipients of anthropogenically produced microplastic that disperse globally via broad air and ocean. Furthermore, the numerous people traveling glaciers for conducting investigation, tourism, fishing, and activities such as fuel burning, inadvertent gasoline or sewage spills, waste generation, and the unintended spread of invasive alien species represent a direct risk to the glacial environment. As of the previous few years, there has been enhanced concern about the prevalence of plastic debris in marine ecosystems connected with glaciers, as it is one of the most common and extensively distributed kind of manmade junk in the globe. Microplastic pollution has undoubtedly reached the Polar Regions. Variability in microplastic collecting and analyzing techniques, the expansive and difficult-to-reach character of Polar Regions, and the absence of integrated studies all impede reliable assessment of microplastic contamination in these areas. More variations in study findings are produced by the decision to incorporate or omit fibers, which are commonly present and may account for most cases of microparticles found in glacier research. However, the expansion of research confirming the ubiquity of microplastics in the glacier area broadens our comprehension of the widespread nature of contamination of microplastic.

References

Absher TM, Ferreira SL, Kern Y, Ferreira AL, Christo SW, Ando RA (2019) Incidence and identification of microfibers in ocean waters in Admiralty Bay, Antarctica. Environ Sci Pollut Res 26(1):292–298. https://doi.org/10.1007/s11356-018-3509-6

Adams JK, Dean BY, Athey SN, Jantunen LM, Bernstein S, Stern G, Diamond ML, Finkelstein SA (2021) Anthropogenic particles (including microfibers and microplastics) in marine sediments of the Canadian Arctic. Sci Total Environ 784:147155. https://doi.org/10.1016/j.scitotenv.2021.147155

Aliko V, Qirjo M, Sula E, Morina V, Faggio C (2018) Antioxidant defense system, immune response and erythron profile modulation in gold fish, Carassius auratus, after acute manganese treatment. Fish Shellfish Immunol 76:101–109. https://doi.org/10.1016/j.fsi.2018.02.042

Allen S, Allen D, Phoenix VR, Le Roux G, Durántez Jiménez P, Simonneau A, Binet S, Galop D (2019) Atmospheric transport and deposition of microplastics in a remote mountain catchment. Nat Geosci 12(5):339–344. https://doi.org/10.1038/s41561-019-0335-5

Ambrosini R, Azzoni RS, Pittino F, Diolaiuti G, Franzetti A, Parolini M (2019) First evidence of microplastic contamination in the supraglacial debris of an alpine glacier. Environ Pollut 253:297–301. https://doi.org/10.1016/j.envpol.2019.07.005

Amélineau F, Bonnet D, Heitz O, Mortreux V, Harding AMA, Karnovsky N, Walkusz W, Fort J, Grémillet D (2016) Microplastic pollution in the Greenland Sea: Background levels and selective contamination of planktivorous diving seabirds. Environmental Pollution, 219, 1131–1139. https://doi.org/10.1016/j.envpol.2016.09.017

Anfuso G, Bolívar-Anillo HJ, Asensio-Montesinos F, Portantiolo Manzolli R, Portz L, Villate Daza DA (2020) Beach litter distribution in Admiralty Bay, King George Island, Antarctica. Mar Pollut Bull 160:111657. https://doi.org/10.1016/j.marpolbul.2020.111657

Antacli JC, Di Mauro R, Rimondino GN, Alurralde G, Schloss IR, González GA, Morales S, Ottero A, Vodopivez C (2024) Microplastic pollution in waters of the Antarctic coastal environment of Potter Cove (25 de Mayo Island/King George Island, South Shetlands). Sci Total Environ 915:170155. https://doi.org/10.1016/j.scitotenv.2024.170155

Aves AR, Revell LE, Gaw S, Ruffell H, Schuddeboom A, Wotherspoon NE, LaRue M, McDonald AJ (2022) First evidence of microplastics in Antarctic snow. Cryosphere 16(6):2127–2145. https://doi.org/10.5194/tc-16-2127-2022

Barnes DKA, Galgani F, Thompson RC, Barlaz M (2009) Accumulation and fragmentation of plastic debris in global environments. Philos Trans R Soc B Biol Sci 364(1526):1985–1998. https://doi.org/10.1098/rstb.2008.0205

Bergami E, Manno C, Cappello S, Vannuccini ML, Corsi I (2020a) Nanoplastics affect moulting and faecal pellet sinking in Antarctic krill (Euphausia superba) juveniles. Environ Int 143:105999. https://doi.org/10.1016/j.envint.2020.105999

Bergami E, Rota E, Caruso T, Birarda G, Vaccari L, Corsi I (2020b) Plastics everywhere: first evidence of polystyrene fragments inside the common Antarctic collembolan. Biol Lett 16(6):20200093. https://doi.org/10.1098/rsbl.2020.0093

Bergmann M, Mützel S, Primpke S, Tekman MB, Trachsel J, Gerdts G (2019) White and wonderful? Microplastics prevail in snow from the Alps to the Arctic. Sci Adv 5(8):eaax1157. https://doi.org/10.1126/sciadv.aax1157

Bessa F, Ratcliffe N, Otero V, Sobral P, Marques JC, Waluda CM, Trathan PN, Xavier JC (2019) Microplastics in gentoo penguins from the Antarctic region. Sci Rep 9(1):14191. https://doi.org/10.1038/s41598-019-50621-2

Carlsson P, Singdahl-Larsen C, Lusher AL (2021) Understanding the occurrence and fate of microplastics in coastal Arctic ecosystems: The case of surface waters, sediments and walrus (Odobenus rosmarus). Science of The Total Environment, 792, 148308. https://doi.org/10.1016/j.scitotenv.2021.148308

Caruso G, Bergami E, Singh N, Corsi I (2022) Plastic occurrence, sources, and impacts in Antarctic environment and biota. Water Biol Secur 1(2):100034. https://doi.org/10.1016/j.watbs.2022.100034

Cózar A, Martí E, Duarte CM, García-de-Lomas J, van Sebille E, Ballatore TJ, Eguíluz VM, González-Gordillo JI, Pedrotti ML, Echevarría F, Troublè R, Irigoien X (2017) The Arctic Ocean as a dead end for floating plastics in the North Atlantic branch of the thermohaline circulation. Sci Adv 3(4):e1600582. https://doi.org/10.1126/sciadv.1600582

Cunningham EM, Ehlers SM, Dick JTA, Sigwart JD, Linse K, Dick JJ, Kiriakoulakis K (2020) High abundances of microplastic pollution in Deep-Sea sediments: evidence from Antarctica and the Southern Ocean. Environ Sci Technol 54(21):13661–13671. https://doi.org/10.1021/acs.est.0c03441

Faggio C, Tsarpali V, Dailianis S (2018) Mussel digestive gland as a model tissue for assessing xenobiotics: an overview. Sci Total Environ 636:220–229. https://doi.org/10.1016/j.scitotenv.2018.04.264

Feng S, Lu H, Yao T, Tang M, Yin C (2023) Analysis of microplastics in soils on the high-altitude area of the Tibetan Plateau: multiple environmental factors. Sci Total Environ 857:159399. https://doi.org/10.1016/j.scitotenv.2022.159399

Geilfus NX, Munson KM, Sousa J, Germanov Y, Bhugaloo S, Babb D, Wang F (2019) Distribution and impacts of microplastic incorporation within sea ice. Mar Pollut Bull 145:463–473. https://doi.org/10.1016/j.marpolbul.2019.06.029

Golubev S (2020) Macroplastic in seabirds at Mirny, Antarctica. Birds 1(1):13–18. https://doi.org/10.3390/birds1010003

Golubeva E, Gradova M (2024) Numerical study of the riverine microplastic distribution in the Arctic Ocean. Water 16(3):441. https://doi.org/10.3390/w16030441

Gosai HG, Mankodi P (2024a) Evaluation of coastal sediments for heavy metal contamination, Bhavnagar coast, gulf of Khambhat, Gujarat, India. Soil Sediment Contam Int J, 1–26. https://doi.org/10.1080/15320383.2024.2319857

Gosai HG, Mankodi P (2024b) Evaluation of surface water from the Western coast Bhavnagar, gulf of Khambhat, Gujarat, India. Thalassas 40:669. https://doi.org/10.1007/s41208-023-00656-0

Gosai HG, Jadeja F, Sharma A, Jain S (2024) Occurrence and toxicity of organic microcontaminants in agricultural perspective: an overview. In: Bhadouria R, Tripathi S, Singh P, Singh R, Singh HP (eds) Organic micropollutants in aquatic and terrestrial environments. Springer Nature, Switzerland, pp 107–126. https://doi.org/10.1007/978-3-031-48977-8_6

Hamilton BM, Harris LN, Provencher JF, Rochman CM (2024) The utility of monitoring snow for microplastics in the Arctic: a pilot study from Iqaluktuuttiaq, Nunavut. Arctic Sci. https://doi.org/10.1139/as-2023-0041

Jones-Williams K, Galloway TS, Peck VL, Manno C (2021) Remote, but Not Isolated—Microplastics in the Sub-surface Waters of the Canadian Arctic Archipelago. Frontiers in Marine Science, 8:666482. https://doi.org/10.3389/fmars.2021.666482

Kanhai LDK, Gårdfeldt K, Lyashevska O, Hassellöv M, Thompson RC, O'Connor I (2018) Microplastics in sub-surface waters of the Arctic Central Basin. Mar Pollut Bull 130:8–18. https://doi.org/10.1016/j.marpolbul.2018.03.011

Kanhai LDK, Gardfeldt K, Krumpen T, Thompson RC, O'Connor I (2020) Microplastics in sea ice and seawater beneath ice floes from the Arctic Ocean. Sci Rep 10(1):5004. https://doi.org/10.1038/s41598-020-61948-6

Kim S-K, Lee H-J, Kim J-S, Kang S-H, Yang E-J, Cho K-H, Tian Z, Andrady A (2021) Importance of seasonal sea ice in the western Arctic Ocean to the Arctic and global microplastic budgets. J Hazard Mater 418:125971. https://doi.org/10.1016/j.jhazmat.2021.125971

Leistenschneider C, Burkhardt-Holm P, Mani T, Primpke S, Taubner H, Gerdts G (2021) Microplastics in the Weddell Sea (Antarctica): a forensic approach for discrimination between environmental and vessel-induced microplastics. Environ Sci Technol 55(23):15900–15911. https://doi.org/10.1021/acs.est.1c05207

Malygina N, Mitrofanova E, Kuryatnikova N, Biryukov R, Zolotov D, Pershin D, Chernykh D (2021) Microplastic pollution in the surface waters from plain and Mountainous Lakes in Siberia, Russia. Water 13(16):2287. https://doi.org/10.3390/w13162287

Mancuso M, Conti Nibali V, Porcino N, Branca C, Natale S, Smedile F, Azzaro M, D'Angelo G, Bottari T (2023) Monitoring of anthropogenic microplastic pollution in antarctic fish (emerald rockcod) from the Terranova Bay after a quarter of century. Sci Total Environ 904:167244. https://doi.org/10.1016/j.scitotenv.2023.167244

Maria G (2019) Anthropogenic microlitter in wastewater and marine samples from Ny-Ålesund, Barentsburg and Signehamna, Svalbard

Materić D, Kjær HA, Vallelonga P, Tison J-L, Röckmann T, Holzinger R (2022) Nanoplastics measurements in northern and southern polar ice. Environ Res 208:112741. https://doi.org/10.1016/j.envres.2022.112741

McQuaid CD (2023) Understanding the effects of climate change on Southern Ocean ecosystems. Afr J Mar Sci 45(3):155–163. https://doi.org/10.2989/1814232X.2023.2239871

Morgana S, Ghigliotti L, Estévez-Calvar N, Stifanese R, Wieckzorek A, Doyle T, Christiansen JS, Faimali M, Garaventa F (2018) Microplastics in the Arctic: A case study with sub-surface water and fish samples off Northeast Greenland. Environmental Pollution, 242, 1078–1086. https://doi.org/10.1016/j.envpol.2018.08.001

Mountford AS, Morales Maqueda MA (2021) Modeling the accumulation and transport of microplastics by sea ice. J Geophys Res Oceans 126(2):e2020JC016826. https://doi.org/10.1029/2020JC016826

Mu J, Qu L, Jin F, Zhang S, Fang C, Ma X, Zhang W, Huo C, Cong Y, Wang J (2019) Abundance and distribution of microplastics in the surface sediments from the northern Bering and Chukchi seas. Environ Pollut 245:122–130. https://doi.org/10.1016/j.envpol.2018.10.097

Parolini M, De Felice B, Lamonica C, Cioccarelli S, Crosta A, Diolaiuti G, Ortenzi MA, Ambrosini R (2021) Macroplastics contamination on glaciers from Italian Central-Western Alps. Environ Adv 5:100084. https://doi.org/10.1016/j.envadv.2021.100084

Peeken I, Primpke S, Beyer B, Gütermann J, Katlein C, Krumpen T, Bergmann M, Hehemann L, Gerdts G (2018) Arctic Sea ice is an important temporal sink and means of transport for microplastic. Nat Commun 9(1):1505. https://doi.org/10.1038/s41467-018-03825-5

Riboni N, Ribezzi E, Nasi L, Mattarozzi M, Piergiovanni M, Masino M, Bianchi F, Careri M (2024) Characterization of small micro-and nanoparticles in Antarctic snow by electron microscopy and Raman micro-spectroscopy. Appl Sci 14(4):1597. https://doi.org/10.3390/app14041597

Rosso B, Scoto F, Hallanger IG, Larose C, Gallet JC, Spolaor A, Bravo B, Barbante C, Gambaro A, Corami F (2024) Characteristics and quantification of small microplastics (<100 μm) in seasonal svalbard snow on glaciers and lands. J Hazard Mater 467:133723. https://doi.org/10.1016/j.jhazmat.2024.133723

Sharma A, Jadeja F, Gosai HG, Jain S (2024) Organic micropollutants in the freshwater ecosystem: environmental effects, potential treatments, and limitations. In: Bhadouria R, Tripathi S, Singh P, Singh R, Singh HP (eds) Organic micropollutants in aquatic and terrestrial environments. Springer Nature, Switzerland, pp 203–224. https://doi.org/10.1007/978-3-031-48977-8_10

Stefánsson H, Peternell M, Konrad-Schmolke M, Hannesdóttir H, Ásbjörnsson EJ, Sturkell E (2021) Microplastics in glaciers: first results from the Vatnajökull ice cap. Sustainability 13(8). https://doi.org/10.3390/su13084183

Sureda A, Capó X, Busquets-Cortés C, Tejada S (2018) Acute exposure to sunscreen containing titanium induces an adaptive response and oxidative stress in Mytillus galloprovincialis. Ecotoxicol Environ Saf 149:58–63. https://doi.org/10.1016/j.ecoenv.2017.11.014

Thomas DN, Dieckmann GS (2003) Sea ice: an introduction to its physics, chemistry, biology and geology. Wiley-Blackwell, Hoboken

Tirelli V, Suaria G, Lusher AL (2022) Microplastics in polar samples. In: Rocha-Santos T, Costa MF, Mouneyrac C (eds) Handbook of microplastics in the environment. Springer International Publishing, pp 281–322. https://doi.org/10.1007/978-3-030-39041-9_4

von Friesen LW, Granberg ME, Pavlova O, Magnusson K, Hassellöv M, Gabrielsen GW (2020) Summer sea ice melt and wastewater are important local sources of microlitter to Svalbard waters. Environ Int 139:105511. https://doi.org/10.1016/j.envint.2020.105511

Waluda CM, Staniland IJ, Dunn MJ, Thorpe SE, Grilly E, Whitelaw M, Hughes KA (2020) Thirty years of marine debris in the Southern Ocean: annual surveys of two island shores in the Scotia Sea. Environ Int 136:105460. https://doi.org/10.1016/j.envint.2020.105460

Wen H, Xu H, Ma Y, Zhang C, Zhang D, Wang X (2024) Diverse and high pollution of microplastics in seasonal snow across northeastern China. Sci Total Environ 907:167923. https://doi.org/10.1016/j.scitotenv.2023.167923

Zarfl C, Matthies M (2010) Are marine plastic particles transport vectors for organic pollutants to the Arctic? Mar Pollut Bull 60(10):1810–1814. https://doi.org/10.1016/j.marpolbul.2010.05.026

Zhang K, Hamidian AH, Tubić A, Zhang Y, Fang JKH, Wu C, Lam PKS (2021) Understanding plastic degradation and microplastic formation in the environment: a review. Environ Pollut 274:116554. https://doi.org/10.1016/j.envpol.2021.116554

Zhang S, Zhang W, Ju M, Qu L, Chu X, Huo C, Wang J (2022) Distribution characteristics of microplastics in surface and subsurface Antarctic seawater. Sci Total Environ 838:156051. https://doi.org/10.1016/j.scitotenv.2022.156051

Zhang F, Guo X, Zhang J, Zhang Z (2024) Microplastics in Qinghai-Tibet Plateau and Yunnan-Guizhou Plateau lakes, China. Sci Total Environ 915:169978. https://doi.org/10.1016/j.scitotenv.2024.169978

Microplastics in Crops

7

Laxmi Singh and Ajay Kumar

Abstract

Microplastics (MPLs), or plastic particles smaller than 5 mm, have become a serious issue because of their amassing in the soil in recent years. These particles have substantial effects on ecosystems and organisms alike. As plants are at the base of the food chain and essential to food security, the effect of microplastics on them is very important. This presents an emerging multifaceted food safety and physiological stress issue that needs to be approached strategically. Therefore, it is imperative to look at the consequences of microplastics because of the possible risks to human health through contamination of the food supply chain. The constituents of these MPLs in consumable plants, the contamination of soils, the process of centralization of MPLs, methods for identifying them, the adverse effects of microplastics on a variety of food commodities, and deleterious effects on ecosystem by them are all covered in this chapter. Few research that has been done thus far suggests that microplastics may influence the physicochemical and biological characteristics of soil, inclusive of porosity, aggregate size, pH, evaporation, water retention capacity, and the accessibility of nutrients. The studies that are currently available on the incidental effects in agriculture and their cumulative effect on mixing with more contaminants are included in this chapter.

Keywords

Microplastics · Environment · Contamination · Biological · Nutrients

L. Singh · A. Kumar (✉)
Department of Biotechnology, Faculty of Engineering and Technology, Rama University, Kanpur, Uttar Pradesh, India

A. Kumar, V. Singh (eds.), *Microplastics Pollution and its Remediation*,
https://doi.org/10.1007/978-981-97-4068-0_7

7.1 Introduction

One of the main issues we face today is plastic waste. Both scientists and the general public are becoming more aware of the ongoing impact and perseverance of pollution from plastic in our natural surroundings (Alimi et al. 2018; Karthik et al. 2018; Choi et al. 2020; Rochman and Hoellein 2020). Humans have utilized plastic products and litter extensively over the past few decades, particularly by the mid-twentieth century. Nevertheless, most of the plastic culminates in junkheap or the surroundings, and barely any of it is recycled (Geyer et al. 2017), making it among the primary sources of human-caused environmental damage (Zhang and Chen 2020). The environment is contaminated with microplastics attributable to the growing usage and inappropriate disposal of plastics (Prata et al. 2020a, b; Ogunola et al. 2018). Microplastics are minuscule particles of plastic. They are described as plastics with a diameter of smaller than 5 mm (Takahashi et al. 2023). Numerous habitats, including the soil, water, lake, sea, and atmosphere, have been found to be contaminated by microplastics (Tang 2023) (Fig. 7.1). Smaller plastic debris causes more problems because it is more bioavailable, more toxic, and more conducive to the transport of plastic particles through ecosystems. This can happen accidentally during production or as a result of abiotic and biotic processes that break down

Fig. 7.1 Plastic pollution in agricultural farmlands. (The picture shows the contaminated fields. These pictures were taken by Prabhakar Singh at Honwad, District Vijayapura, Karnataka, India)

larger plastic debris, such as atmospheric oxidation, photooxidation (UV radiation), thermo-oxidation, mechanical action, and microbial activity.

The public's health is impacted when consuming products containing microplastics, such as seafood, crustaceans, fish, fruits, and vegetables. Globally, microplastics have been found in a wide range of agricultural soils, from 10 to nearly 13,000 items kg^{-1} (Uwamungu et al. 2022; Ya et al. 2021). As a result, studies on the ecological toxicology of MPLs to soil ecosystems are growing in popularity and are thought to be essential in understanding the safety of microplastics for the environment (de SouzaMachado et al. 2018). Over three million species of plants are edible, meaning that plants are the primary source of food. Plant variety is a crucial characteristic of terrestrial ecosystems (Li et al. 2016). Numerous species have been studied regarding the effects of microplastics inclusive of broad beans. A crucial component of risk assessments is determining how phytotoxic microplastics are to food crops due to their ability to reduce crop yield and transfer trophic levels to people through the food chain.

7.2 Chemical Composition of Microplastics

Agricultural operations introduce microplastics into cropland soils, which are classified as primary or secondary components according to where they originate from. The primary microplastics are made of tiny plastic granules that are utilized as the foundation for commodity products such as cosmetics (Islam 2019) and stray and eliminated from materials such as fish traps. Secondary MPLs are created during the breaking down of larger plastic materials, such as commercial products (Fig. 7.2). Secondary microplastics mostly consist of fibers from tire abrasion, synthetic clothing laundering, and plastic coating particles (Lehtiniemi et al. 2018). These all work together to cause MPLs contamination (Koelmans et al. 2019). Exposure to external elements, primarily solar radiation and ocean waves, is the source of this breakdown. According to Tang et al. (2018), polystyrene is a significant commercial polymer that finds extensive application in everyday life, such as the manufacturing of plastics, plastic packaging, and insulation for buildings, although due to inappropriate disposal methods, polystyrene products are becoming a significant agent contributing to MPLs pollution (Hidalgo-Ruz et al. 2012; Gong et al. 2022).

The farmland soil is exposed to microplastics through a variety of methods, including irrigation with sewage, the application of compost and biosolids, and the fragmentation of other agrarian instruments (Henseler et al. 2019). When agricultural plastics are left in the soil, in the form of trickle pipes and plasticulture, they break down and release secondary microplastics (Wu et al. 2019). Plastics are continuously weathered once discharged into the environment, producing microplastics among other tiny fragments (Abdolahpur Monikh et al. 2022). Changes in the physicochemical characteristics of the soil and the activity of soil animals may be the cause of the detrimental effects on crop growth. Microplastics will become more prevalent as long as plastic is continuously introduced into agriculture and continues to degrade there. It's yet unclear and is an emerging

Fig. 7.2 Commonly found microplastics

hazard that how microplastics may affect crop development and the condition of soil. The downside of microplastics is that they take an extremely long time to break down into innocuous molecules. MPLs have the ability to alter the soil properties, affecting various lifeforms, that involve plants as well (Wang et al. 2022). Literature has demonstrated that plant roots and leaves can absorb microplastics. Consequently, trophic transmission from soil to organisms and humans along the food chain has become a matter of worry (Dar et al. 2019; Nair and Perumal 2022).

7.3 Sources of MPLs in Soil Ecosystems

Plastic pollution was initially identified in marine habitats. Consequently, plenty of studies suggest the source, prevalence, and ultimate destiny of plastic in the aquatic ecosystem, including seawater, seafloor deposits, and coastal areas. The origins, distribution, and ultimate destiny of MPLs in soil are largely unknown (Akhtar et al.

2021). Research in this area evince the abundance of plastic in soil ecosystems is emerging these days, around 0.03–6.7% of plastic in the soils of factory zones (Fuller and Gautam 2016). Plastic eventually mixes in with a complicated blend of organic material and mineral substitutes in the soil after accumulation. Soil organic matter can extremely stabilize and last for many hundred years as a result of interactions with organic minerals (Deng and Dixon 2002; Li et al. 2023). However, soil organic matter comes from a variety of sources, primarily leftovers from different stages of decomposition in plants and microbes. Because of the great variety of soil organic matter elements, identifying plastic residues in soil chemically presents a unique difficulty. It is challenging to identify plastic materials, especially those that are very tiny, based solely on their chemical properties because of the wide variety of functional groups found in soil organic matter. According to research, incessant practices in intensive agriculture, such as the use of mulch sheets, the addition of manure, compost, scraps, municipal biosolids or septage, using irrigation water, atmospheric sedimentation, and other practices, were closely associated with the abundance of MPLs in soil ecosystems (Fig. 7.3).

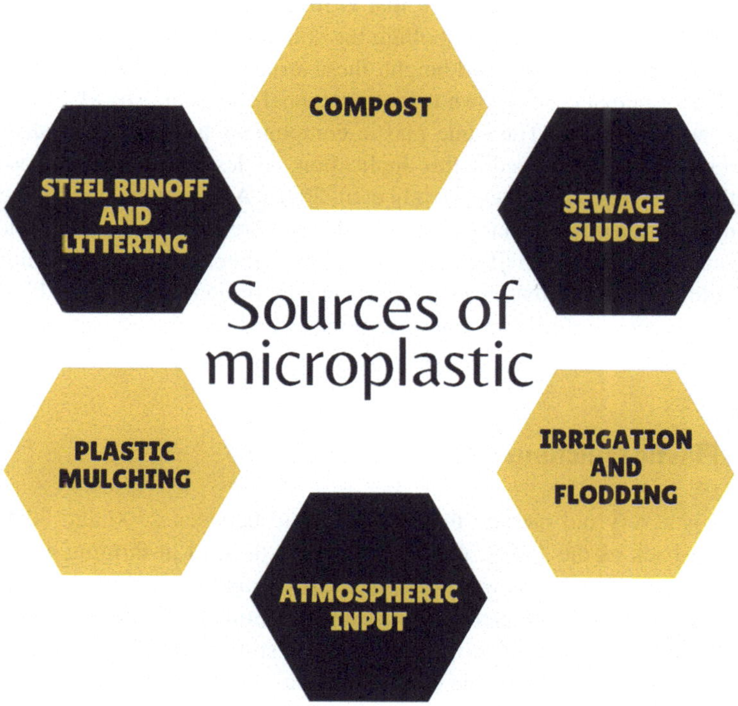

Fig. 7.3 Sources of microplastics in soil

7.3.1 Greenhouse Covering and Films

Greenhouses are long-lasting cultivation constructs primarily coated in plastic films that let in solar radiation while also lowering soil radiation and gas emissions, as well as the possibility of pest infection. Furthermore, these structures maintain particular environmental parameters, such as humidity and temperature, that are necessary for the best possible progression and maturation of plants (Huynh et al. 2023; Salimi et al. 2023). While several polymers may be utilized to produce greenhouse agricultural sheets, three types of polymers account for over 80% of global production: LDPE, vinyl and ethyl ethanoate, and PVC (Cao et al. 2017). The half-life of plastic films, which varies from 6 to 45 months based on UV stabilizers in their chemical makeup, geographical atmosphere, and pesticide usage, along with other considerations, is one of the constraints of employing them in greenhouses (Espi et al. 2006). Consequently, it is necessary to replace greenhouse plastic films with new ones, which adds to the ongoing production of plastic trash. This waste can be burned, recycled, or, in the very pathetic case scenario, dumped to the midden, or left in croplands itself (Wright et al. 2020).

Another polymer-based substance that is frequently used in agriculture is low tunnel films, having the ability to pollute the soil with plastic waste, the only dissimilarity is their complexity and height, these structures work similarly to greenhouses. The principal crops grown using low tunnel sheets are strawberries, melon varieties, and asparagus. The same plastic components used in the production of greenhouse films are also used in this application, but less plastic waste is produced than with greenhouse coverings (Wright et al. 2020). A semicircular construction is more suited for absorbing solar radiation and greatly reduces the risk of wind damage, replacing air, the tunnel film's sealing reduces temperature loss. In addition, the coating minimizes water loss as in comparison with ordinary unprotected land. The tunnel film is more affordable than greenhouses and has the advantage of being simple to install.

7.3.2 Plastic Mulching

Mulching is a very old method that dates back to between 2500 and 3000 years B.C. As far back as the 1950s, plastic mulch was utilized in farming as a crucial technology for increasing the warmth of lands, lowering loss of water, inhibiting the proliferation of grasses, enhancing fertility and structure, facilitating early intake of nutrients and development of roots, prevention of soil erosion, harvesting of crops more quickly, and more profitable cultivation (Nair 2010). It involves covering the soil's surface with organic material sections such as crop debris, crop remnants, fallen leaf, intact fibers, wood chips, ash, or biological waste, which are the materials typically used to create mulching covers in a classical way (Lal 2015). Notwithstanding, compared to the most conventional approaches, the advent of films which are polymer derived in mulching led about considerable amelioration as it was simpler, less expensive, and more widely applicable (Mansoor et al. 2022). The predominant polymer used in mulching is low-density polyethylene (LDPE), which has a thickness of 12–80 μm. Its half-life is exceptionally quick about 8–18 weeks, and it can be biodegraded and lower the expenses related to removal at the end of each crop season. The primary benefit of mulch film that is biodegradable (BDM) is the way it may be plowed underneath the soil post using and will break down naturally, because of the activity of soil organisms (Qi et al. 2018). Nevertheless, completely biodegradable plastic has not yet been discovered (Steinmetz & Schröder 2022; Lwanga et al. 2017; Horton et al. 2017).

7.3.3 Solar Films

Solarization films are one kind of mulching technique. FAO states that a kind of hydrothermal operation occurs in damp soil that gets wrapped in plastic films and subjected to bright sunlight for 4–5 weeks during the summer season. By warming and raising the soil's calefaction at various deepness, this strategy aims to sterilize the soil and lower or perhaps eradicate plant diseases that may have developed through subsequent cropping periods (Qi et al. 2020). Moreover, traces of pesticides in soil may degrade due to solar films. Because of their affordability, efficiency, ease of installation, and ability to produce unique technological alternatives, in irrigation and silage plastic films play a crucial role in storage and irrigation systems. To preserve nutritional benefits, these films are crucial when making silage from grasses or crops for cattle to consume during colder months. Furthermore, plastic is widely used in vegetable and fruit conveying systems (Hussain and Hamid 2003). With the exception of polyvinyl chloride used for irrigation, the various additional usage typically requires low- or high-density polyethylene.

7.4 Concerns About Environmental Risks Associated with Microplastics in Croplands

Although polymer-based products have many advantages in agriculture, their extensive usage has resulted in large volumes of plastic leftovers that stay in the soil (Liu et al. 2014; Jiang et al. 2020). The quality of soil, terrestrial biota, and eventually food webs may be seriously affected by these resistant residues. Exposure to microplastics has been linked to a number of reported negative effects, such as reduction in the species and animal's behavior and soil microbiota, damage to soil texture and aggregation, which can lead to changes in the chemicophysical characteristics of soil (Prata et al. 2020a, b), delay in the rise of roots and shoots of plants and modification of the carbon cycle and greenhouse gas emissions (Boots et al. 2019) (Fig. 7.4).

Despite these negative consequences, some researchers continue to contend that using polymer-based products in agriculture has more advantages than

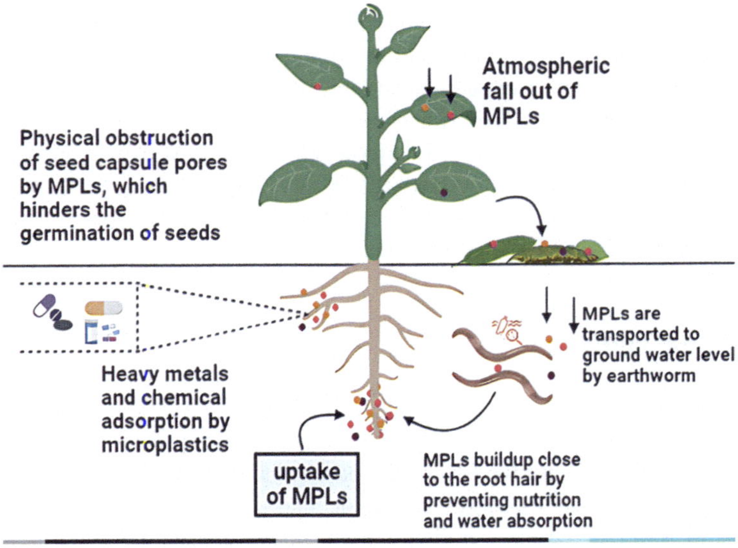

Fig. 7.4 Impact of microplastic in the environment

disadvantages (Jiang et al. 2020). Several researchers found that exposure to microplastics doesn't harm plants, soil animals, or soil; in fact, it may have improved the conditions in which these organisms flourish (Chae and An 2020). For instance, the biological quality and function of soil are not significantly affected in the short term by applying pure polyethylene microplastic in the field (Brown et al. 2023). However, the nitrogen content and biomass of plant leaves are greatly increased by polyamide microplastics (de SouzaMachado et al. 2019). Additionally, by enhancing photosynthesis, polystyrene microplastics have been shown to lengthen roots and boost plant growth in wheat (Zong et al. 2021). However, they have been shown to inhibit these same processes in onions (Allium cepa) and broad beans (V. faba) (Jiang et al. 2019) because of oxidative damage and genotoxicity to roots, especially at high doses (Giorgetti et al. 2020). Moreover, microplastics may be ingested, inhaled, or absorbed by organs, whereupon they may have adverse effects on health, including immunological reactions, inflammation, and cell damage (Sixto et al. 2021).

7.4.1 Impact of MPLs on Soil

The chemicophysical qualities of soil may be negatively impacted by the buildup of plastic trash on the soil's facade and topsoil. These effects are:

- Alterations in the mobility of nutrients and moisture due to the possibility that plastic detritus might hinder water entry and movement rippling, which will eventually lead to the salinity of secondary.

- Decrease in soil pH due to immoderate mineralization.
- Modifications in apparent density (an indicator of soil compaction), stabilization of aggregates, pore space, and aerification.
- Impact on seedlings development, length of roots, and surface area of roots.
- Rise in temperature of the soil.
- Polymeric additives, like esters of phthalic acid, furfurans, polychlorinated dibenzofurans (PCDFs), or metals, can contaminate soil.

The release of poisonous and endocrine compounds like bisphenol A is one way that plastic endangers ecosystems after it finds its way into the soil and surrounding environment. Polychlorinated biphenyls and heavy metals are further harmful pollutants that plastic can absorb. Microplastics are very small plastic particles that pose a threat to the environment. When taken up by biota, these plastic items act as carriers for pollutants (Koelmans et al. 2013), and so, adsorbed contaminants might be incorporated into the food chain.

7.4.2 Effect on Soil Animals

The disposal of microplastics in agricultural or waste soils poses a major risk to other terrestrial creatures. For instance, the consumption of microplastics has been extensively documented in several species of seabirds, with plastic residues identified in their digestive system. Birds living in agricultural ecosystems may perhaps be poisoned by ingesting microplastics. Birds are additionally vulnerable under food chain transfers, a pathway that has been shown in marine food webs. Leed and Smithson (2019) suggested that carnivorous birds may absorb microscopic human-caused litter from plastic waste that their prey has already eaten. According to the literature, microplastics have an inimical effect on the growth, reproduction, and death of many earthworm species that live in soil. In addition, when released from plastic, chemicals like phthalates and plasticizing agents can damage soil-dwelling organisms. These additives may be washed out since they are only irregularly integrated into the polymer backbone of many polymers. Certain phthalates, such as diisooctyl phthalate, have been shown to suppress soil microbiological activity (Das et al. 2021). They are regarded as dangerous soil contaminants because they may have mutagenic, and carcinogenic qualities, or may act as hormonally active agents (Wang et al. 2017). Phthalates can be absorbed by plants, and subsequently pass on to humans and affect them after they are discharged into the environment (Horton & Dixon 2018) (Fig. 7.5).

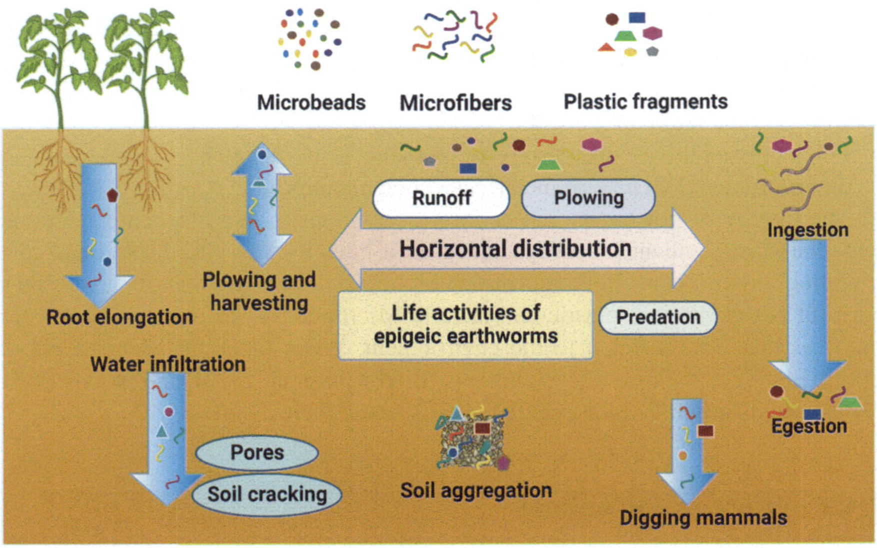

Fig. 7.5 Common microplastics and their effect in agroecosystems

7.4.3 Impact on Plants

There are several ways that microplastics might impact plants and crops. Microplastics can cause oxidative damage by changing the activity of enzymes within cells. Cells in plants and protoplasts can ingest microplastics (Tang et al. 2023). Microplastics have an inimical impact on the development of seeds and roots, although it is unclear how long these effects will continue (Fig. 7.6). Microplastics can enter plants through their roots, build up, and migrate throughout the plant as a result of physiological processes like transpiration. Additionally, it has been proposed that microplastics may impact plants by altering the biophysical characteristics of soil. Microplastics generally possess the ability to change plant variety via modifying soil texture and microbiota (Malinconico 2017).

7.4.3.1 On Overall Development of Plants

Throughout the life cycle of plants, including various stages, continued presence and movement of MPLs may promptly influence the production and growth of crops (Fageria et al. 2002). Based on studies, the terrestrial ecosystem is believed to contain 4–23 more instances of microplastics over marine environment (Wang et al. 2019). Additionally, croplands may hold bounteous microplastics than ocean basins (Hageneder 2022). Nonetheless, new research indicates that plants can absorb microplastics and that the cell wall's pores may have pliability, allowing the granules to pass through more readily. Microplastics' harmful effects might hinder growth, result in anomalies, and lower output altogether (Fig. 7.7). However, the

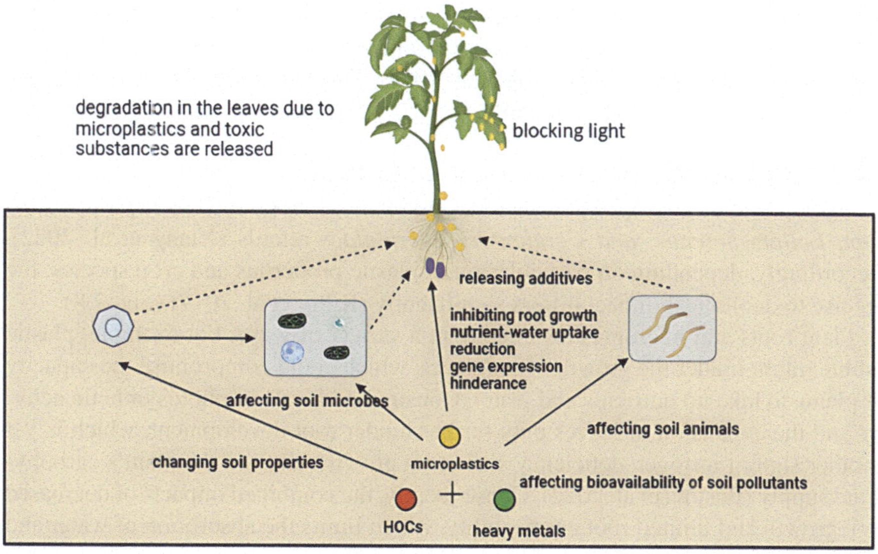

Fig. 7.6 Consequences of MPLs in the environment

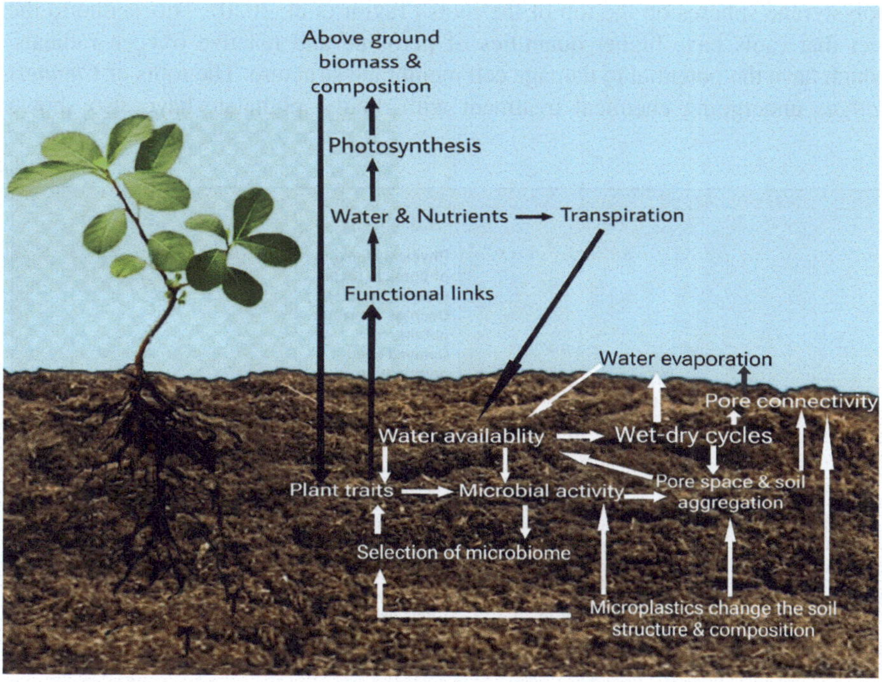

Fig. 7.7 Impact of microplastics in soil and plants

detrimental backlashes of MPLs depend on their specific attributes based on the polymer, quantity, dimensions, appearance, and deterioration, as well as the type of plant and development phase (Rillig et al. 2019). This is because different crops are responsive to microplastics differently.

7.4.3.2 On Seed Germination and Root Growth

Being exposed to microplastics, seed development in crops such as *Lepidium sativum, Lolium perenne, and Ceratopteris pteridoides* retards (Xiang et al. 2022). Accordingly, depending on particular microplastic properties and crop species, the precise toxicological impact differs significantly (Rillig et al. 2017) (Fig. 7.8).

Plant roots and microplastics interact in a variety of ways. For example, plastic rubble might hinder the growth of root hairs, which could compromise the capacity of plants to take up nutrients and water. Constrained levels of photosynthetic activity and the expansion of leaves may further hinder root development, which is yet another sign of nitrogen deficiency since root growth relies on the plant's carbohydrate supply (Postma et al. 2022). Consequently, the combined impacts of decreased root growth and limited root permeability, which limits the absorption of water and nutrients, may demonstrate the potent detrimental consequences of polyethylene buildup. It has been documented that MPLs can attach externally to the roots of duckweed (Dovidat et al. 2020); similarly, *Arabidopsis thaliana* has accumulated polystyrene spheres on the top of the roots (Taylor et al. 2020). This is due to the fact that roots have higher quantities of peroxide and reactive oxygen radicals, which have the potential to damage cell membrane structure. The roots of *Cucumis sativus* undergoing chemical treatment with dibutyl phthalate have also shown

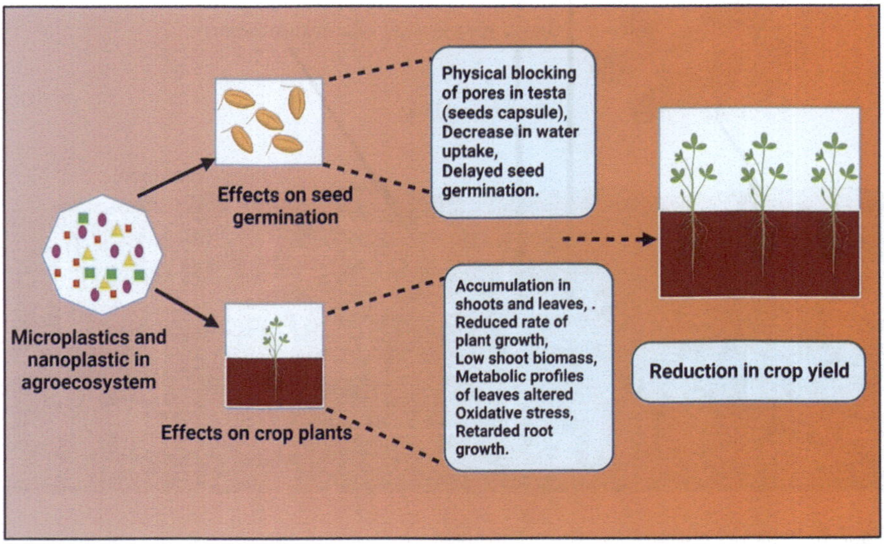

Fig. 7.8 This figure shows the effect of microplastic in seed germination and in the overall growth of the plant

a similar kind of accumulation of MPLs (Zhang et al. 2015). Moreover, these may have a toxic effect on the cytosol and at gene level on roots. The deterioration and abnormalities in the ECM and membranes within cells of roots may result from adhered microplastics (Lian et al. 2021; Li et al. 2021; Wang et al. 2022). In the root cells, it has been found to affect the rates of chromosomal and nuclear anomaly, decrease the mitotic rate, and increase the micronucleus frequency. Comparably, MPLs between 30 and 600 μm have the potential to decrease the viability of the root cells of common duckweed. This could result in damage and modifications to the production of ROS in roots causing alteration in gene expression, lipid peroxidation, and phenylpropanoid manufacture in cell organelles. and their subsequent effects on the root's oxidative stress and antioxidant defense systems (Wang et al. 2022). Additionally, the MPLs that are absorbed may weaken the cell walls and create larger pores, which would enable larger particles to get through the walls and worsen the injury (Chen et al. 2020a). Particularly for root crops, these detrimental effects on root development may ultimately result in a decrease in yield. Microplastics also have several negative effects on root functions, such as reducing transpiration and obstructing the intake of water and nutrients, which may affect radicle structure or root dispersion, influencing nutrient intake; agricultural roots may be harmed by microplastics through effects on their biomass, size, action survival, and development of secondary roots (Gao et al. 2019; Lian et al. 2021; Li et al. 2021; Liu et al. 2020; Lopez et al. 2022).

7.4.3.3 On Plant Growth and Tissue Differentiation

Crops including *Lactuca sativa*, *Vicia faba*, *Glycine max*, *Allium cepa*, *Triticum aestivum*, *Zea mays*, and *Oryza sativa* are adversely affected by MPLs in terms of their growth and tissue development (Zhang et al. 2022). MPLs absorbed at the root can move by apoplastic and symplastic pathways to the aerial portions of plants, where they can agglomerate within the vascular system. Microplastic particles can attach to leaves through atmospheric deposition and possibly impact stomatal functions, which could affect photosynthesis and transpiration. Furthermore, microplastics can enter a crop's root system and hinder tissue development and crop growth (Pignattelli et al. 2020). Additionally, microplastics may limit photosynthesis by influencing the leaves' and shoots' ability to produce chlorophyll. Additionally, their accumulation in leaf veins and stem vascular bundles may inhibit the absorption as well as the transfer of nutrients and water, which would impede the development of stems and tissues (Liu et al. 2020; Dong et al. 2020; Chen et al. 2020a). Vascular plants have the capacity to engross and collect microplastics in larger concentrations, as evidenced by studies, and as such, they may be employed in the remediation of microplastics from soil. This method can be applied to lessen the buildup of MPLs on land and in water. Nevertheless, there is still a lack of quantitative data regarding microplastic absorption, which could help to prevent phytoaccumulation of these newly developing pollutants. Additionally, pathogenic bacteria can be easily transmitted by MPLs to the soil, build up, and suppress plant growth

(Oberbeckmann and Labrenz 2020). Beyond MPLs' entry and transfer into crop tissues, the physiological conditions unique to those tissues may cause the released chemical compounds from MPLs to be significantly elevated. This could lead to the crop tissue sorbing the MPLs internally or peripherally, which could result in toxicity for cell (Maity et al. 2021). However, depending on the crop species and microplastic properties these specific effects on plants may differ. For instance, the chlorophyll content of corn leaves was not significantly affected by polyethylene but was dramatically reduced by polylactic acid at greater concentrations as 10% (Wang et al. 2021).

7.5 Translocation of Microplastics Inside the Plant

Different plant species have different rates of uptake, transport, and building up organic pollutants based on their characteristics (Ng et al. 2018). The surface area, density, volume, potential of roots, xylem, rate of growth, water loss and exchange, potential of tonoplast, and cell membrane are some significant features of plants that probably influence the absorption of organic pollutants. Depending on their dimension and kind, microplastics can enter plant cells and penetrate various parts of the plant entailing copious tissues. Under regulated conditions, plants have been found to absorb very small MPLs (0.2 μm). Vegetables' roots can absorb and store polystyrene microplastics, which can then move in an upward direction (Li et al. 2020; Zeb et al. 2024). In locations where the lateral roots start to appear, the particles penetrated in a way similar to the gap entryway. Very small plastics of micro sizes are effectively absorbed as a result of entrance and unique properties of the microplastics. MPLs start moving toward the shoot once they enter. Elevated rates of transpiration facilitated the assimilation of plastic particles, suggesting that transpiration pull served as the principal factor driving their migration. Elevated rates of transpiration facilitated the assimilation of plastic particles, suggesting that transpiration pull served as the principal factor driving their migration. This route tracks the apoplast passage of water radially through the cell walls of the root's endodermal layer, which contains the suberin, and the cells of undeveloped root regions that have yet to establish Casparian strip and band (Fig. 7.9).

Agricultural processes such as plowing and harvesting, water infiltration, runoff, earthworm and other animal harrowing activity, and root elongation are thought to affect microplastic movement and distribution once they are incorporated into the soil. The process of physically combining soil particles in tilled strata, such as soil microplastics, would result from soil tillage and ploughing. Water penetration in the wet-dry cycles of soil may hasten the vertical movement of microplastics downhill in sandy soil, particularly for smaller microplastic particles (O'Connor et al. 2020). Since crop roots' elongation and expansion can produce bioturbation, which can have a major impact on soil particle movements, it is assumed that crop roots would also have an impact on the distribution of microplastics. Numerous reports have shown that microplastics can build in soil through a variety of channels. They are

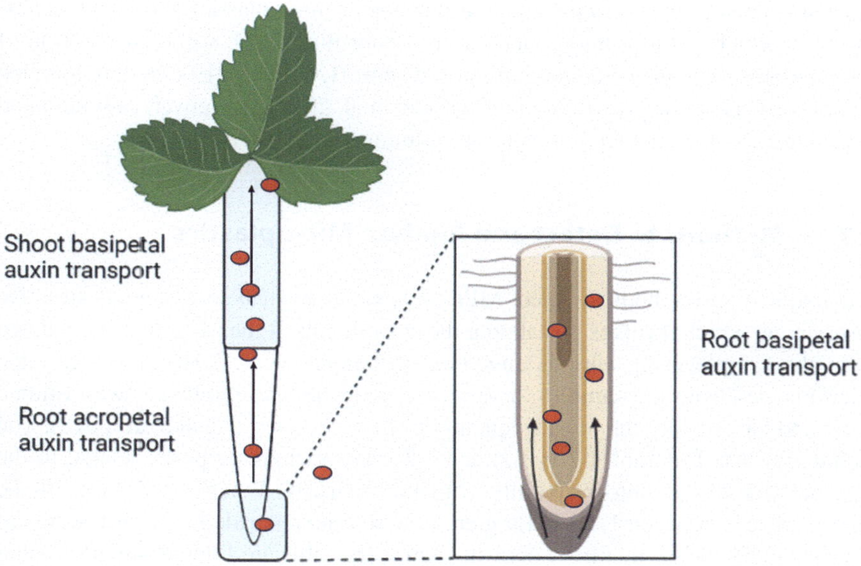

Fig. 7.9 Translocation of microplastics inside plants

thought to travel vertically up the soil profile as a result of plant root growth, soil fauna action, and water penetration. It is yet unknown how MPLs are absorbed by plant roots from the soil, directly deposited onto shoots from the atmosphere, or irrigated with wastewater. It is unclear from the research that is currently available whether there is a risk to humans as well as food chain components from the MPLs exposure route. Comparably, the weathering and degradation of these developing pollutants in the environment may change their characteristics and cause leachates to be released

7.6 Microplastics Interaction with Other Environmental Pollutants

In particular, the impact of soil microplastics (MPLs) on crops when mixed with other contaminants, as well as their direct lethal features affects the growth of crops drastically. Pollutants such as heavy metals, antibiotics, agrochemicals, and other organic contaminants may be adsorbed to the surfaces of microplastics, and microplastic residues may act as external transporters for these contaminants (Xu et al. 2020). Literature has revealed the presence of medications in biosolids, such as ciprofloxacin, llotycin, lincomycin, biaxin, vibramycin-D, doxycycline, monistat, and many more. These materials can withstand soil erosion and are absorbed by plants (de Carvalho and Neto 2016). These drugs then gather in various parts of these plants (Goldstein et al. 2012) and can stunt plant growth (da Costa et al. 2018).

Several studies looked into the coupled impacts of microplastics and heavy metals together, which led to a bigger decline in the fertility of the soil, deterioration of microorganisms in the soil, and nutrient transport (Lv et al. 2019), leading to much notable effects on the growth of plants (Kader et al. 2017). Moreover, cadmium and microplastics may interact and increase cadmium toxicity.

7.7 Methods to Detect and Analyze Microplastics

The majority of methods to detect MPLs rely on a physical sort of nondestructive assessment, which involves visualizing the constituents of the MPL that are isolated from the sample using various analytical techniques (Fig. 7.10). To extract the microplastics from the samples, a variety of reagents are employed, with diluted acids and bases being the most frequently utilized (Akhbarizadeh et al. 2018). The techniques listed in Tables 7.1, 7.2, and 7.3 serve various purposes related to the polymer that comes into proximity with the matrices. When the density of the microplastic is between 1.2 and 1.9 g/cm^3, the sodium chloride in solution serves to separate it. The MPL that have densities lower than this are filtered out after being separated (Bläsing & Amelung 2018). The 50% KI solution, the 96% v/v C_2OH_6l solution, and the 90% v/v $ZnCl_2$ solution are additional reagents that serve a similar purpose (Zobkov and Esiukova 2017). H_2O_2 is primarily utilized as an oxidizing agent, dissolving organic particles that differ from microplastics in order to clean the sample as much as feasible for measurement techniques (Herrera et al. 2019).

In particular, for sample matrices containing tissues together with samples that have a lot of organic matter, diluted acid and base solutions like hydrochloric acid, nitric acid, sodium hypochlorite, sodium hydroxide, and potassium hydroxide are

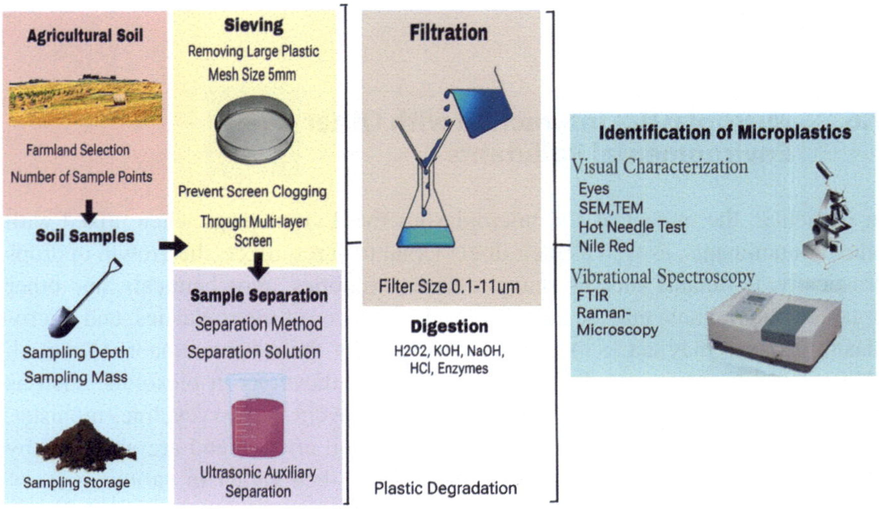

Fig. 7.10 Various methods to analyze and detect MPLs

Table 7.1 Methods to detect MPLs by using sample purification

Method	Advantages	Disadvantages	References
Separation of densities	• High separation efficiency • Ease of use • Low cost • Capacity for handling several samples	• High-density solutions are costly or hazardous • They take a lot of time to separate • High-density plastics can't be separated by a low-density solution	Quinn et al. (2017)
Electrostatic separation	• Good effectiveness in separating • Easy to use • Quick • Basic sample management	• Challenging to distinguish between tiny MPLs • Samples must be spread out and allowed to dry	Felsing et al. (2018)
Oil separation	• Effective separation • Easy to use • Cheap • Quick	• Determination of interfering MPLs	
Froth floatation	• Large-scale sample separation • Easy sample management	• Complicated procedures • Tedious • Weak efficacy of separation	Goethals (2023)
Magnetic extraction	• Distinct small-sized MPLs • Reasonably priced	• Tedious • Difficult processes • Harm to sample	Grbic et al. (2019)

Table 7.2 Methods to detect MPLs using digestion

Method	Advantages	Disadvantages	References
H_2O_2 treatment	• It is suitable for biological samples • Gentle method • It doesn't break down plastic	• Destructive to MPLs • Tedious • Generate high temperature	de Aragão Belé et al. (2021)
Alkaline treatment	• Reasonably priced • Appropriate for biological samples • Powerful ability to digest	• Effectiveness differs depending on the MPL type	Dong et al. (2023)
Acid treatment	• Removal of organic materials is efficient • Economically beneficial	• It is impossible to remove carbon black	Zou et al. (2021)
Enzyme treatment	• Appropriate for biological samples • Doesn't damage plastic	• Costlier • The degree of soil organic matter removal effectiveness remains uncertain	Mbachu et al. (2021)

Table 7.3 Techniques to identify and quantify MPLs

Method	Advantages	Disadvantages	References
Visual characterization	• Quick, simple, and convenient • Low-cost • Recognizes colors, sizes, and shapes • Not harmful	• Insufficient knowledge about the composition of plastics	Chen et al. (2020b)
Fourier transform infrared spectroscopy	• The technique is quick and is not harmful, and can identify the type of plastic. It is also automated	• Costlier • Complicated to examine nanoplastics (NPs) • Severe water interference	Corcoran et al. (2009)
Raman spectroscopy	• Detect plastic composition • High-resolution images are generated • Nanoplastics can be analyzed • Simple to prepare sample • Not harmful	• Costlier • Tedious	Araujo et al. (2018)
Ultrafiltration	• Low-cost • Accessible • High quantities can be used	• No separation from the particle-matrix • Low flow rates • Membrane with tiny holes	Enfrin et al. (2020)
Ultracentrifugation	• Easy to use • Affordable • Basic technique	• Slow • Little amounts • Damage to the sample • Sample deterioration • Absence of separation from the particle-matrix	Li et al. (2021)
Thermal extraction desorption, -Gas chromatography, Mass spectrometry	• Greater amounts of the samples • Quickly locates MPLs of any size and quantifies them	• Detrimental to MPLs • A dry sample is required	Bao et al. (2021)

employed in the process of digestion. They target the organic matter in order to liberate MPLs from the framework so that they can subsequently be separated by filtration (Nuelle et al. 2014) (Fig. 7.11). When K-proteinase is utilized in enzymatic digestion, a consequential portion of the organic matter containing the microplastic can be broken down, resulting in damage to this particular type of microplastic (Cole et al. 2014). Along with density separation, some other techniques are

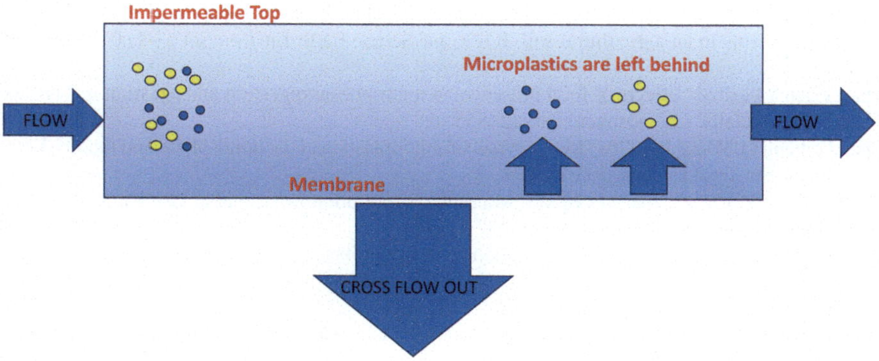

Fig. 7.11 Diagram for ultrafiltration process

employed like electrostatic separation, oil separation, froth flotation, magnetic extraction, filtration, and digestion (Junhao et al. 2021) as shown in Tables 7.1 and 7.2. These microplastics can be visually characterized or can be analyzed by vibrational spectroscopy (Table 7.3).

Advantages and limitations of different methods used for analysis and detection of MPLs.

References

Abdolahpur Monikh F, Darbha GK, Vijver MG, Peijnenburg WJ (2022) An overview of methodologies for tracing and quantifying microplastics in environmental samples. In: Advances in toxicology and risk assessment of nanomaterials and emerging contaminants, pp 21–46

Akhbarizadeh R, Moore F, Keshavarzi B (2018) Investigating a probable relationship between microplastics and potentially toxic elements in fish muscles from northeast of Persian Gulf. Environ Pollut 232:154–163

Akhtar N, Syakir Ishak MI, Bhawani SA, Umar K (2021) Various natural and anthropogenic factors responsible for water quality degradation: a review. Water 13(19):2660

Alimi OS, Farner Budarz J, Hernandez LM, Tufenkji N (2018) Microplastics and nanoplastics in aquatic environments: aggregation, deposition, and enhanced contaminant transport. Environ Sci Technol 52(4):1704–1724

Araujo CF, Nolasco MM, Ribeiro AM, Ribeiro-Claro PJ (2018) Identification of microplastics using Raman spectroscopy: latest developments and future prospects. Water Res 142:426–440

Bao ZZ, Chen ZF, Lu SQ, Wang G, Qi Z, Cai Z (2021) Effects of hydroxyl group content on adsorption and desorption of anthracene and anthrol by polyvinyl chloride microplastics. Sci Total Environ 790:148077

Bläsing M, Amelung W (2018) Plastics in soil: analytical methods and possible sources. Sci Total Environ 612:422–435

Boots B, Russell CW, Green DS (2019) Effects of microplastics in soil ecosystems: above and below ground. Environ Sci Technol 53(19):11496–11506

Brown E, MacDonald A, Allen S, Allen D (2023) The potential for a plastic recycling facility to release microplastic pollution and possible filtration remediation effectiveness. J Hazard Mater Adv 10:100309

Cao D, Wang X, Luo X, Liu G, Zheng H (2017) Effects of polystyrene microplastics on the fitness of earthworms in an agricultural soil. IOP Conf Series: Earth Environ Sci 61(1):012148. IOP Publishing

Chae Y, An YJ (2020) Effects of food presence on microplastic ingestion and egestion in Mytilus galloprovincialis. Chemosphere 240:124855

Chen G, Feng Q, Wang J (2020a) Mini-review of microplastics in the atmosphere and their risks to humans. Sci Total Environ 703:135504

Chen Y, Leng Y, Liu X, Wang J (2020b) Microplastic pollution in vegetable farmlands of suburb Wuhan, central China. Environ Pollut 257:113449

Choi JS, Hong SH, Park JW (2020) Evaluation of microplastic toxicity in accordance with different sizes and exposure times in the marine copepod Tigriopus japonicus. Mar Environ Res 153:104838

Cole M, Webb H, Lindeque PK, Fileman ES, Halsband C, Galloway TS (2014) Isolation of microplastics in biota-rich seawater samples and marine organisms. Sci Rep 4(1):4528

Corcoran PL, Biesinger MC, Grifi M (2009) Plastics and beaches: a degrading relationship. Mar Pollut Bull 58(1):80–84

da Costa JP, Paço A, Santos PS, Duarte AC, Rocha-Santos T (2018) Microplastics in soils: assessment, analytics and risks. Environ Chem 16(1):18–30

Dar MI, Green ID, Khan FA (2019) Trace metal contamination: transfer and fate in food chains of terrestrial invertebrates. Food Webs 20:e00116

Das MT, Kumar SS, Ghosh P, Shah G, Malyan SK, Bajar S et al (2021) Remediation strategies for mitigation of phthalate pollution: challenges and future perspectives. J Hazard Mater 409:124496

de Aragão Belé TG, Neves TF, Cristale J, Prediger P, Constapel M, Dantas RF (2021) Oxidation of microplastics by O3 and O3/H2O2: surface modification and adsorption capacity. J Water Process Eng 41:102072

de Carvalho DG, Neto JAB (2016) Microplastic pollution of the beaches of Guanabara Bay, Southeast Brazil. Ocean Coastal Manag 128:10–17

de Souza Machado AA, Kloas W, Zarfl C, Hempel S, Rillig MC (2018) Microplastics as an emerging threat to terrestrial ecosystems. Glob Chang Biol 24(4):1405–1416

de Souza Machado AA, Lau CW, Kloas W, Bergmann J, Bachelier JB, Faltin E et al (2019) Microplastics can change soil properties and affect plant performance. Environ Sci Technol 53(10):6044–6052

Deng Y, Dixon JB (2002) Soil organic matter and organic-mineral interactions. In: Soil mineralogy with environmental applications, vol 7. pp 69–107

Dong M, Luo Z, Jiang Q, Xing X, Zhang Q, Sun Y (2020) The rapid increases in microplastics in urban lake sediments. Sci Rep 10(1):848

Dong H, Wang X, Niu X, Zeng J, Zhou Y, Suona Z et al (2023) Overview of analytical methods for the determination of microplastics: current status and trends. TrAC Trends Anal Chem 167:117261

Dovidat LC, Brinkmann BW, Vijver MG, Bosker T (2020) Plastic particles adsorb to the roots of freshwater vascular plant Spirodela polyrhiza but do not impair growth. Limnol Oceanogr Lett 5(1):37–45

Enfrin M, Lee J, Le-Clech P, Dumee LF (2020) Kinetic and mechanistic aspects of ultrafiltration membrane fouling by nano- and microplastics. J Membr Sci 601:117890

Espi E, Salmerón A, Fontecha A, García Y, Real AI (2006) Plastic films for agricultural applications. J Plastic Film Sheet 22(2):85–102

Fageria NK, Baligar VC, Clark RB (2002) Micronutrients in crop production. Adv Agron 77:185–268

Felsing S, Kochleus C, Buchinger S, Brennholt N, Stock F, Reifferscheid G (2018) A new approach in separating microplastics from environmental samples based on their electrostatic behavior. Environ Pollut 234:20–28

Fuller S, Gautam A (2016) A procedure for measuring microplastics using pressurized fluid extraction. Environ Sci Technol 50(11):5774–5780

Gao M, Liu Y, Song Z (2019) Effects of polyethylene microplastic on the phytotoxicity of di-n-butyl phthalate in lettuce (Lactuca sativa L. var. ramosa Hort). Chemosphere 237:124482
Geyer R, Jambeck JR, Law KL (2017) Production, use, and fate of all plastics ever made. Sci Adv 3(7):e1700782
Giorgetti L, Spanò C, Muccifora S, Bottega S, Barbieri F, Bellani L, Castiglione MR (2020) Exploring the interaction between polystyrene nanoplastics and Allium cepa during germination: internalization in root cells, induction of toxicity and oxidative stress. Plant Physiol Biochem 149:170–177
Goethals B (2023) Development of a model to predict the recovery of microplastics from sediment by Froth flotation (Doctoral dissertation, Ghent University)
Goldstein MC, Rosenberg M, Cheng L (2012) Increased oceanic microplastic debris enhances oviposition in an endemic pelagic insect. Biol Lett 8(5):817–820
Gong X, Wang Y, Huang D, Zhang J (2022) Effects of microplastics of different sizes on the Chlorella vulgaris-Ganoderma lucidum co-pellets formation processes. Sci Total Environ 820:153266
Grbic J, Nguyen B, Guo E, You JB, Sinton D, Rochman CM (2019) Magnetic extraction of microplastics from environmental samples. Environ Sci Technol Lett 6(2):68–72
Hageneder F (2022) Earth Spirit: healthy planet: global meltdown or global healing. John Hunt Publishing, Winchester
Henseler M, Brandes E, Kreins P (2019) Microplastics in agricultural soils: a new challenge not only for agro-environmental policy?
Herrera A, Ŝtindlová A, Martínez I, Rapp J, Romero-Kutzner V, Samper MD et al (2019) Microplastic ingestion by Atlantic chub mackerel (Scomber colias) in the Canary Islands coast. Mar Pollut Bull 139:127–135
Hidalgo-Ruz V, Gutow L, Thompson RC, Thiel M (2012) Microplastics in the marine environment: a review of the methods used for identification and quantification. Environ Sci Technol 46(6):3060–3075
Horton AA, Dixon SJ (2018) Microplastics: an introduction to environmental transport processes. Wiley Interdiscip Rev Water 5(2):e1268
Horton AA, Walton A, Spurgeon DJ, Lahive E, Svendsen C (2017) Microplastics in freshwater and terrestrial environments: evaluating the current understanding to identify the knowledge gaps and future research priorities. Sci Total Environ 586:127–141
Hussain I, Hamid H (2003) Plastics in agriculture. In: Plastics and the environment, pp 185–209
Huynh HX, Tran LN, Duong-Trung N (2023) Smart greenhouse construction and irrigation control system for optimal Brassica juncea development. PLoS One 18(10):e0292971
Islam S (2019) Environmental governance towards microplastic pollution: the case of personal care and cosmetics products in Bangladesh (Master's thesis)
Ji X, Ma Y, Zeng G, Xu X, Mei K, Wang Z et al (2021) Transport and fate of microplastics from riverine sediment dredge piles: implications for disposal. J Hazard Mater 404:124132
Jiang C, Yin L, Li Z, Wen X, Luo X, Hu S et al (2019) Microplastic pollution in the rivers of the Tibet Plateau. Environ Pollut 249:91–98
Jiang X, Chang Y, Zhang T, Qiao Y, Klobučar G, Li M (2020) Toxicological effects of polystyrene microplastics on earthworm (Eisenia fetida). Environ Pollut 259:113896
Junhao C, Xining Z, Xiaodong G, Li Z, Qi H, Siddique KH (2021) Extraction and identification methods of microplastics and nanoplastics in agricultural soil: a review. J Environ Manag 294:112997
Kader MA, Senge M, Mojid MA, Ito K (2017) Recent advances in mulching materials and methods for modifying soil environment. Soil Tillage Res 168:155–166
Karthik R, Robin RS, Purvaja R, Ganguly D, Anandavelu I, Raghuraman R et al (2018) Microplastics along the beaches of southeast coast of India. Sci Total Environ 645:1388–1399
Koelmans AA, Besseling E, Wegner A, Foekema EM (2013) Plastic as a carrier of POPs to aquatic organisms: a model analysis. Environ Sci Technol 47(14):7812–7820

Koelmans AA, Nor NHM, Hermsen E, Kooi M, Mintenig SM, De France J (2019) Microplastics in freshwaters and drinking water: critical review and assessment of data quality. Water Res 155:410–422

Lal R (2015) Restoring soil quality to mitigate soil degradation. Sustainability 7:5875–5895

Leed R, Smithson M (2019) Ecological effects of soil microplastic pollution. Science Insights 30, 70–84

Lehtiniemi M, Hartikainen S, Näkki P, Engström-Öst J, Koistinen A, Setälä O (2018) Size matters more than shape: ingestion of primary and secondary microplastics by small predators. Food Webs 17:e00097

Li J, Qu X, Su L, Zhang W, Yang D, Kolandhasamy P et al (2016) Microplastics in mussels along the coastal waters of China. Environ Pollut 214:177–184

Li C, Busquets R, Campos LC (2020) Assessment of microplastics in freshwater systems: a review. Sci Total Environ 707:135578

Li J, Wang B, Chen Z, Ma B, Chen JP (2021) Ultrafiltration membrane fouling by microplastics with raw water: behaviors and alleviation methods. Chem Eng J 410:128174

Li Q, Wang L, Fu Y, Lin D, Hou M, Li X et al (2023) Transformation of soil organic matter subjected to environmental disturbance and preservation of organic matter bound to soil minerals: a review. J Soils Sediments 23(3):1485–1500

Lian J, Liu W, Meng L, Wu J, Zeb A, Cheng L et al (2021) Effects of microplastics derived from polymer-coated fertilizer on maize growth, rhizosphere, and soil properties. J Clean Prod 318:128571

Liu EK, He WQ, Yan CR (2014) 'White revolution' to 'white pollution'—agricultural plastic film mulch in China. Environ Res Lett 9(9):091001

Liu P, Zhan X, Wu X, Li J, Wang H, Gao S (2020) Effect of weathering on environmental behavior of microplastics: properties, sorption and potential risks. Chemosphere 242:125193

Lopez MD, Toro MT, Riveros G, Illanes M, Noriega F, Schoebitz M et al (2022) Brassica sprouts exposed to microplastics: effects on phytochemical constituents. Sci Total Environ 823:153796

Lv X, Dong Q, Zuo Z, Liu Y, Huang X, Wu WM (2019) Microplastics in a municipal wastewater treatment plant: fate, dynamic distribution, removal efficiencies, and control strategies. J Clean Prod 225:579–586

Lwanga EH, Gertsen H, Gooren H, Peters P, Salánki T, van der Ploeg M et al (2017) Incorporation of microplastics from litter into burrows of Lumbricus terrestris. Environ Pollut 220:523–531

Maity S, Biswas C, Banerjee S, Guchhait R, Adhikari M, Chatterjee A, Pramanick K (2021) Interaction of plastic particles with heavy metals and the resulting toxicological impacts: a review. Environ Sci Pollut Res 28:60291–60307

Malinconico M (ed) (2017) Soil degradable bioplastics for a sustainable modern agriculture. Springer

Mansoor Z, Tchuenbou-Magaia F, Kowalczuk M, Adamus G, Manning G, Parati M et al (2022) Polymers use as mulch films in agriculture—a review of history, problems and current trends. Polymers 14(23):5062

Mbachu O, Jenkins G, Pratt C, Kaparaju P (2021) Enzymatic purification of microplastics in soil. MethodsX 8:101254

Mbachu O, Kaparaju P, Pratt C (2023) Plastic pollution risks in bioretention systems: a case study. Environ Technol 44(17):2525–2538

Nair KP (2010) The agronomy and economy of important tree crops of the developing world

Nair HT, Perumal S (2022) Trophic transfer and accumulation of microplastics in freshwater ecosystem: risk to food security and human health. Int J Ecol 2022:1234078

Ng EL, Lwanga EH, Eldridge SM, Johnston P, Hu HW, Geissen V, Chen D (2018) An overview of microplastic and nanoplastic pollution in agroecosystems. Sci Total Environ 627:1377–1388

Nuelle MT, Dekiff JH, Remy D, Fries E (2014) A new analytical approach for monitoring microplastics in marine sediments. Environ Pollut 184:161–169

O'Connor JD, Murphy S, Lally HT, O'Connor I, Nash R, O'Sullivan J et al (2020) Microplastics in brown trout (Salmo trutta Linnaeus, 1758) from an Irish riverine system. Environ Pollut 267:115572

Oberbeckmann S, Labrenz M (2020) Marine microbial assemblages on microplastics: diversity, adaptation, and role in degradation. Annu Rev Mar Sci 12:209–232
Ogunola OS, Onada OA, Falaye AE (2018) Mitigation measures to avert the impacts of plastics and microplastics in the marine environment (a review). Environ Sci Pollut Res 25:9293–9310
Pignattelli S, Broccoli A, Renzi M (2020) Physiological responses of garden cress (L. sativum) to different types of microplastics. Sci Total Environ 727:138609
Postma JV (2022) An inexpensive atmospheric microplastic collector for use in remote areas. Atmos Pollut Res 13(10):101550
Prata JC, da Costa JP, Lopes I, Duarte AC, Rocha-Santos T (2020a) Environmental status of (micro) plastics contamination in Portugal. Ecotoxicol Environ Saf 200:110753
Prata JC, da Costa JP, Lopes I, Duarte AC, Rocha-Santos T (2020b) Environmental exposure to microplastics: an overview on possible human health effects. Sci Total Environ 702:134455
Qi Y, Yang X, Pelaez AM, Lwanga EH, Beriot N, Gertsen H et al (2018) Macro-and micro-plastics in soil-plant system: effects of plastic mulch film residues on wheat (Triticum aestivum) growth. Sci Total Environ 645:1048–1056
Qi R, Jones DL, Li Z, Liu Q, Yan C (2020) Behavior of microplastics and plastic film residues in the soil environment: a critical review. Sci Total Environ 703:134722
Quinn B, Murphy F, Ewins C (2017) Validation of density separation for the rapid recovery of microplastics from sediment. Anal Methods 9(9):1491–1498
Rillig MC, Ziersch L, Hempel S (2017) Microplastic transport in soil by earthworms. Sci Rep 7(1):1362
Rillig MC, Lehmann A, de Souza Machado AA, Yang G (2019) Microplastic effects on plants. New Phytol 223(3):1066–1070
Salimi M, Channab BE, El Idrissi A, Zahouily M, Motamedi E (2023) A comprehensive review on starch: structure, modification, and applications in slow/controlled-release fertilizers in agriculture. Carbohydr Polym 322:121326
Sixto A, El-Morabit B, Trujillo-Rodríguez MJ, Carrasco-Correa EJ, Miró M (2021) An automatic flow-through system for exploration of the human bioaccessibility of endocrine disrupting compounds from microplastics. Analyst 146(12):3858–3870
Steinmetz Z, Schröder H (2022) Plastic debris in plastic-mulched soil—a screening study from western Germany. PeerJ 10:e13781
Takahashi Y, Tanaka K, Kajiwara T, Suzuki G, Osako M, Kuramochi H (2023) Cross-sectional microstructural analysis to evaluate the crack growth pattern of weathered marine plastics. Chemosphere 331:138794
Tang G, Liu M, Zhou Q, He H, Chen K, Zhang H et al (2018) Microplastics and polycyclic aromatic hydrocarbons (PAHs) in Xiamen coastal areas: implications for anthropogenic impacts. Sci Total Environ 634:811–820
Tang L, Feng JC, Li C, Liang J, Zhang S, Yang Z (2023) Global occurrence, drivers, and environmental risks of microplastics in marine environments. J Environ Manag 329:116961
Taylor SE, Pearce CI, Sanguinet KA, Hu D, Chrisler WB, Kim YM et al (2020) Polystyrene nano- and microplastic accumulation at Arabidopsis and wheat root cap cells, but no evidence for uptake into roots. Environ Sci Nano 7(7):1942–1953
Uwamungu JY, Wang Y, Shi G, Pan S, Wang Z, Wang L, Yang S (2022) Microplastic contamination in soil agro-ecosystems: a review. Environ Adv 9:100273
Vethaak AD, Legler J (2021) Microplastics and human health. Science 371(6530):672–674
Wang W, Ndungu AW, Li Z, Wang J (2017) Microplastics pollution in inland freshwaters of China: a case study in urban surface waters of Wuhan, China. Sci Total Environ 575:1369–1374
Wang J, Liu X, Li Y, Powell T, Wang X, Wang G, Zhang P (2019) Microplastics as contaminants in the soil environment: a mini-review. Sci Total Environ 691:848–857
Wang C, Yu J, Lu Y, Hua D, Wang X, Zou X (2021) Biodegradable microplastics (BMPs): a new cause for concern? Environ Sci Pollut Res 28:66511–66518
Wang F, Wang Q, Adams CA, Sun Y, Zhang S (2022) Effects of microplastics on soil properties: current knowledge and future perspectives. J Hazard Mater 424:127531

Wright SL, Ulke J, Font A, Chan KLA, Kelly FJ (2020) Atmospheric microplastic deposition in an urban environment and an evaluation of transport. Environ Int 136:105411

Wu X, Pan J, Li M, Li Y, Bartlam M, Wang Y (2019) Selective enrichment of bacterial pathogens by microplastic biofilm. Water Res 165:114979

Xiang Y, Jiang L, Zhou Y, Luo Z, Zhi D, Yang J, Lam SS (2022) Microplastics and environmental pollutants: key interaction and toxicology in aquatic and soil environments. J Hazard Mater 422:126843

Xu B, Liu F, Cryder Z, Huang D, Lu Z, He Y et al (2020) Microplastics in the soil environment: occurrence, risks, interactions and fate—a review. Crit Rev Environ Sci Technol 50(21):2175–2222

Ya H, Jiang B, Xing Y, Zhang T, Lv M, Wang X (2021) Recent advances on ecological effects of microplastics on soil environment. Sci Total Environ 798:149338

Zeb A, Liu W, Ali N, Shi R, Wang Q, Wang J et al (2024) Microplastic pollution in terrestrial ecosystems: global implications and sustainable solutions. J Hazard Mater 461:132636

Zhang Z, Chen Y (2020) Effects of microplastics on wastewater and sewage sludge treatment and their removal: a review. Chem Eng J 382:122955

Zhang K, Gong W, Lv J, Xiong X, Wu C (2015) Accumulation of floating microplastics behind the three Gorges Dam. Environ Pollut 204:117–123

Zhang Y, Wang K, Chen W, Ba Y, Khan K, Chen W et al (2022) Effects of land use and landscape on the occurrence and distribution of microplastics in soil, China. Sci Total Environ 847:157598

Zobkov M, Esiukova E (2017) Microplastics in Baltic bottom sediments: quantification procedures and first results. Mar Pollut Bull 114(2):724–732

Zong X, Zhang J, Zhu J, Zhang L, Jiang L, Yin Y, Guo H (2021) Effects of polystyrene microplastic on uptake and toxicity of copper and cadmium in hydroponic wheat seedlings (Triticum aestivum L.). Ecotoxicol Environ Saf 217:112217

Zou Y, Xu Q, Zhang G, Li F, Li F (2021) Influence of six digestion methods on the determination of polystyrene microplastics in organisms using the fluorescence intensity. Adv Anal Sci 2(2)

Microplastics in Foods

8

Amit Joshi, Nahid Akhtar, and Ajay Kumar

Abstract

Microplastic pollution has emerged as a pressing environmental and public health issue, with widespread contamination observed in various ecosystems and food chains worldwide. This chapter provides a comprehensive overview of microplastics in foods, addressing their sources, types, detection methods, occurrence in different food types, human health implications, regulatory frameworks, mitigation strategies, and future research directions. Microplastics originate from diverse sources, including the breakdown of larger plastic debris, microbeads in personal care products, synthetic textiles, and industrial processes. These minute particles pose significant risks to ecosystems and human health due to their persistence and potential to accumulate toxic contaminants. Detection and analysis of microplastics in foods present challenges due to their small size and complex composition. Various techniques, such as microscopy, spectroscopy, and chemical analysis, are employed to identify and quantify microplastics in food samples accurately. Microplastics have been detected in a wide range of food types, including seafood, drinking water, salt, agricultural products, and processed foods. Ingestion of microplastics may lead to physical injury, chemical exposure, inflammatory responses, and hormonal disruption, raising concerns about potential health impacts. Future research directions include investigating the environmental fate and transport of microplastics, assessing their ecological

A. Joshi
Department of Biochemistry, Kalinga University, Naya Raipur, Chhattisgarh, India

N. Akhtar
School of Bioengineering and Biosciences, Lovely Professional University, Jalandhar, Punjab, India

A. Kumar (✉)
Department of Biotechnology, Faculty of Engineering and Technology, Rama University, Kanpur, Uttar Pradesh, India

A. Kumar, V. Singh (eds.), *Microplastics Pollution and its Remediation*, https://doi.org/10.1007/978-981-97-4068-0_8

and human health impacts, developing analytical methods and detection techniques, and exploring innovative mitigation and remediation technologies. Overall, addressing microplastic pollution requires interdisciplinary collaboration, stakeholder engagement, and evidence-based decision-making. By implementing effective strategies and advancing scientific knowledge, we can mitigate the impacts of microplastic pollution and create a cleaner and healthier environment for current and future generations.

Keywords

Microplastics · Pollution · Food chain · Health implications · Detection methods

8.1 Introduction

Just over a hundred years after the inception of the initial modern plastic polymer, plastics have evolved from a scientific marvel to one of the most significant global environmental dilemmas of our time. Plastic possesses the capacity to serve as a definitive indication of human-induced pollution in the current Anthropocene epoch. Plastics are broadly categorized into thermoplastics and thermosetting plastics (thermosets), differing in their ability to be remelted; thermosets cannot be remelted, while thermoplastics can. Examples of thermoplastics include polyethylene (PE), polypropylene (PP), polyvinyl chloride (PVC), polyethylene terephthalate (PET), polystyrene (PS), polycarbonate (PC), and polyamide (PA), whereas common types of thermosets encompass polyurethane (PU), epoxy resins, vinyl esters, and silicones (Pouzada and Sergio Pouzada 2021; Jagadeesh et al. 2022). Among the most prevalent types of plastic encountered in microplastics are thermoplastics, notably PE, PP, PVC, PS, and PET. Microplastics are characterized as a diverse blend of variously shaped materials, denoted as fragments, fibers, spheres, granules, pellets, flakes, or beads, ranging in size from 0.1 to 5000 μm (Quilis et al. 2024; Kumari et al. 2024). They originate from various sources, including the breakdown of larger plastic debris, microbeads from personal care products, fibers from synthetic clothing, and industrial processes. These minute particles pose a significant threat to ecosystems and human health due to their widespread distribution and persistence in the environment. The prevalence of microplastics in various environmental compartments, including oceans, rivers, soil, and even the atmosphere, highlights their global impact (Rathore et al. 2024). However, recent research has also shed light on the presence of microplastics in the food chain, raising concerns about their potential implications for human health. Understanding the sources and types of microplastics in foods is crucial for assessing exposure risks and implementing effective mitigation strategies (Protyusha et al. 2024; Bharti et al. 2024). Furthermore, accurate detection and analysis methods are essential for quantifying microplastic contamination levels in food samples and identifying potential sources. This chapter will explore the sources and types of microplastics found in foods, discuss detection and analysis methods, examine the occurrence of microplastics in different food types, evaluate the human health implications, propose mitigation and prevention strategies, and suggest future research directions to address this crucial environmental and public health issue.

8.2 Sources and Types of Microplastics in Foods

Microplastics in foods originate from multiple sources, reflecting the widespread contamination of the environment by plastic debris. The primary sources of microplastics in foods (Olmo-Gilabert et al. 2024; Saeedi 2024) include:

- Environmental Pollution: Plastic waste, including larger items like bottles and bags, undergoes degradation over time due to environmental factors such as sunlight, heat, and mechanical forces. This process results in the formation of microplastics that can contaminate soil, water bodies, and air, eventually finding their way into the food chain.
- Industrial Processes: Certain industrial activities, such as the manufacturing and processing of plastics, can release microplastics into the environment. These may include microbeads used in personal care products, plastic pellets (nurdles) used as raw materials in plastic production, and microfibers generated during the production and washing of synthetic textiles (Jasso-Salcedo et al. 2024).
- Agricultural Practices: The application of plastic mulches in agriculture for weed control and moisture retention can lead to the fragmentation of these films, resulting in the release of microplastics into the soil (Tian et al. 2024). Additionally, the use of plastic-based fertilizers and pesticides can contribute to microplastic contamination in agricultural products.
- Atmospheric Deposition: Microplastics can be transported over long distances through the atmosphere and subsequently deposited onto terrestrial and aquatic environments (Zhu et al. 2024). This atmospheric deposition may result from the fragmentation of larger plastic items, as well as the release of microplastic particles from vehicle tires and road markings.

The types of microplastics found in foods vary depending on their sources and environmental pathways (Milne et al. 2024). Common types of microplastics detected in food samples include:

- Fibers: Microscopic plastic fibers shed from synthetic clothing, textiles, and fishing gear.
- Fragments: Small fragments of plastic resulting from the breakdown of larger plastic items, such as bottles, packaging materials, and fishing nets.
- Microbeads: Tiny spherical plastic particles used in personal care products like exfoliating scrubs and toothpaste.
- Pellets: Small, pellet-shaped plastic resin pellets (nurdles) used as raw materials in plastic manufacturing processes.
- Films: Thin plastic films used in agriculture as mulches or wraps for food packaging.

- Others: Various other forms of microplastics, including microspheres, granules, and foam particles, may also be present in food samples.

Understanding the diverse sources and types of microplastics in foods is essential for assessing exposure risks and implementing effective mitigation strategies to safeguard food safety and human health.

8.3 Detection and Analysis Methods

Detecting and analyzing microplastics in various matrices, including food samples, present significant challenges due to their small size, diverse shapes, and complex compositions (Martín-Gómez et al. 2024). However, researchers have developed several techniques and methodologies to identify and quantify microplastics accurately. These detection and analysis methods can be broadly categorized into physical, chemical, and spectroscopic techniques, each with its advantages and limitations.

8.3.1 Visual Inspection and Microscopy

Optical Microscopy: Optical microscopes, including stereomicroscopes and polarized light microscopy, are commonly used for visual identification and characterization of microplastics based on their size, shape, and color (Fritz et al. 2024).

Scanning Electron Microscopy (SEM): SEM provides high-resolution imaging of microplastics, allowing for detailed morphological analysis and elemental identification using energy-dispersive X-ray spectroscopy (EDS) (Li et al. 2024).

8.3.2 Chemical Analysis

Fourier Transform Infrared Spectroscopy (FTIR): FTIR spectroscopy is utilized to identify the chemical composition of microplastics by analyzing their molecular vibrations (Circelli et al. 2024). This technique enables the differentiation of various polymer types present in microplastic samples.

Raman Spectroscopy: Raman spectroscopy provides information about the molecular structure and composition of microplastics based on their unique spectral fingerprints (Guo et al. 2024). It offers rapid and nondestructive analysis of microplastic samples.

8.3.3 Thermal Analysis

Thermogravimetric Analysis (TGA): TGA measures the change in weight of microplastic samples as a function of temperature, allowing for the determination of their thermal stability and decomposition characteristics (Enyoh et al. 2024). This technique aids in distinguishing between different polymer types based on their thermal properties.

8.3.4 Spectroscopic Techniques

Atomic Absorption Spectroscopy (AAS) and Inductively Coupled Plasma Mass Spectrometry (ICP-MS): These techniques are employed for the elemental analysis of microplastics, particularly metal contaminants associated with plastic particles (Bacon et al. 2024; Chovancová et al. 2024; Kadadou et al. 2024).

Fluorescence Spectroscopy: Fluorescence spectroscopy is used to detect fluorescently labeled microplastics or plastic additives in complex matrices, offering high sensitivity and selectivity.

8.3.5 Microplastic Extraction and Isolation Methods

Density Separation: Density-based methods, such as density gradient centrifugation and sink-float separation, are employed to isolate microplastics from organic matter and debris in food samples (Mushtak et al. 2024).

Filtration and Sieving: Filtration through pore-sized membranes and sieving using mesh filters are commonly employed to separate microplastics from food matrices based on their size.

Integration of multiple complementary techniques, such as combining microscopy with spectroscopic analysis or thermal characterization, enhances the accuracy and reliability of microplastic detection and analysis in food samples (Mushtak et al. 2024). Moreover, standardization of methods and protocols is essential for ensuring consistency and comparability of results across studies, facilitating a comprehensive understanding of microplastic contamination in the food chain.

8.4 Occurrence of Microplastics in Different Food Types

Microplastics have been detected in various food items, raising concerns about their potential impact on human health. The contamination of different food types with microplastics is attributed to various factors, including environmental pollution, processing techniques, packaging materials, and agricultural practices. Understanding the prevalence and distribution of microplastics in different food types is crucial for assessing exposure risks and implementing mitigation strategies. Seafood, particularly fish and shellfish, is prone to microplastic contamination due to the extensive pollution of marine environments. Microplastics can be ingested by marine organisms through direct exposure to contaminated water or through the consumption of prey organisms. Studies have reported the presence of microplastics in a wide range of seafood products, including fish fillets, shellfish, and crustaceans.

Drinking water sources, such as tap water and bottled water, have also been found to contain microplastic particles. Microplastics can enter drinking water supplies through various pathways, including atmospheric deposition, surface runoff, and wastewater discharge (Srisiri et al. 2024; la Cecilia et al. 2024). Additionally, the leaching of microplastics from plastic pipes and packaging materials during water treatment and distribution processes contributes to their presence in drinking water. Salt, obtained from seawater through evaporation or mining of rock salt deposits, has been found to contain microplastic contaminants. Seawater serves as a major sink for microplastics, with plastic debris accumulating in marine sediments and saltwater bodies. Consequently, salt harvested from these environments may contain microplastic particles derived from marine sources. Microplastics can also contaminate agricultural products through various pathways, including the use of plastic mulches, irrigation with microplastic-contaminated water, and the application of plastic-based fertilizers and pesticides (Zhou et al. 2024). Fruits, vegetables, grains, and other agricultural products may accumulate microplastics from soil, water, and atmospheric deposition, posing potential risks to human consumers (see Table 8.1). Processed foods, including packaged snacks, convenience meals, and beverages, may contain microplastics originating from packaging materials and processing equipment. Plastic packaging materials, such as bottles, containers, and wrappers, can release microplastic particles into food products during storage, transportation, and handling processes. Overall, the occurrence of microplastics in different food types highlights the pervasive nature of plastic pollution and its potential implications for human dietary exposure. Efforts to mitigate microplastic contamination in the food chain require interdisciplinary collaboration among scientists, policymakers, industry stakeholders, and consumers to develop effective solutions and promote sustainable practices.

Table 8.1 Microplastics related to health hazards

Microplastic type	Size range (μm)	Potential health hazards	Examples
Fibers	5–1000	Respiratory irritation, digestive tract blockage	Polyester, nylon, acrylic
Fragments	1–1000	Physical injury, chemical leaching	PET, PE, PP, PS
Microbeads	1–100	Ingestion, gastrointestinal irritation	Polyethylene, polypropylene
Pellets	1–5	Ingestion, toxic chemical absorption	Nurdles
Films	Varies	Choking hazard, chemical migration	PVC, LDPE, HDPE
Spheres	1–500	Inhalation, ingestion	Polystyrene, acrylic
Granules	1–1000	Ingestion, gut irritation	Polyethylene, polypropylene
Foam particles	1–1000	Ingestion, respiratory issues	Expanded polystyrene

8.5 Human Health Implications

The presence of microplastics in the environment and food chain has raised growing concerns about their potential impacts on human health. While research on this topic is still emerging, several potential health implications associated with exposure to microplastics have been identified (Mondal et al. 2024). Ingestion and Absorption: Microplastics present in food and drinking water may be ingested by humans. These particles can range in size from micrometers to millimeters, making them small enough to be consumed unintentionally. Once ingested, microplastics can potentially be absorbed by the gastrointestinal tract and enter systemic circulation. Physical Damage: Larger microplastics or plastic fragments may cause physical damage to the gastrointestinal tract, leading to abrasions, inflammation, or obstruction (Puskic et al. 2024). Additionally, sharp edges or irregular shapes of microplastics could potentially puncture or damage the intestinal lining. Chemical Exposure: Microplastics can adsorb and accumulate chemical contaminants from the environment, including persistent organic pollutants (POPs), heavy metals, and plastic additives. When ingested, these contaminants may leach from microplastics into the body, potentially causing toxicity and adverse health effects. Inflammatory Responses: Exposure to microplastics has been associated with inflammatory responses in the body. Studies have suggested that microplastics may trigger immune reactions and inflammatory processes, which could contribute to the development or exacerbation of inflammatory-related diseases. Hormonal Disruption: Some chemicals found in microplastics, such as phthalates and bisphenol A (BPA), have endocrine-disrupting properties (Kim et al. 2024). These chemicals can interfere with hormonal signaling pathways in the body, potentially leading to reproductive disorders, developmental abnormalities, and other hormonal imbalances (He and Yin 2024). Microbial Interactions: Microplastics in the gastrointestinal tract may interact with gut microbiota, influencing microbial composition and function. Disruption of the gut microbiome balance could have implications for digestive health, immune function, and overall well-being. Translocation and Bioaccumulation: There is growing evidence to suggest that microplastics may translocate from the gastrointestinal tract to other tissues and organs in the body. Additionally, microplastics and associated contaminants may bioaccumulate in various organs over time, potentially increasing health risks with prolonged exposure. While the full extent of the health implications of microplastic exposure is still being elucidated, precautionary measures and further research are warranted to better understand and mitigate potential risks to human health. Strategies to address microplastic pollution in the environment and food chain, along with regulations to limit exposure to harmful chemicals associated with plastics, are essential for safeguarding public health. Additionally, promoting sustainable practices and reducing plastic consumption can help minimize the release of microplastics into the environment (Gayathri et al. 2024), ultimately benefiting both human health and the planet.

8.6 Microplastics in Food Chain

Plastic contamination has become ubiquitous across the planet, with evidence of its presence found in diverse and remote environments, from Arctic snow to desolate deserts. Of particular concern are microplastics, minute particles smaller than 5 mm, which pervade oceans, air, and soil as a result of plastic degradation (Saikumar et al. 2024). Studies have shown that microplastics pose significant risks to human health, damaging cells, disrupting reproductive systems, and interfering with the endocrine system. Moreover, they serve as carriers for harmful chemicals, potentially leading to conditions like cancer and weakened immune responses. The pervasiveness of microplastic contamination means that they can easily infiltrate the human body (Saha and Saha 2024). Inhalation of airborne microplastics is a constant risk, with every breath potentially introducing plastic particles and associated additives into the respiratory system. Additionally, research has revealed that micro- and nanoplastics can penetrate the skin barrier, meaning that even clothing, largely made from synthetic fibers, can contribute to internal plastic exposure. Further compounding the issue is the presence of microplastics in our food supply. Agricultural practices involving the use of plastic-based fertilizers and packaging contribute to the contamination of crops and seafood. Consequently, consuming these items exposes individuals to additional microplastic intake, with certain foods exhibiting higher levels of contamination. Seafood, for instance, often contains significant quantities of microplastics due to pollution in marine environments. Consumption of seafood results in the ingestion of these particles, with estimates suggesting an annual intake of tens of thousands of microplastic pieces per person (Jasso-Salcedo et al. 2024). Similarly, beverages like tea, which frequently utilize plastic-based tea bags, can release billions of microplastic particles into brewed liquids. Even seemingly innocuous items like fruits and vegetables are not immune, as they can absorb microplastics from soil and water sources. This phenomenon is particularly pronounced in certain produce such as apples and carrots, which can contain over 100,000 microplastics per gram. Alarming levels of microplastic contamination extend to everyday consumables like beer and salt, highlighting the pervasive nature of this issue (Damaj et al. 2024). With approximately 28 microplastic particles per liter of beer and significant microplastic presence in common table salt, it is evident that plastic pollution has infiltrated various facets of our lives.

Biomagnification of microplastics in food chains refers to the process by which microplastic particles accumulate and increase in concentration as they move up the food chain. This phenomenon occurs when organisms at lower trophic levels ingest microplastics along with their food and are subsequently consumed by organisms at higher trophic levels. As a result, microplastics become more concentrated in the tissues of predators at the top of the food chain (McMullen et al. 2024). The biomagnification of microplastics can have significant ecological and health implications. For example, predatory species such as large fish, marine mammals, and birds may accumulate high levels of microplastics in their tissues, leading to potential adverse effects on their health and reproductive success. Additionally, humans consuming seafood or other organisms from contaminated food chains may also be exposed to higher concentrations of microplastics, raising concerns about potential risks to human health. Several factors influence the biomagnification of microplastics in food chains, including the size and type of microplastic particles, the feeding behavior of organisms, and the efficiency of digestion and excretion. Smaller microplastic particles, such as fibers and fragments, are more readily ingested and absorbed by organisms, increasing their potential for biomagnification. Additionally, organisms with longer lifespans and higher trophic levels tend to accumulate greater concentrations of microplastics due to the cumulative effects of bioaccumulation over time.

Microplastics enter the food chain through various pathways, with primary sources including:

- Marine Pollution: Plastic debris in oceans and water bodies degrade over time into smaller particles, leading to the ingestion of microplastics by aquatic organisms such as fish, shellfish, and plankton.
- Agricultural Practices: The use of plastic-based mulches, fertilizers, and irrigation systems in agriculture contributes to soil and water contamination, resulting in the uptake of microplastics by crops and livestock.
- Atmospheric Deposition: Microplastics suspended in the air settle onto land and water surfaces, where they can be ingested by terrestrial animals and aquatic organisms.

8.6.1 Pathways of Microplastics in the Food Chain

Once introduced into the environment, microplastics follow complex pathways through the food chain:

- Bioaccumulation: Microplastics accumulate in the tissues of organisms at lower trophic levels, such as algae and small invertebrates, through ingestion or absorption. As predators consume these organisms, microplastics biomagnify, reaching higher concentrations in the tissues of larger predators.
- Trophic Transfer: Microplastics move up the food chain as predators feed on prey containing microplastics, leading to the transfer of plastic particles from one trophic level to another.

8.6.2 Implications for Human Health and the Environment

The presence of microplastics in the food chain raises significant concerns for both human health and the environment:

Human Health Risks: Consuming seafood and agricultural products contaminated with microplastics exposes humans to potential health risks, including the ingestion of plastic particles and associated chemicals (He and Yin 2024). Studies have linked microplastic exposure to adverse effects on reproductive health, immune function, and metabolic processes.

Ecological Impacts: Microplastics can disrupt ecosystems and biodiversity by interfering with nutrient cycling, altering feeding behaviors, and causing physical harm to organisms. Additionally, plastic ingestion can lead to internal injuries, reduced growth rates, and reproductive abnormalities in wildlife.

Poly-fluoroalkyl substances (PFAS) are a group of synthetic chemicals characterized by their strong carbon-fluorine bonds, which impart unique properties such as water and oil resistance, heat stability, and chemical inertness. These properties make PFAS valuable in a wide range of industrial and consumer applications, including waterproof textiles, nonstick cookware, firefighting foams, and food packaging. PFAS compounds are highly persistent in the environment and resist degradation through natural processes, leading to their accumulation in soil, water, air, and living organisms (Thapa et al. 2024). Their widespread use and environmental persistence have raised concerns about potential adverse effects on human health and the environment. Exposure to PFAS can occur through various pathways, including:

- Drinking Water: PFAS can leach into groundwater and surface water from industrial sites, landfills, and wastewater treatment plants, contaminating drinking water sources. Individuals can ingest PFAS by consuming contaminated water or food grown with contaminated water.
- Food: PFAS can accumulate in the food chain through bioaccumulation, with higher levels found in fish, meat, dairy products, and crops irrigated with contaminated water or treated with PFAS-containing pesticides.
- Air: PFAS can be released into the air during manufacturing, use, and disposal of PFAS-containing products, leading to inhalation exposure for nearby communities.

Research suggests that exposure to PFAS may be associated with various adverse health effects, including:

- Toxicity: Some PFAS compounds have been linked to liver damage, immune system suppression, thyroid disorders, and developmental effects in laboratory animals and human studies.
- Carcinogenicity: Certain PFAS compounds, such as perfluorooctanoic acid (PFOA) and perfluorooctanesulfonate (PFOS), have been classified as potential carcinogens based on animal studies and limited evidence in humans.

Reproductive and Developmental Effects: Exposure to PFAS during pregnancy may be associated with reduced birth weight, altered hormone levels, and developmental delays in infants and children. PFAS contamination poses risks to ecosystems and wildlife, including:

- Bioaccumulation: PFAS can accumulate in the tissues of aquatic and terrestrial organisms, leading to biomagnification in food webs and potential harm to wildlife populations.
- Persistence: PFAS compounds can persist in the environment for extended periods, leading to long-term contamination of soil, water, and wildlife habitats.

PFAS contamination represents a significant environmental and public health challenge, requiring coordinated efforts to mitigate exposure, regulate use, and remediate contaminated sites. By addressing PFAS contamination comprehensively, we can work toward safeguarding human health and the environment from the harmful effects of these persistent and widespread chemicals.

8.7 Regulatory Framework and Guidelines

As the awareness of microplastic pollution grows, regulatory bodies around the world are increasingly recognizing the need to address this environmental and public health concern (Baechler et al. 2024). While comprehensive regulations specifically targeting microplastics are still evolving, existing regulatory frameworks and guidelines play a crucial role in managing and mitigating microplastic pollution.

8.7.1 International Initiatives

International organizations such as the United Nations Environment Programme (UNEP) and the World Health Organization (WHO) have started to address microplastic pollution through various initiatives and reports. These efforts aim to raise awareness, promote research collaboration, and develop guidelines for monitoring and managing microplastics in the environment and food chain.

8.7.2 Regional Regulations

Some regions have implemented specific regulations or guidelines to address microplastic pollution. For example, the European Union has adopted measures to reduce plastic pollution, including bans on single-use plastics and restrictions on microplastics in certain products. Similarly, countries in Asia, such as China and Japan, have introduced initiatives to tackle plastic waste and microplastic contamination in water bodies.

8.7.3 Food Safety Standards

Regulatory agencies responsible for food safety, such as the Food and Drug Administration (FDA) in the United States and the European Food Safety Authority

(EFSA) in Europe, are actively assessing the potential risks associated with microplastics in food. These agencies conduct risk assessments, establish safety standards, and provide guidance on monitoring and mitigating microplastic contamination in food products.

8.7.4 Product Regulations

Regulations governing the use of plastic materials in consumer products, cosmetics, and packaging also indirectly contribute to addressing microplastic pollution. Bans or restrictions on microbeads in personal care products, for example, help reduce the release of microplastics into the environment.

8.7.5 Research and Monitoring Programs

Many countries have established research programs and monitoring networks to assess the prevalence and impacts of microplastic pollution. These initiatives provide valuable data and scientific evidence to support regulatory decision-making and policy development.

Despite these efforts, challenges remain in developing comprehensive regulatory frameworks to effectively address microplastic pollution (Jani et al. 2024). Key challenges include the lack of standardized methods for measuring and monitoring microplastics, limited understanding of their long-term impacts on ecosystems and human health, and the global nature of plastic pollution requiring coordinated international action.

Moving forward, continued research, collaboration among stakeholders, and strengthened regulatory measures are essential for reducing microplastic pollution and protecting both the environment and public health. This includes the development of standardized protocols for sampling and analysis, the establishment of safe exposure limits for microplastics in food and water, and the promotion of sustainable practices to prevent further plastic contamination.

8.8 Mitigation and Prevention Strategies

Addressing the issue of microplastic pollution requires a multifaceted approach that involves both mitigating existing contamination and preventing future sources of pollution (Rangel-Buitrago et al. 2023). Here are some key strategies for mitigating and preventing microplastic pollution:

8.8.1 Reduction of Plastic Use

One of the most effective ways to mitigate microplastic pollution is to reduce the overall production and consumption of plastic (Prata et al. 2019). This can be achieved through measures such as promoting the use of reusable products, implementing plastic bans or restrictions on single-use plastics, and encouraging the adoption of alternative materials.

8.8.2 Proper Waste Management

Improving waste management systems is essential for preventing plastic waste from entering the environment (Hahladakis 2020). This includes expanding recycling infrastructure, implementing effective waste collection and disposal programs, and raising public awareness about the importance of proper waste disposal.

8.8.3 Pollution Control Measures

Implementing pollution control measures, such as installing litter traps in rivers and waterways, can help intercept plastic debris before it enters the oceans (Helinski et al. 2021). Additionally, wastewater treatment plants can be upgraded to remove microplastics from effluent before it is discharged into the environment.

8.8.4 Microplastic Filtration and Removal

Developing and deploying filtration technologies to remove microplastics from water sources can help reduce contamination levels (Lenz and Labrenz 2018). These technologies may include mesh filters, sediment traps, and advanced filtration systems capable of capturing microplastic particles.

8.8.5 Sustainable Product Design

Encouraging the development of eco-friendly products and packaging materials can help minimize the release of microplastics into the environment. This involves promoting the use of biodegradable and compostable materials, as well as designing products with minimal plastic content.

8.8.6 Education and Awareness

Increasing public awareness about the impacts of microplastic pollution and the importance of responsible consumption and disposal practices is crucial for driving behavioral change. Educational campaigns, outreach programs, and environmental initiatives can help raise awareness and empower individuals to take action.

8.8.7 Research and Innovation

Investing in research and innovation is essential for developing new technologies and solutions to address microplastic pollution. This includes funding research projects to better understand the sources, pathways, and impacts of microplastics, as well as supporting the development of innovative cleanup technologies and materials.

8.8.8 Policy and Regulation

Enacting and enforcing policies and regulations at the local, national, and international levels is critical for tackling microplastic pollution (Hettiarachchi and Meegoda 2023). This may involve implementing bans or restrictions on plastic products, setting standards for microplastic levels in water and food, and establishing incentives for sustainable practices. By implementing these mitigation and prevention strategies in a coordinated manner, it is possible to reduce the prevalence of microplastics in the environment and protect ecosystems, wildlife, and human health from the harmful impacts of plastic pollution. Collaboration among governments, industries, nongovernmental organizations, and the public is essential for achieving meaningful progress in the fight against microplastic pollution.

8.9 Future Research Directions

As the understanding of microplastic pollution continues to evolve, there are several key research directions that warrant further investigation to address knowledge gaps and inform effective mitigation and management strategies. These future research directions include: Environmental Fate and Transport: Research is needed to better understand the fate and transport of microplastics in various environmental compartments, including oceans, rivers, soils, and the atmosphere (Akdogan and Guven 2019). This includes studying the mechanisms of microplastic transport, deposition, and accumulation in different ecosystems, as well as assessing the long-term persistence and potential for bioaccumulation in biota. Ecological Impacts: Further research is needed to elucidate the ecological impacts of microplastic pollution on aquatic and terrestrial ecosystems. This includes investigating the effects of microplastics on wildlife health, behavior, reproduction, and ecosystem functioning, as well as assessing the potential for trophic transfer and biomagnification of microplastics through food webs. Human Exposure and Health Effects: Research on the human health effects of microplastic exposure is still in its infancy and requires further investigation. This includes assessing the sources, pathways, and levels of human exposure to microplastics through food, water, air, and consumer products, as well as elucidating the potential health impacts, including toxicity, inflammation, and carcinogenicity. Analytical Methods and Detection Techniques: Developing standardized methods and techniques for the detection, quantification, and characterization of microplastics in environmental samples is essential for advancing research in this field. This includes improving the sensitivity, accuracy, and reliability of analytical techniques, as well as developing novel approaches for identifying microplastic sources and tracking their spatial and temporal distribution. Microplastic Risk Assessment: Further research is needed to develop robust risk assessment frameworks for evaluating the environmental and human health risks associated with microplastic pollution. This includes integrating data on microplastic exposure, toxicity, and ecological impacts to assess risk levels and inform risk management decisions at local, regional, and global scales. Mitigation and Remediation Technologies: Research on the development and implementation of cost-effective and scalable technologies for mitigating and remediating microplastic pollution is essential. This includes exploring innovative approaches for removing microplastics from the environment, such as advanced filtration systems, biodegradable materials, and nature-based solutions. Socioeconomic and Policy Dimensions: Understanding the socioeconomic drivers and implications of microplastic pollution, as well as the effectiveness of policy interventions and governance mechanisms, is critical for fostering sustainable solutions (Joshi et al. 2023; Chaudhry and Sachdeva 2021). This includes conducting interdisciplinary research to assess the social, economic, and cultural factors influencing plastic consumption, waste management practices, and regulatory policies. By addressing these future research directions, scientists, policymakers, and stakeholders can work together to advance our understanding of microplastic pollution and develop evidence-based strategies to mitigate its environmental and human health impacts (Wilhelm et al. 2024). Collaboration and cooperation among researchers from diverse disciplines, along with engagement with communities and decision-makers, are essential for achieving meaningful progress in addressing this complex and multifaceted challenge.

8.10 Conclusion

Microplastic pollution represents a significant environmental and public health challenge that requires urgent attention and concerted action. As our understanding of the sources, distribution, and impacts of microplastics continues to evolve, it is evident that comprehensive strategies are needed to mitigate and manage this pervasive problem. From the contamination of marine ecosystems to the potential risks posed to human health through food and water consumption, the consequences of microplastic pollution are far-reaching and multifaceted. However, by implementing a combination of prevention, mitigation, and remediation measures, it is possible to reduce the prevalence of microplastics in the environment and safeguard ecosystems, wildlife, and human well-being. Key elements of effective microplastic pollution management include reducing plastic consumption, improving waste management practices, developing innovative technologies for microplastic detection and removal, and fostering international collaboration and policy coordination. Moreover, raising public awareness and promoting behavioral changes are essential for fostering a culture of sustainability and responsible consumption. While significant progress has been made in recent years, much remains to be done to address the complex and interconnected challenges posed by microplastic pollution. Continued research, innovation, and collaboration across disciplines and sectors will be essential for advancing our understanding of microplastics and developing evidence-based solutions to protect the environment and human health for generations to come. In essence, by working together and taking decisive action, we can mitigate the impacts of microplastic pollution and create a cleaner, healthier, and more sustainable future for all.

References

Akdogan Z, Guven B (2019) Microplastics in the environment: a critical review of current understanding and identification of future research needs. Environ Pollut 254:113011

Bacon JR, Butler OT, Cairns WR, Cavoura O, Cook JM, Davidson CM, Mertz-Kraus R (2024) Atomic spectrometry update—a review of advances in environmental analysis. J Anal At Spectrom 39:11

Baechler BR, De Frond H, Dropkin L, Leonard GH, Proano L, Mallos NJ (2024) Public awareness and perceptions of ocean plastic pollution and support for solutions in the United States. Front Mar Sci 10:1323477

Bharti S, Sharma VP, Kumar R, Das P (2024) Scenario of microplastics waste in the ecosystem: strategies toward monitoring and management. In: Remediation of plastic and microplastic waste. CRC Press, Boca Raton, pp 141–159

Chaudhry AK, Sachdeva P (2021) Microplastics' origin, distribution, and rising hazard to aquatic organisms and human health: socio-economic insinuations and management solutions. Reg Stud Mar Sci 48:102018

Chovancová K, Michalides NM, Košľabová M, Končálová M, Halko R (2024) Green direct ultrasonic slurry sampling and electrothermal atomic absorption spectrometry method for metal analysis of solid plastics samples. Anal Methods 16:990

Circelli L, Cheng Z, Garwood E, Yuksel K, Di Iorio E, Angelico R, Colombo C (2024) Comparison of ATR-FTIR and NIR spectroscopy for identification of microplastics in biosolids. Sci Total Environ 916:170215

Damaj S, Trad F, Goevert D, Wilkesmann J (2024) Bridging the gaps between microplastics and human health. Microplastics 3(1):46–66

Enyoh CE, Ovuoraye PE, Rabin MH, Qingyue W, Tahir MA (2024) Thermal degradation evaluation of polyethylene terephthalate microplastics: insights from kinetics and machine learning algorithms using non-isoconversional TGA data. J Environ Chem Eng 12(2):111909

Fritz M, Deutsch LF, Wijaya KP, Götz T, Fischer CB (2024) An image-processing tool for size and shape analysis of manufactured irregular polyethylene microparticles. Microplastics 3(1):124–146

Gayathri NP, Priya V, Prabhakaran V, Prasad G (2024) Understanding the impact of microplastic contamination on soil quality and eco-toxicological risks in horticulture: a comprehensive review. Case Stud Chem Environ Eng 9:100633

Guo X, Dai H, He L (2024) Migration testing of microplastics from selected water and food containers by Raman microscopy. J Hazard Mater 462:132798

Hahladakis JN (2020) Delineating and preventing plastic waste leakage in the marine and terrestrial environment. Environ Sci Pollut Res 27(11):12830–12837

He Y, Yin R (2024) The reproductive and transgenerational toxicity of microplastics and nanoplastics: a threat to mammalian fertility in both sexes. J Appl Toxicol 44(1):66–85

Helinski OK, Poor CJ, Wolfand JM (2021) Ridding our rivers of plastic: a framework for plastic pollution capture device selection. Mar Pollut Bull 165:112095

Hettiarachchi H, Meegoda JN (2023) Microplastic pollution prevention: the need for robust policy interventions to close the loopholes in current waste management practices. Int J Environ Res Public Health 20(14):6434

Jagadeesh P, MavinkereRangappa S, Siengchin S, Puttegowda M, Thiagamani SMK, Hemath Kumar M et al (2022) Sustainable recycling technologies for thermoplastic polymers and their composites: a review of the state of the art. Polym Compos 43(9):5831–5862

Jani V, Wu S, Venkiteshwaran K (2024) Advancements and regulatory situation in microplastics removal from wastewater and drinking water: a comprehensive review. Microplastics 3(1):98–123

Jasso-Salcedo AB, Díaz-Cruz CA, Rivera-Vallejo CC, Jiménez-Regalado EJ, Aguirre-Loredo RY (2024) Human consumption of microplastics via food type and habits: recent review. Water Air Soil Pollut 235(2):1–22

Joshi A, Vuree S, Nguyen TL, Kaushik V (2023) Computational toxicity analysis of industrial dyes and inhibitory effect of wheat bran associated enzymes. AIP Conf Proc 2800:020018. AIP Publishing

Kadadou D, Tizani L, Alsafar H, Hasan SW (2024) Analytical methods for determining environmental contaminants of concern in water and wastewater. MethodsX 12:102582

Kim JA, Kim MJ, Choi JY, Park YS, Kim JH, Choi CY (2024) Exposure to bisphenol A and fiber-type microplastics induce oxidative stress and cell damage in disk abalone Haliotis discus hannai: bioaccumulation and toxicity. Fish Shellfish Immunol 144:109277

Kumari A, Pawar D, Saini S, Kumar A, Budhwar S, Sharma SP (2024) Challenges in biodegradation of plastic waste from food packaging. In: Advances in sustainable food packaging technology. Apple Academic Press, pp 251–269

la Cecilia D, Philipp M, Kaegi R, Schirmer M, Moeck C (2024) Microplastics attenuation from surface water to drinking water: impact of treatment and managed aquifer recharge–and identification uncertainties. Sci Total Environ 908:168378

Lenz R, Labrenz M (2018) Small microplastic sampling in water: development of an encapsulated filtration device. Water 10(8):1055

Li Y, Zhu Y, Huang J, Ho YW, Fang JKH, Lam EY (2024) High-throughput microplastic assessment using polarization holographic imaging. Sci Rep 14(1):2355

Martín-Gómez B, Elmore JS, Valverde S, Ares AM, Bernal J (2024) Recent applications of chromatography for determining microplastics and related compounds (bisphenols and phthalate esters) in food. Microchem J 197:109903

McMullen K, Vargas FH, Calle P, Alavarado-Cadena O, Pakhomov EA, Alava JJ (2024) Modelling microplastic bioaccumulation and biomagnification potential in the Galápagos penguin ecosystem using Ecopath and Ecosim (EwE) with Ecotracer. PLoS One 19(1):e0296788

Milne MH, De Frond H, Rochman CM, Mallos NJ, Leonard GH, Baechler BR (2024) Exposure of US adults to microplastics from commonly-consumed proteins. Environ Pollut 343:123233

Mondal C, Mondal N, Moon U, Biswas SJ, Hossain A (2024) Contamination of microplastics in the marine food web with special reference to seafood: present status and future concern. In: Spatial modeling of environmental pollution and ecological risk. Woodhead Publishing, Sawston, pp 175–207

Mushtak F, Prakash J, Katoch SS (2024) Microplastics in complex soil matrix: recovery, identification and removal using micro nano techniques. Micro Nano Eng 22:100237

Olmo-Gilabert R, Fagiano V, Alomar C, Rios-Fuster B, Compa M, Deudero S (2024) Plastic webs, the new food: dynamics of microplastics in a Mediterranean food web, key species as pollution sources and receptors. Sci Total Environ 918:170719

Pouzada AS, Sergio Pouzada A (2021) Selection of thermoplastics. In: Design and manufacturing of plastics products, pp 87–140

Prata JC, Silva ALP, Da Costa JP, Mouneyrac C, Walker TR, Duarte AC, Rocha-Santos T (2019) Solutions and integrated strategies for the control and mitigation of plastic and microplastic pollution. Int J Environ Res Public Health 16(13):2411

Protyusha GB, Kavitha B, Robin RS, Nithin A, Ineyathendral TR, Shivani SS et al (2024) Microplastics in oral healthcare products (OHPs) and their environmental health risks and mitigation measures. Environ Pollut 343:123118

Puskic PS, Slocombe R, Ploeg R, Roman L, Lea MA, Hutton I, Bridle AR (2024) Exploring the pathology of liver, kidney, muscle, and stomach of fledgling seabirds associated with plastic ingestion. J Hazard Mater 465:133306

Quilis SA, Hernández-Martínez AM, Arribas AM, Pérez JG, García-March JR, Medialdea JT (2024) High prevalence of microplastics in the digestive tract of Scyliorhinus canicula (Linneaus, 1758) shows the species biomonitoring potential. Mar Pollut Bull 200:116051

Rangel-Buitrago N, Neal WJ, Galgani F (2023) Plastics in the anthropocene: a multifaceted approach to marine pollution management. Mar Pollut Bull 194:115359

Rathore C, Saha M, de Boer J, Desai A, Gupta P, Naik A, Subha HY (2024) Unraveling the land-based discharge of microplastics from sewers to oceans—A comprehensive study and risk assessment in wastewaters of Goa, India. Sci Total Environ 913:169621

Saeedi M (2024) How microplastics interact with food chain: a short overview of fate and impacts. J Food Sci Technol 61(3):403–413

Saha SC, Saha G (2024) Effect of microplastics deposition on human lung airways: a review with computational benefits and challenges. Heliyon 10:e24355

Saikumar S, Mani R, Ganesan M, Dhinakarasamy I, Palanisami T, Gopal D (2024) Trophic transfer and their impact of microplastics on estuarine food chain model. J Hazard Mater 464:132927

Srisiri S, Haetrakul T, Dunbar SG, Chansue N (2024) Microplastic contamination in edible marine fishes from the upper Gulf of Thailand. Mar Pollut Bull 198:115785

Thapa BS, Pandit S, Mishra RK, Joshi S, Idris AM, Tusher TR (2024) Emergence of per-and poly-fluoroalkyl substances (PFAS) and advances in the remediation strategies. Sci Total Environ 916:170142

Tian H, Zheng C, Huang X, Qi C, Li B, Du Z et al (2024) Effects of farmland residual mulch film-derived microplastics on the structure and function of soil and earthworm Metaphireguillelmi gut microbiota. Sci Total Environ 915:170094

Wilhelm K, Woor S, Jackson M, Albini D, Young N, Karamched P et al (2024) Microplastic pollution on historic facades: hidden 'sink' or urban threat? Environ Pollut 343:123128

Zhou R, Dong Z, Li Z, Zhou W, Li Y, Xing L et al (2024) Adsorption-desorption behavior of florpyrauxifen-benzyl on three microplastics in aqueous environment as well as its mechanism and various influencing factors. Ecotoxicol Environ Saf 272:116066

Zhu J, Xu A, Shi M, Su Y, Liu W, Zhang Y et al (2024) Atmospheric deposition is an important pathway for inputting microplastics: insight into the spatiotemporal distribution and deposition flux in a mega city. Environ Pollut 341:123012

Microplastics in Personal Care Products and Cosmetics

9

Shruti Chaudhari, Himakshi Parmar, and Prakash Samnani

Abstract

The environmental predominance of microplastics has grown to be a significant worldwide concern. Personal care and cosmetic products (PCCPs) with microbeads are single of the important causes of environmental microplastics. Solid primary microplastics with a diameter of less than five millimeters, known as microbeads, are utilized in cosmetic products to either exfoliate or cleanse the skin. Several products like soap, shampoo, deodorant, toothpaste, moisturizer, wrinkle treatment, sunscreen lotion, face mask, lipstick, eye shadow, kids' bubble bath, and more are known to contain microplastics, which can be spherical or irregular in shape. Since the quantities of plastic materials used in the formulations of personal care products and cosmetics globally are not publicly disclosed, it is impossible to estimate the true extent of the damage caused by microplastics. Once disposed of, full biodegradation of these plastics is not possible due to their stable chemical nature coupled with environmental constraints and the lack of an effective recovery mechanism. Scientists have proposed a prohibition on goods containing microbeads, mostly due to their potential effects on human health and the environment, and the fact that there were always choices available at comparable costs. Due to growing public concerns, some European and other countries around the world have taken governmental action against these microbeads and, to a lesser extent, other microplastics used in cosmetics. This book chapter offers an in-depth overview of the literature on microbeads in cosmetics up to this

S. Chaudhari · H. Parmar
Department of Environmental Studies, Faculty of Science, The Maharaja Sayajirao University of Baroda, Vadodara, India

P. Samnani (✉)
Department of Chemistry, Faculty of Science, The Maharaja Sayajirao University of Baroda, Vadodara, India
e-mail: prakash.samnani-chem@msubaroda.ac.com

A. Kumar, V. Singh (eds.), *Microplastics Pollution and its Remediation*,
https://doi.org/10.1007/978-981-97-4068-0_9

point, emphasizing the efforts taken to discourage their use and raising concerns about whether upcoming contamination from microbeads and/or microplastics in cosmetics has been effectively addressed.

Keywords

Microplastics · Cosmetics (Personal care products) · Microplastic beads · Pollution · Toxicity

9.1 Introduction

"Plastic era" is what we currently live in because of the widespread use, flexibility, durability, and affordability of plastics due to their unique properties and chemical structure (Anagnosti et al. 2021). The term Plastic has been used here in a broader sense to includes all the synthetic polymeric materials converted into a variety of useful products as single polymer or combinations like copolymers, block copolymers, blends, composites, mixtures, etc. Plastics pollution is a well-known phenomenon, and all countries are making attempts to contain the menace. The use of plastics has resulted in to a newer form of pollution, namely, microplastics pollution. In general terms, microplastics are depicted as plastic particles with dimension range involving 100 nm to 5 mm; however, the term is also occasionally used to refer to nanoplastics, which are smaller forms of plastics that are less than 100 nm (Zhou et al. 2023). Microplastics can originate from various sources and can possess a variety of impacts on the natural environment, animals, and humans. Due to radiation, oxidation, and heat, larger plastics on coasts and in water break down, which serves as the major source of microplastics release into the aquatic environment. These are termed as secondary microplastics. Some plastics, known as primary microplastics, are purposefully made in tiny sizes, and added to goods like encapsulates, cleaners, cosmetics, detergents, abrasives, and fertilizers to offer a desired feature. Municipal wastewater is purportedly among the principal contributors of microplastics pollution in the environment among other sources. In terms of the components of municipal wastewater, ordinary household activities including washing dishes, doing laundry, taking a bath, and using the toilet are some of the sources of microplastics in municipal wastewater (Jessieleena et al. 2023).

Environmental problems due to microplastics pollution have become more prevalent, with many public outcries recently focusing on specific microbeads, category of microplastics originating from personal care and cosmetic products (PCCPs). Because Microbeads are ubiquitous in nearly all environments, they are today the subject of extraordinary attention on a global scale. Cosmetic microplastics are currently under investigation by the public, legislators, the media, business, and the international scientific community. It has been reported that microbeads can carry hazardous inorganic and organic persistent pollutants and trace chemicals. It has also been demonstrated that hazardous heavy metals, like chromium, can travel through PCCP microbeads and into aquatic environments (Rubin and Zucker 2022).

With their mechanical peeling effect, microbeads are highly desired exfoliating and abrasive substances that eliminate superficial blemishes, makeup flaws, and dead skin cells (Kozlowska et al. 2019). They have many other applications in skin exfoliating and washing (microbeads), opacity control, smooth and silky texture, skin lighting, and viscosity control (Tagg and Ivar do Sul 2019; UNEP 2016; van Leeuwenhoeklaan 2018; Yurtsever 2019a, 2019b). Microbeads are included in both "leave-on" and "rinse-off" cosmetics. While rinse-off cosmetics containing these particles directly enter sewage, leave-on cosmetics after their use are either removed with cotton pad, tissues, or washed off, nonetheless not right away like rinse-off cosmetics because their purpose is to stay on the body for a while. Thus, some of these particles end up as solid waste. Shampoos, shower gels, soaps, and products for coloring and nourishing hair as well as treatments for body and hair bleaching are examples of "wash off" cosmetics containing microplastics. Microplastics in "leave-on" cosmetics are incorporated in make-up (powders, foundations, concealers, eyeliners, eye shadows, mascaras, eye pencils), sun care, skin care (like body lotion), lip care, deodorants, nail care products, hair care (like leave-on dry shampoos, conditioners). For example, sparkle in the form of microplastic particles may be included in eye shadows and mascara (Guerranti et al. 2019).

An important category of microplastics is glitter used in cosmetics as a decorative element, and is made of metalized polyethylene terephthalate (PET). Despite their prevalent use in cosmetics, these microplastics have not received much attention. This is because most of the research collects samples of microplastics from near surface, while these particles are expected to sink in sediments. Glitter is easily distinguished from other primary and secondary microplastics and can be just as damaging as any other microplastic. Such metallized polymers also have a variety of other applications on large scale, for example, decoration, packaging, insulation, and electronics manufacture. Scarce data is available on how they behave in environments after disposal.

Cosmetics that are left-on have more complicated formulae than those that are rinsed off; they can contain up to three or four types of polymers. Microplastics are sometimes used in the formulations of leave-on and rinse-off cosmetics as scent encapsulates. Cosmetic companies manufacturing such products assert that most of the polymers in their recipes are in liquid state and are therefore safe, with the exception of solid microplastics (Dauvergne 2018). Concerns are, however, being voiced against the use of such "liquid microplastics," many of which are nonbiodegradable, and not much information is available on their other harmful effects. Also, their behavior in various environment segments is not known, there being strong possibility of the "liquid microplastics" getting converted to solid microbeads. Important examples of liquid microplastics used in cosmetics are carbomer (polyacrylic acids), cyclopentasiloxane (decamethylcyclopentasiloxane), and acrylic(co) polymers.

A prior study by Gouin et al. (2011) calculated that everyone in the USA consumes about 2.4 mg of microplastics per day through cosmetics, resulting in 263 tons of emissions annually in the USA only. Further, it was also calculated that a single application of these problematic products would emit between 4594 and

94,500 microbeads, the majority of which were polyethylene-based compounds (Napper et al. 2015).

9.2 Types of Plastic Used in Cosmetics and Personal Care Products

Plastic materials have been utilized as PCCP constituents for many years; the first patents on the subject date back to the 1960s. Even now, plastic materials are a major source of invention for modern PCCPs (Lochhead 2007; *Patil, A.* 2013). The plastics used as constituents in PCCP preparations that are covered at this point fall into two key groups: thermoset plastics, such as polyester resins and polyurethanes; and thermoplastics, such as polypropylene, polyethylene, polystyrene, polytetrafluoroethylene (Teflon), polymethylacrylate, and polyamide. Thermosets become hard on heating; the process is irreversible. Thermoplastics, on the other hand, becomes pliable or moldable on heating. Thermoset plastics tend to be more brittle than thermoplastics. Further, the constituents of the plastic PCCP comprise many copolymers in addition to homopolymers. Polyethylene terephthalate is created through the polymerization of ethylene terephthalate monomers, homopolymers, which are polymer chains of a single monomer type, are created. Poly(butylene terephthalate)/PET and other copolymers are gradually replacing PET, the homopolymer (Becker et al. 2014). According to Dąbrowska et al. (2022) polyethylene makes up more than four-fifths of the microplastics found in cosmetics, while some of these tiny plastic particles have been shown to contain nylon, polyethylene terephthalate, polypropylene, and polymethyl methacrylate, among other plastic polymers.

9.3 Function of Microplastics in Product Formulations

Items such as shampoo, shower gel, toothpaste, cream, blush powder, eye shadow, makeup, skin cream, foundation, hairspray, deodorant, liquid makeup, nail polish, eye color, mascara, shaving cream, facial cleanser, lotion, bubble bath, baby products, sunscreen, hair color, and insect repellents are just a few of the PCCPs that contain plastic ingredients in their formulation. Different items contain plastic materials in varying amounts; in some situations, this ranges from less than 1% to over 90%.

The plastic materials have been used in preparations for an extensive scope of purposes, provisional on the type, composition, shape, and size of the polymers used. These purposes include emulsifiers, viscosity regulators, opacifying agents, bulking agents, liquid absorbents, film formers, binders for an "optical blurring" effect (e.g., of wrinkles), glitters, skin conditioning, oral care, exfoliants, abrasives, such as, gellants, tooth polishing in denture adhesives, for measured period release of several active ingredients, and sportive phase for the delivery of vitamins, oils, fragrances, insect repellents, moisturizers, sun filters, and a variability of other vigorous ingredients. It is evident that these polymers have uses beyond the well-known and widely reported scouring action of microbeads (Leslie 2014).

Microbeads have also emerged as essential components of the preparation for a range of PCCPs, such as body creams, baby products, lotions and creams, bubble baths, deodorants, face powders, eye shadows, hair colorings, hair sprays, toothpastes, shower gels, shampoos, mascara, nail polishes, sunscreens, and shaving creams (Leslie et al. 2022). Various PCCPs now utilize these microplastics for purposes beyond their original purpose of scrubbing. They consist of the following: controlled time release of various active elements, for gellants in denture adhesives, for bulking mediators, emulsifiers, ingredients, binders, film formers, exfoliants, sparkles, opacifying agents, viscosity controllers, tooth polishing in oral care, skin conditioning, insect repellents, conditioners, sun strainers, adsorptive phase for the delivery of other ingredients and stabilizers (Tian et al. 2022)Hung et al. (2022) estimate that polyethylene accounts for 90% of the microbeads used in PCCPs for cosmetic purposes, with the remaining 10% consisting of different plastic polymers like nylon, polymethyl methacrylate, polypropylene, and polyethylene terephthalate. Table 9.1 lists several functions of various polymers used as microbeads in different cosmetic formulations.

9.4 Sources, Routes, and Fate of Microplastics from Cosmetics

Primary and secondary microplastics are the two general categories of microplastics observed in the environment. Primary microplastics are made specifically and entirely for a variety of industrial applications, such as cosmetic microbeads, whereas secondary microplastics are created when bigger plastic particles break (Amobonye et al. 2021). Mixing processes of compressed plasticizer plastic, polymers, and additional required additives; extrusion of the molten mixture with micro-area extruders; shaping the blend into the wanted form; and ultrafine crushing in a grinder are described as part of the industrial production of these microbeads (Wainaina et al. 2020; Grimmer 1996). The big cosmetic companies in the world have been under fire for using a lot of microbeads in their various PCCPs over the past several years, and they have received a large share against the effects of cosmetic microplastic on the environment (Goyal and Jerold 2023). The first study by Zitko and Hanlon (1991), which showed the ability for cosmetic microplastics to pollute the environment and predicted their accumulation in various ecosystems, initiated the learning of the environmental effects of cosmetic microbeads. Many scientific efforts have now been directed toward the isolation, identification, characterization of these microbeads in the environment.

Current studies have attempted to calculate the number of cosmetic-derived primary microplastics in the environment, although confirming the existence, distribution and amount of microbeads in the marine environment has proven extremely difficult due to their small size (Liu et al. 2022a; Park and Kim 2019; Rondoni et al. 2021), and also due to simultaneous presence of secondary microplastics. Estimates

Table 9.1 Examples of various functions of different polymer microbeads in cosmetics

Sr. No	Polymer name	Chemical formula	Examples of functions in personal care and cosmetic products (PCCP) formulation
1.	Nylon-6	$(C_6H_{11}NO)_n$	Viscosity-controlling agent and bulker
2.	Nylon-12 (polyamide-12)	$(C_{12}H_{23}NO)_n$	Opacifying, controlling viscosity, and bulking (e.g., antiaging products)
3.	Poly (butylene terephthalate)	$(C_{12}H_{12}O_4)_n$	Regulating viscosity and creation of films
4.	Poly (ethylene isoterephthalate)	–	Bulking agent
5.	Poly (ethylene terephthalate)	$(C_{10}H_8O_4)_n$	Adhesives, film creation, hair fixing agent; viscoelastic regulating, aesthetic agent, (e.g., glitters in makeup and bubble bath)
6.	Poly (methyl methacrylate)	$(C_5H_8O_2)_n$	Sorbent for the transportation of active components
7.	Poly (pentaerythrityl terephthalate)	–	Creation of films
8.	Poly (propylene terephthalate)	$(C_{10}H_8O_4)_n$	Dermal conditioned behavior, emulsion stabilization
9.	Polyethylene	$(C_2H_4)_n$	Stabilizer for powdered substances, viscoelastic management, film formation, and abrasive
10.	Polypropylene	$(C_3H_6)_n$	Agents for bulking and raising viscosity
11.	Polystyrene	$(C_8H_8)_n$	Film formation
12.	Polytetrafluoroethylene (Teflon)	$(C_2F_4)_n$	Skin shampoo and conditioner, slip modification, binding agent, and thickening ingredient
13.	Polyurethane	$(C_{27}H_{36}N_2O_{10})_n$	Film formation (e.g., facial masks, sunscreens, mascara)
14.	Polyacrylate	$(C_3H_3NaO_2)_n$	Regulating viscosity
15.	Acrylates copolymer	$(C_3H_4O_2)_n$	Binding agent, film-forming agent, hair fixing agent, and suspension agent
16.	Allyl stearate/vinyl acetate copolymers	$(C_{21}H_{40}O_2)_n$	Hair fixing agent and film formation
17.	Ethylene/propylene/styrene copolymer	$(C_{12}H_{14})_n$	Regulating viscosity
18.	Ethylene/methyl acrylate copolymer	$(C_6H_{10}O_2)_n$	Film formation
19.	Ethylene/acrylate copolymer	$(C_5H_8O_2)_n$	Film formation in gellant (such as lipstick, stick cosmetics, and hands lotions) and water-resistant protection
20.	Butylene/ethylene/styrene copolymer	$(C_{13}H_{16})_n$	Viscosity controlling
21.	Styrene acrylates copolymer	$(C_{11}H_{11}O_2)_n$	Aesthetic, colorful microspheres (e.g., makeup)
22.	Trimethylsiloxysilicate (silicone resin)	$(C_6H_{18}O_5Si_3)_n$	Film formation (e.g., color cosmetics, skin care, sun care)

have been made available based on the amount of microplastics present in source materials.

Therefore, it was calculated that the concentration of microplastics in face cleansers was 2450 particles/g or 0.05 g/g (geometric mean), whereas shower gels had an estimated 2.15 particles/g or 0.02 g/g (geometric mean) of the particles. Based on this, it was projected that 1.2 × 10^4 tons of microplastics originating from cosmetics are released into the atmosphere annually. Moreover, it is projected that throughout the past 50 years, 3.0 × 10^5 tons of microplastics created from PCCP have accumulated in the environment (Sun et al. 2020).

Many of the cosmetic microbeads used to create PCCPs are produced and processed in Europe, the Americas, and certain parts of Asia, according to research. The USA is reported to have the greatest relative microplastics concentration in PCCPs consumed, followed by Asia and Europe, which includes China, Japan, Thailand, and South Korea. The trend in production, however, differs from the microbeads being emitted into the environment. This is thought to be partially caused by regional differences in the volume of these items used as well as regional differences in the technological sophistication of wastewater treatment systems.

Estimates place the world's microbead emissions at 15.9% from Europe and Central Asia, 15.8% from China, 17.2% from North America, 15% from Oceania and East Asia, and 18.3% from South Asia. Furthermore, it was found that the Middle East and South America/Africa contributed 8.7% and 9.1%, respectively, to the global microbead pollution. A compilation of literature showing microplastics (amount, shape, color, characterization by infrared (IR) spectroscopy, etc.) found in personal care products in some countries is shown in Table 9.2.

PP polypropylene, *PE* polyethylene, *PA* polyacrylate, *PVC* polyvinyl chloride, *CP* Cellophane, *AC* acrylates copolymers, *PS* polystyrene.

Multinational cosmetics companies have spent millions on nanotechnology throughout the years (Raj et al. 2012) mainly to improve the active ingredients' intradermal delivery and utilize them as existent UV absorbers in sunscreen formulations (Zacharopoulou and Varvaresou 2012). In cosmetics, nanoparticles—ranging in size from 1 nm to 100 nm—mostly take the form of inorganic substances including silica, ZnO, and TiO_2. Cosmetics contain purposefully added nanoscale plastics, or nanoplastics, mostly in the form of nanocapsules (Pastrana et al. 2018; Fytianos et al. 2020). Solid polymeric nanoparticles are not meant to be used for a long duration in rinse-off cosmetic products because they would be too little to offer the necessary benefits (Sherrington et al. 2016; Anagnosti et al. 2021). However, Hernandez et al. discovered an unexpected scrub in 2017 that included PE nanoparticles (Hernandez et al. 2017). The formation of the nanoparticles most likely resulted from the disintegration of microplastic beads while the mixture was being stirred and the formula was being prepared.

Table 9.2 Microplastics in personal care products from different countries

Country	Product	Amount of microplastics	Shape	Color	IR result of microbeads	Reference
Slovenia	Body scrub	4.82 g/100 mL	Irregular	White	PE	Kalčíková et al. (2017)
	Facial scrub	0.74 g/100 mL	Irregular	White	PE	
New Zealand	Hand soap	0.19 g/ 100 g	–	–	–	Gregory (1996)
	Facial scrub	6.91 g/ 100 g	–	–	PE	
China	Face wash	0.031–0.040 g/10 g	Slightly granular, irregular	Pink, dark green	PE	Bhavsar and Gore (2023)
	Face scrub	0.077–0.352 g/10 g	Granular, irregular	Granular, Brown	PE	
Malaysia	Six samples of PCP (toothpaste, facial, and body scrub)	–	Granular, spherical, Irregular	Colorless, White, pink, Purple, Orange	PE	Suardy et al. (2020)
UK	Toothpaste	2500 item/g	Irregular	Blue	PE	Carr et al. (2016)
Vietnam	Facial cleanser/scrubs Toothpaste Body wash/scrubs	–	–	–	PE AC PS	(Hoang et al. (2022)
–	Wrinkle creams. Glitter in bubble bath or make up	–	–	–	–	Detraz (2017)
India	Toothpaste	327–832 Particles/g	Fibers, Fragments	–	CP, PVC, PA, PP	Madhumitha et al. (2022)
Turkey	Toothpaste	0.25–4.17 g Particles/10 g	Opaque fragment	–	PE	Ustabasi, (2019)

9.5 Cosmetics Microplastics: Ecological and Biological Interactions

Plastic microbeads created from silicones, functionalized polymers, hydrophilic agents (polystyrene, polyethylene, and polypropylene), and film-forming agents (Lochhead 2017) generate a roller-bearing effect that will provide a smooth surface and the capacity to disperse, making them an essential component of makeup businesses. Though mostly spherical, these microbeads also come in a variety of forms and colors (thread, ellipse, and irregularly frayed). When washed away, the microbeads travel via the water to other bodies of freshwater streams, canals, land, sewage, and lastly the oceanic atmosphere. Cosmetics are therefore seen to be a potentially dangerous source of microplastics (MPs). Ecosystems are negatively impacted by their presence (Yuan et al. 2022). According to Li et al. (2021b) MPs migrate vertically on soil surfaces due to chemical, biological, and gravitational processes, and they eventually seep into deep soils. Most notably, MPs can more easily adsorb contaminants from the environment due to their huge identifiable surface characteristics and surface area (Atugoda et al. 2020). Sewage treatment facilities can reduce the microbeads load but not all such processes are 100% efficient in removing them. Thus, many of the microbeads enter the natural water systems through treated sewage water. Furthermore, the removed particles become part of the sludge, which gets transferred to landfills or agriculture fields where the dried sludge is often applied as soil conditioner.

MPs adsorb organic pollutants through a variety of noncovalent processes, including hydrogen bonding, ionic interactions, interactions caused by van der Waals forces, and hydrophobic bonds. Moreover, the ability of pollutants, both inorganic and organic, to be adsorbed is also affected by the chemical characteristics and topography of the MPs, including the large specific surface area, size of the particles, advancing age, the total number of surface sites for adsorption, hydrophobicity, crystallization, functional structures, pore-filling of polymers structure, and polarity. Limiting parameters affecting MPs' capacity to adsorb organic contaminants include their metal cationic concentration, chlorine replacement number, hydrogen deposits, charges of organic contaminants, and environmental elements like moisture, temperature, pH, and ionic strength (Li et al. 2021a, 2021b). Antibiotics (cephalexin, tetracycline, amoxicillin, fluoroquinolones and ciprofloxacin hydrochloride) are important hydrophilic organic pollutants that can interact with MPs (Fu et al. 2021). There have been recent reports that microplastics in soil and aquatic environments serve as heavy metal vectors. The MPs may be beneficial for attracting heavy metals such as lead, copper, chromium, and cadmium by adsorption as the ecosystem's temperature rises and its salinity falls, provided the particles can be separated from rest of the system subsequently. Research indicates that biofilm buildup and intra-particle diffusion have an impact on the adsorption ability of heavy metals on microplastics (Liu et al. 2022b; Cao et al. 2021). Because of their higher concentration than in the sea environment, in the lentic and lotic ecology, the MPs emitted by personal care items are extremely active chemically, physiologically, and biologically. These MPs are pseudo-persistent in nature.

Because of dilution and degradation, the concentration of MPs in the marine environment is lower than in the freshwater ecosystem.

9.6 Microplastics Toxicity in Humans

Microplastics have been reported in human blood. They can enter the human bodies through a variety of exposures, viz., skin contact, inhalation, and ingestion. The human skin barrier function keeps microplastics from entering skin layers or the circulatory system, making exposure to microplastics through the skin less important. Regarding the various forms of toxicity studies, there are few data on the effects of microplastics on bioaccumulation, neurotoxicity, translocation, and reproductive toxicity in human cells. However, studies in mouse models involving PS microplastics and nanoplastic particles show effect on neurotransmission, leading to imbalance and neurotoxicity in enzymes because of the microplastics' bioaccumulation and translocation after absorption. It has been noted that being exposed to polystyrene microplastics causes fatal losses because the microplastics cause an overactive immunological response. Therefore, it is hypothesized that ingesting microplastics could put humans at comparable risk. Another recent study has shown presence of microplastics in the plaques removed from the patients suffering from cardiovascular diseases, indicating increased chances of heart-related problems due to microplastics and nanoplastics. Working with environmental microplastics in real time is therefore crucial to observing the impact of toxic effects on human health. More studies are emerging about the toxic effects of these particles.

9.7 Determination of Microplastics in PCCPs

The following section describes some of the literature methods for determining microplastics in some PCCPs. Substantial literature exists for determination of microplastics in aquatic environments and soil/sediments (Chaudhari and Samnani 2022).

9.7.1 From Face Scrub

A procedure for extraction was created to identify and measure the MP in face wash. Acetone was utilized in the extraction procedure, and the approach is described briefly here. One gram of the face scrub sample and five milliliters of acetone were added to a clean glass beaker. After shaking, the liquid was transferred to a second glass beaker that held 50 mL of deionized water. Using a stainless-steel spoon, microplastic particles floating over the surface of the water were gathered and moved to a glass petri dish. The gathered particles were left to dry up at ambient temperature for a whole night before being weighed. The influences of multiple parameters (sample to acetone proportion, acetic acid proportion, and extraction time) on the efficacy of the abstraction technique were also examined through spiking experiments in pre-obtained polyethylene MPs (average size: 242 ± 74.9 μm) at a concentration of 20 mg per g of the face scrub sample. There was also a blank sample with no MPs added. By splitting the number of MPs extracted (in mg) by the number of MPs spiked (20 mg), the method's efficiency (Recovery %) was determined (Shalumon, 2023).

9.7.2 From Face Cleanser

Ten grams of face cleanser was suspended in 100 mL of warm distilled water (around 45–50 °C) to extract microbeads. The combination above was then added to the beaker with 2 mL of concentrated HNO_3. After that, the contents in the beakers were magnetically stirred at room temperature for roughly 15–20 min at a measured rate. After that, the solution was filtered via a grade 42 Whatman filter paper to make sure all the microplastics had been removed. To lessen the number of water-soluble contaminants in the sample, all the microbeads were cleaned with distilled water (around 45–50 °C) following the filtration process. After being left to dry in the lab for roughly 2 days, the microbeads on the funnel's filter were carefully gathered and placed in the sample container (Bhavsar and Gore 2023).

9.7.3 From Shower Gel

There were three identical samples of every product with a wet weight of 1.5–2.0 g (that is, microplastics and the quantity of substances containing solution) that were homogenized with 200–500 mL of ultra cleaned water, then boiled and continuously stirred till the soluble constituents completely vanished. The resultant mixture was passed through a membrane filter made of nitrocellulose that was warmed up to 50 °C for a whole day to extract all insoluble particles (mostly MPs) larger than 10 μm. The filter was allowed to dry out in an incubator set at 50 °C. The weight of every particle in the filter was ascertained by measuring the weight difference between the filter pre- and post-filtration. The weight ratios of the elements to the product were then computed, which can be expressed as weight of piece or weight of byproducts × 100%. Each particle was put into a glass vial (Lei, 2017).

9.7.4 From Toothpaste

Ten grams of toothpaste was combined with 500 mL of deionized water and heated along with magnetic stirring. Filtration was started right away when the whole sample disappeared. Samples were allowed to filter naturally with the aid of gravity. Filters were dried up in an oven at 50 °C temperature for 7–8 h after the filtration process. The sample's percentage of undissolved solids was determined by weighing the samples after they had cooled to room temperature. After the filtering process was finished, the solids were combined with 50 mL of distilled water. Because polyethylene is less dense than water, the MPs that were present were predicted to float. The combinations were given 2 h to settle. Following the particle settling process, the liquid portion was transferred to glass flasks to produce floating particles, and the mixture was further dried at 75 °C temperature. The dried-out solids' weights were noted (Ustabasi, 2019).

9.8 Risk Assessment of Microplastics from Cosmetics

To determine the extent to which toothpaste is contributing to microplastics pollution, a risk assessment was conducted. The following formula is used to compute the estimated microplastics emissions values (Madhumitha, 2022).

$YME = DU \times POP_{ind} \times TP_{use} \times OR \times M.P. \times N_{days}$.

where,

1. (YME) = Yearly Microplastics Emission from any source.
2. (DU)—The quantity of daily usage.
3. (POP_{Ind})—Indian population.
4. (TP_{use})—The quantity of toothpaste needed for a single application.
5. (OR)—Occurrence ratio of microplastics in the total number of samples.
6. (MP)—A percentage of microplastics in the sample.
7. (N_{days})—Everyday number for a year.

Similar relations can be developed for estimating the microplastics emissions from other cosmetics.

9.9 The Current Regulations for Mitigating and Controlling Microplastics Pollution

According to one estimate, the 320 million tons of plastic waste created annually now will increase three times by 2050 (Kumar et al. 2021). Jiang (2018) has studied the stages taken to decrease the contamination brought on by microplastics, such as policies, laws, and levies, as well as the roles played by international administrations, NGOs, and governments in preventing adverse consequences and preserving the environment. To safeguard the environment against plastic pollution and to keep an eye on its disadvantageous effects on humans and aquatic living, several academic institutions, government bodies, and research groups have worked together on mitigation and management strategies. As stated by Luís and Spínola (2010), higher prices for plastic bag are imposed in Germany, as well as Netherlands, Norway, Sweden, and Switzerland. After imposing a fee on polybags in 2015, Portugal's usage of plastic has decreased to 74% of total usage (Mitrano and Wohlleben 2020). Although research found that microbeads were present in many cosmetics, several countries have outlawed the use of these tiny particles in cosmetic items, even if MPs make up very little of them. In California, sales of cosmetics and personal hygiene products containing microbeads were banned in 2015

(Syberg et al. 2018) that went into force on January 1, 2020. Plastic bags and products containing microbeads are prohibited in the following five US states: the state of New York, Minnesota, Illinois, Maine, and California (Casebeer 2015; Auta et al. 2017). The prohibition on microbeads has likewise been maintained by other countries. To support the ban, the trade association and Dutch government, for instance, urged the industry to cease manufacturing microbeads and published an announcement titled "Beat the Microbeads, 2016" (Xanthos and Walker 2017). Similarly, the UK banned the manufacturing of cosmetics containing microplastics in January 2018 (Leadsom 2016). So far, while 127 different nations have laws controlling the use of plastics, 115 countries have put in place various regulations and laws to ban both national and international plastic waste. The use and manufacture of certain types of plastics have been banned in some countries, or fines are being levied.

In addition to bans and taxes, other measures including recycling and ecolabeling have also been employed. Recycling is a common practice, with rates varying across countries. For example, they are 50% in the Netherlands, Germany, the Czech Republic, Slovenia, and Sweden, and 20% in France (Ogunola et al. 2018), and 7% in India (Singh et al. 2017). Eco-labeling is one tactic for lessening the pollution caused by plastic in the oceanic water (Auta et al. 2017; Pettipas et al. 2016). Eco-labeling has been successfully applied in numerous industrialized nations. In order to reduce the amount of plastic waste in the ocean, the EU and the UK, for instance, have modified eco-labeling (Ogunola et al. 2018). Numerous Nordic countries, like Finland, Sweden, Norway, Denmark, and Iceland have also embraced eco-labeling (Kristanti et al. 2023).

In January 2018, the European Authority passed a guideline urging the recycling and reuse of all plastic packaging in addition to decreasing the usage of single-use plastics and microplastics. The initial agreement to address plastic limits was the 1973 Marpol agreement, also known as the International Convention for the Prevention of Pollution from Ships. On December 31, 1988, the convention's Annex V—which addresses the absolute ban on disposing of any sort of plastic waste—went into force (Parker 2021).

Additionally, in 2015, The Microbeads-Free Water Act, enacted by the US federal government, prohibits the production and sale of items that include microbeads (Lam et al. 2018). Moreover, the Environmental Protection Act was permitted in Canada to restrict the import, distribution, and production of microbeads, particularly those utilized in cosmetics (Pettipas et al. 2016; Jiang 2018). The year 2015 saw the signing of a microbead phase-out agreement between the state of South Australia and New South Wales, which are two of Australia's six states, and the "national industry organization of makers and marketers of hygiene, cosmetic and specialized products" (ACCORD) (EPA, 2016). In 2018, the environment minister of South Africa stated that the country's administration is consulting with the cosmetics sector over the elimination of microbeads in makeup (Molewa, 2018). Table 9.3 gives existing government policies on controlling pollution due to microbeads.

As an alternative, some biomaterials have been suggested to be useful in cosmetics applications. For example, chitin and chitosan-based materials, seaweed

Table 9.3 The present state of microbead and microplastic regulation and prohibition

Country	Year	Policies	References
The USA	2015	The 2015 Microbeads-Free Waters Act	(Lamichhane et al. 2023; McDevitt et al. 2017)
		Prohibition of cosmetic products containing microbeads	(FDA, U. 2015)
Canada	2016	The production, distribution, and importation of personal hygiene items containing microbeads are all forbidden by the microbeads act	(Pettipas et al. 2016)
California	2015	By 2020, microbeads and plastic will not be used in personal care products	(Syberg et al. 2018)
The UK	2016	Outlawing cosmetics that include microbeads	(Rist and Hartmann 2018)
	2017	To outlaw the usage of microplastics in personal care items and cosmetics	(Lamichhane et al. 2023; Xanthos and Walker 2017)
Italy	2020	Products like cosmetics with microbeads cannot be marketed	(Guerranti et al. 2019)
India	2021	Plastic bags that are single-use and less than 50 μm thickness banned	(Lamichhane et al. 2023; Laskar and Kumar 2019)
New Zealand	2018	Banning the production and distribution of personal hygiene products with microbeads	–
Sweden	2019	Amendments to acts prohibiting the usage of microbeads in cosmetic products	–
Korea	2017	Microbeads are classified as raw materials that are not appropriate for use in cosmetic products	(Hurley, 2019)
	2019	Korea's Ministry of Food and Drug Safety is developing a test procedure to find microbeads in cosmetic items	–
	2021	Guidelines regarding the amount of microplastics in some household chemical goods, such as fabric softener, laundry detergent, deodorant, and cleaning solutions	–
Taiwan	2020	Act changes pertaining to the use of microbeads in cosmetics announced	(Tsai, 2021)

polysaccharide-based microbeads, etc., are considered green substitutes of microplastics.

9.10 Conclusion

Understanding the path, destiny, and transport of cosmetic microplastics to soils and water bodies has undergone enormous effort in recent decades. Cleaning up plastics contamination requires additional strategies for eliminating microplastics or MPs. Finding every possible combination of microplastic type, form, and overall morphology is the first challenge in eliminating tiny plastic particles. The two primary problems have following issues with handling microplastics: (1) there isn't a precise definition of what microplastics are, which makes it hard to compare the outcomes of various studies; and (2) the difficulty in comparing the results because different studies on microplastics have used different methodologies.

Uncontrolled processes including abrasion and deterioration of bigger plastic products, or secondary microplastics, are considered as an important sources of MPs pollution. Despite making up a very minor portion of the anticipated total load of environmental MPs, MPs originating from PCCPs have a significant influence-With the ongoing attepmts, by several countries, to control use of these particles in PCCPs, researchers expect that in near future, contribution to overall MPs pollution due to these microbeads will reduce substantially.

References

Amobonye A, Bhagwat P, Raveendran S, Singh S, Pillai S (2021) Environmental impacts of microplastics and nanoplastics: a current overview. Front Microbiol 12:3728

Anagnosti L, Varvaresou A, Pavlou P, Protopapa E, Carayanni V (2021) Worldwide actions against plastic pollution from microbeads and microplastics in cosmetics focusing on European policies. Has the issue been handled effectively? Mar Pollut Bull 162:111883

Atugoda T, Wijesekara H, Werellagama D, Jinadasa K, Bolan NS, Vithanage M (2020) Adsorptive interaction of antibiotic ciprofloxacin on polyethylene microplastics: implications for vector transport in water. Environmental Technology & Innovation 19:100971

Auta HS, Emenike CU, Fauziah SH (2017) Distribution and importance of microplastics in the marine environment: a review of the sources, fate, effects, and potential solutions. Environ Int 102:165–176

Becker LC, Bergfeld WF, Belsito DV, Hill RA, Klaassen CD, Liebler DC, Marks JG Jr, Shank RC, Slaga TJ, Snyder PW (2014) Safety assessment of modified terephthalate polymers as used in cosmetics. Int J Toxicol 33(3_suppl):36S–47S

Bhavsar PS, Gore AH (2023) Extraction and analysis of microplastic beads from personal care products. Curr Anal Chem 19(2):184–189

Cao Y, Zhao M, Ma X, Song Y, Zuo S, Li H, Deng W (2021) A critical review on the interactions of microplastics with heavy metals: mechanism and their combined effect on organisms and humans. Sci Total Environ 788:147620

Casebeer, T. (2015). Why the ocean clean up project Won't save our seas:"Boyan Slat's ocean cleanup Project's net Array". Planet experts 2015.

Chaudhari S, Samnani P (2022) Determintion of microplastics in pond water. Mater Today Proc 77:91–98. https://doi.org/10.1016/j.matpr.2022.10.141

Dąbrowska A, Mielańczuk M, Syczewski M (2022) The Raman spectroscopy and SEM/EDS investigation of the primary sources of microplastics from cosmetics available in Poland. Chemosphere 308:136407

Dauvergne P (2018) The power of environmental norms: marine plastic pollution and the politics of microbeads. Environmental Politics 27(4):579–597

FDA U (2015) The microbead-free waters act: FAQs. US Food and Drug Administration, US Department of Health and Human Services

Fu L, Li J, Wang G, Luan Y, Dai W (2021) Adsorption behavior of organic pollutants on microplastics. Ecotoxicol Environ Saf 217:112207

Fytianos G, Rahdar A, Kyzas GZ (2020) Nanomaterials in cosmetics: Recent updates. Nanomaterials 10(5):979

Gouin T, Roche N, Lohmann R, Hodges G (2011) A thermodynamic approach for assessing the environmental exposure of chemicals absorbed to microplastic. Environ Sci Technol 45(4):1466–1472

Goyal N, Jerold F (2023) Biocosmetics: technological advances and future outlook. Environ Sci Pollut Res 30(10):25148–25169

Gregory MR (1996) Plastic 'scrubbers' in hand cleansers: a further (and minor) source for marine pollution identified. Mar Pollut Bull 32(12):867–871

Grimmer RA (1996) Process for manufacturing plastic microspheres. Google Patents, In

Guerranti C, Martellini T, Perra G, Scopetani C, Cincinelli A (2019) Microplastics in cosmetics: environmental issues and needs for global bans. Environ Toxicol Pharmacol 68:75–79

Hernandez LM, Yousefi N, Tufenkji N (2017) Are there nanoplastics in your personal care products? Environ Sci Technol Lett 4(7):280–285

Hung C-M, Huang C-P, Hsieh S-L, Chen Y-T, Ding D-S, Hsieh S, Chen C-W, Dong C-D (2022) Exposure of Goniopora columna to polyethylene microplastics (PE-MPs): effects of PE-MP concentration on extracellular polymeric substances and microbial community. Chemosphere 297:134113

Jiang J-Q (2018) Occurrence of microplastics and its pollution in the environment: a review. Sustain Prod Consum 13:16–23

Jessieleena A, Rathinavelu S, Velmaiel KE, John AA, Nambi IM (2023) Residential houses—a major point source of microplastic pollution: insights on the various sources, their transport, transformation, and toxicity behaviour. Environ Sci Pollut Res 30(26):67919–67940

Kalčíková G, Alič B, Skalar T, Bundschuh M, Gotvajn AŽ (2017) Wastewater treatment plant effluents as source of cosmetic polyethylene microbeads to freshwater. Chemosphere 188:25–31

Kozlowska J, Prus W, Stachowiak N (2019) Microparticles based on natural and synthetic polymers for cosmetic applications. Int J Biol Macromol 129:952–956

Kristanti RA, Hadibarata T, Wulandari NF, Sibero MT, Darmayati Y, Hatmanti A (2023) Overview of microplastics in the environment: type, source, potential effects and removal strategies. Bioprocess Biosyst Eng 46(3):429–441

Kumar R, Verma A, Shome A, Sinha R, Sinha S, Jha PK, Kumar R, Kumar P, Shubham, & Das, S. (2021) Impacts of plastic pollution on ecosystem services, sustainable development goals, and need to focus on circular economy and policy interventions. Sustain For 13(17):9963

Lam C-S, Ramanathan S, Carbery M, Gray K, Vanka KS, Maurin C, Bush R, Palanisami T (2018) A comprehensive analysis of plastics and microplastic legislation worldwide. Water Air Soil Pollut 229:1–19

Lamichhane G, Acharya A, Marahatha R, Modi B, Paudel R, Adhikari A, Raut B, Aryal S, Parajuli N (2023) Microplastics in environment: global concern, challenges, and controlling measures. Int J Environ Sci Technol 20(4):4673–4694

Laskar N, Kumar U (2019) Plastics and microplastics: a threat to environment. Environ Technol Inno 14:100352

Leadsom, A. (2016). Microbead ban announced to protect sealife: Department for Environment, food & rural affairs

Leslie H (2014) Review of microplastics in cosmetics. IVM Inst Environ Stud 476:1–33

Leslie HA, Van Velzen MJ, Brandsma SH, Vethaak AD, Garcia-Vallejo JJ, Lamoree MH (2022) Discovery and quantification of plastic particle pollution in human blood. Environ Int 163:107199

Li H, Lu X, Wang S, Zheng B, Xu Y (2021a) Vertical migration of microplastics along soil profile under different crop root systems. Environ Pollut 278:116833

Li Y, Sun Y, Li J, Tang R, Miu Y, Ma X (2021b) Research on the influence of microplastics on marine life. Earth Environ Sci, IOP Conference Series

Liu P, Zheng Y, Yuan Y, Zhang T, Li Q, Liang Q, Su T, Qi Q (2022a) Valorization of polyethylene terephthalate to Muconic acid by engineering pseudomonas putida. Int J Mol Sci 23(19):10997

Liu Q, Wu H, Chen J, Guo B, Zhao X, Lin H, Li W, Zhao X, Lv S, Huang C (2022b) Adsorption mechanism of trace heavy metals on microplastics and simulating their effect on microalgae in river. Environ Res 214:113777

Lochhead R (2017) The use of polymers in cosmetic products. Cosmet Sci Technol 13:171–221
Lochhead, R. Y. (2007). The role of polymers in cosmetics: recent trends
Luís IP, Spínola H (2010) The influence of a voluntary fee in the consumption of plastic bags on supermarkets from Madeira Island (Portugal). J Environ Plan Manag 53(7):883–889
McDevitt JP, Criddle CS, Morse M, Hale RC, Bott CB, Rochman CM (2017) Addressing the issue of microplastics in the wake of the microbead-free waters act a new standard can facilitate improved policy. Environ Sci Technol 51(12):6611–6617
Mitrano D, Wohlleben W (2020) Microplastic regulation should be more precise to incentivize both innovation and environmental safety. *Nat Commun* 11:5324
Napper IE, Bakir A, Rowland SJ, Thompson RC (2015) Characterisation, quantity and sorptive properties of microplastics extracted from cosmetics. Mar Pollut Bull 99(1–2):178–185
OECD, S. (2014). The state of play on extended producer responsibility (EPR): opportunities and challenges
Ogunola OS, Onada OA, Falaye AE (2018) Mitigation measures to avert the impacts of plastics and microplastics in the marine environment (a review). Environ Sci Pollut Res 25:9293–9310
Park SY, Kim CG (2019) Biodegradation of micro-polyethylene particles by bacterial colonization of a mixed microbial consortium isolated from a landfill site. Chemosphere 222:527–533
Parker L (2021) Global treaty to regulate plastic pollution gains momentum. National Geographic. https://www.nationalgeographic.com/environment/article/global-treaty-to-regulate-plastic-pollution-gains-momentum.
Pastrana H, Avila A, Tsai CS (2018) Nanomaterials in cosmetic products: the challenges with regard to current legal frameworks and consumer exposure. NanoEthics 12:123–137
Patil A, Ferritto MS (2013) Polymers for personal care and cosmetics: Overview. In: Polymers for personal care and cosmetics, pp 3–11
Pettipas S, Bernier M, Walker TR (2016) A Canadian policy framework to mitigate plastic marine pollution. Mar Policy 68:117–122
Raj S, Jose S, Sumod U, Sabitha M (2012) Nanotechnology in cosmetics: opportunities and challenges. J Pharm Bioallied Sci 4(3):186
Rist S, Hartmann NB (2018) Aquatic ecotoxicity of microplastics and nanoplastics: lessons learned from engineered nanomaterials. Emerging Environmental Contaminants? Freshwater Microplastics, pp 25–49
Rondoni G, Chierici E, Agnelli A, Conti E (2021) Microplastics alter behavioural responses of an insect herbivore to a plant-soil system. Sci Total Environ 787:147716
Rubin AE, Zucker I (2022) Interactions of microplastics and organic compounds in aquatic environments: a case study of augmented joint toxicity. Chemosphere 289:133212
Sherrington C, Darrah C, Hann S, Cole G, Corbin M (2016) Study to support the development of measures to combat a range of marine litter sources. Report for European Commission DG Environment, p 410
Singh V, Braddick D, Dhar PK (2017) Exploring the potential of genome editing CRISPR-Cas9 technology. Gene 599:1–18
Suardy NH, Tahrim NA, Ramli S (2020) Analysis and characterization of microplastic from personal care products and surface water in Bangi, Selangor. Sains Malaysiana 49(9):2237–2249
Sun Q, Ren S-Y, Ni H-G (2020) Incidence of microplastics in personal care products: an appreciable part of plastic pollution. Sci Total Environ 742:140218
Syberg K, Hansen SF, Christensen TB, Khan FR (2018) Risk perception of plastic pollution: importance of stakeholder involvement and citizen science. In: Freshwater microplastics: emerging environmental contaminants, pp 203–221
Tagg AS, Ivar do Sul JA (2019) Is this your glitter? An overlooked but potentially environmentally-valuable microplastic. Mar Pollut Bull 146:50–53
Tian L, van Putten RJ, Gruter GJM (2022) Plastic pollution. The role of (bio) degradable plastics and other solutions. In: Biodegradable polymers in the circular plastics economy, pp 59–81
UNEP (2016) Plastic in cosmetics: are we polluting the environment through our personal care? Plastic ingredients that contribute to marine microplastic litter, UN
van Leeuwenhoeklaan, A. (2018). Annex XV restriction report proposal for a restriction

Wainaina S, Awasthi MK, Sarsaiya S, Chen H, Singh E, Kumar A, Ravindran B, Awasthi SK, Liu T, Duan Y (2020) Resource recovery and circular economy from organic solid waste using aerobic and anaerobic digestion technologies. Bioresour Technol 301:122778

Xanthos D, Walker TR (2017) International policies to reduce plastic marine pollution from single-use plastics (plastic bags and microbeads): a review. Mar Pollut Bull 118(1–2):17–26

Yuan Z, Nag R, Cummins E (2022) Human health concerns regarding microplastics in the aquatic environment-from marine to food systems. Sci Total Environ 823:153730

Yurtsever M (2019a) Glitters as a source of primary microplastics: an approach to environmental responsibility and ethics. J Agric Environ Ethics 32(3):459–478

Yurtsever M (2019b) Tiny, shiny, and colorful microplastics: are regular glitters a significant source of microplastics? Mar Pollut Bull 146:678–682

Zacharopoulou O, Varvaresou A (2012) Nanotechnology in cosmetology. Epitheorese Klinikes Farmakologias kai Farmakokinetikes 30(1):51–54

Zhou Y, Ashokkumar V, Amobonye A, Bhattacharjee G, Sirohi R, Singh V, Flora G, Kumar V, Pillai S, Zhang Z (2023) Current research trends on cosmetic microplastic pollution and its impacts on the ecosystem: a review. Environ Pollut 320:121106

Zitko V, Hanlon M (1991) Another source of pollution by plastics: skin cleaners with plastic scrubbers. Mar Pollut Bull 22(1):41–42

Microplastics in Air and Their Health Hazard

10

Md Rayhanul Islam Rayhan, Mohammad Toha,
and Md Mostafizur Rahman

Abstract

Microplastics in the air have emerged as a major concern for the global environment because of their ubiquity and potential health implications. This chapter reviewed the occurrences, distributions, sources, and health concerns of airborne microplastics (MPs) in both indoor and outdoor air. Additionally, it reviewed the statistical approaches useful for measuring human health risks from airborne MPs. We found that all studies around the world identified a detectable amount of MPs in the collected samples, indicating an omnipresence of MPs in the air. Various MP sources were identified, and they varied in outdoor and indoor environments. In most cases, fragmented MPs were dominant in indoor and outdoor

M. R. I. Rayhan
Laboratory of Environmental Health and Ecotoxicology, Department of Environmental Sciences, Jahangirnagar University, Dhaka, Bangladesh

Centre for Environmental Research and Development (CERD), Dhaka, Bangladesh

M. Toha
Laboratory of Environmental Health and Ecotoxicology, Department of Environmental Sciences, Jahangirnagar University, Dhaka, Bangladesh

Department of Environmental Science, Bangladesh University of Professionals, Dhaka, Bangladesh

M. M. Rahman (✉)
Laboratory of Environmental Health and Ecotoxicology, Department of Environmental Sciences, Jahangirnagar University, Dhaka, Bangladesh

Centre for Environmental Research and Development (CERD), Dhaka, Bangladesh

Department of Environmental Science, Bangladesh University of Professionals, Dhaka, Bangladesh

Department of Environmental Sciences, Jahangirnagar University, Dhaka, Bangladesh
e-mail: rahmanmm@juniv.edu

A. Kumar, V. Singh (eds.), *Microplastics Pollution and its Remediation*,
https://doi.org/10.1007/978-981-97-4068-0_10

air samples. Polyethylene was found to be the most common polymer in most of the studies. Several studies concluded that MPs in outdoor and indoor air can potentially cause adverse human health conditions, whereas indoor air was identified as more detrimental than outdoor air. However, in-depth studies on human health assessment for MPs are very limited. Several statistical tools can be used to calculate health risks.

Keywords

Indoor Air · Outdoor Air · Fragmented MPs · Polyethylene Polymer · Endocrine Disruption

10.1 Introduction

Plastic production around the world has steadily risen since the mid-nineteenth century, driven by factors such as population growth, improved lifestyles, and economic expansion, reaching 460 million metric tons in 2019. Surprisingly, only 9% of plastic waste is recycled, while the rest of it deposits in the environmental compartments, which causes the expansion of microplastics (MPs) in the built environment (OECD 2022). Additionally, the widespread popularity of plastic is attributed to its durability, lightweight nature, flexibility, ease of production, and low cost, making it highly favored among the general population. These characteristics have contributed to the increased production and utilization of plastic materials (Haque et al. 2023). MPs typically refer to tiny plastic particles whose size does not exceed 5 millimeters in length (Afrin et al. 2020; Rana et al. 2023).

Microplastics are widely distributed across the environment along with other emerging air pollutants like polychlorinated biphenyls (PCBs) (Rayhan et al. 2024) and polyaromatic hydrocarbons (PAHs) (Moorchilot et al. 2023). For example, MPs have been detected in water (Alam et al. 2019; Kirstein et al. 2021; Mercy et al. 2022; Panno et al. 2019), sediment Van Cauwenberghe et al. 2015; Ding et al. 2019; Huang et al. 2021; Osorio et al. 2021), road dust (O'Brien et al. 2023; Patchaiyappan et al. 2021; Su et al. 2020), breast milk and blood (Leslie et al. 2022), human food (Hossain et al. 2019; Khan and Setu 2022; Oliveri Conti et al. 2020), and in human stool (Schwabl et al. 2019; Yan et al. 2022). However, airborne microplastics have not been studied as extensively as their aquatic and terrestrial counterparts (Sharaf Din et al. 2024). Outdoor microplastic pollution, particularly in the form of atmospheric deposition, has received limited attention from researchers. Furthermore, studies investigating indoor microplastic pollution are significantly lacking compared to research conducted on outdoor environments (C. Wang et al. 2023). MPs have been discovered in urban and suburban atmospheres and remote regions far from identifiable microplastic sources. This implies that the air plays a crucial role in transporting microplastics over long distances (Liao et al. 2021).

Air pollution from particulate matter is a critical health issue worldwide. In 2019 alone, the estimated premature deaths of over 6.5 million people worldwide were attributed to the inhalation of air containing fine particulate matter (Murray 2020).

Therefore, it is imperative to understand and monitor our air quality. Microplastic particles are one of the crucial and emerging air pollutants that demand pronounced health impacts. The uncontrolled abundance of MPs in the air can lead to damage to the human respiratory system by inflammation (Di Dong et al. 2020), oxidative stress (Jeon et al. 2022), and genotoxicity (Gasperi et al. 2018). Exposure to microplastics carries various health risks, encompassing physical strain, cellular harm, inflammation, oxidative stress, immune reactions, toxicity over both short and long terms, potential carcinogenic effects, genetic damage, developmental issues, gut microbiome disruption, decreased quality of sperm, lower levels of testosterone, and adverse effects on cognitive functions like memory and learning (Molina and Benedé 2022; Yuan et al. 2022). It is crucial to recognize that our comprehension of the health hazards linked to microplastics is continuously advancing, with notable uncertainties prevailing in this area. Nevertheless, empirical studies have confirmed that microplastics pose a real threat to both ecosystems and human health (Sheraz et al. 2023).

Additionally, Liao et al. 2021 suggested that the health risks linked to airborne microplastics hinge on multiple factors, including their quantity and diverse properties such as size, shape, and chemical composition. Certain shapes of microplastics may pose a higher potential health risk to organisms. For example, fibers are prone to becoming tangled within the intestinal tract, potentially remaining in the organism for extended periods, increasing the likelihood of adverse effects (Rochman 2018). Given these backgrounds, this chapter aims to review the occurrences, distributions, sources, and health concerns of airborne microplastics (MPs) in both indoor and outdoor air. Also, statistical approaches for human health risk assessments have been reviewed.

10.2 Microplastics in Indoor Air

10.2.1 Sources of Microplastics in Indoor Air

Different types of plastics find widespread use indoors owing to their durability, versatility, and cost-effectiveness (Welden and Lusher 2020). These plastics are fragmented into smaller pieces and appear as microplastics in indoor environments. Microplastics in indoor environments primarily originate from household items such as carpets, furniture, clothing, curtains, and floor surfaces (Fig. 10.1). O'Brien et al. (2020) summarized that the usage of household clothes dryers in indoor settings can significantly contribute to the release of MP. On the other hand, Jenner et al. 2021 indicated that there was no significant difference in the abundance of microplastics within households, regardless of whether clothes were hanging or dryers and carpets were employed. The possible sources of fiber microplastics found in indoor environments, composed of polyethylene terephthalate or polypropylene, include carpets, curtains, textiles, torn clothing, and household furnishings (Dris et al. 2017; Zhang et al. 2020). Soltani et al. (2021) observed a doubling in the prevalence of polymers in households with carpets compared to those without.

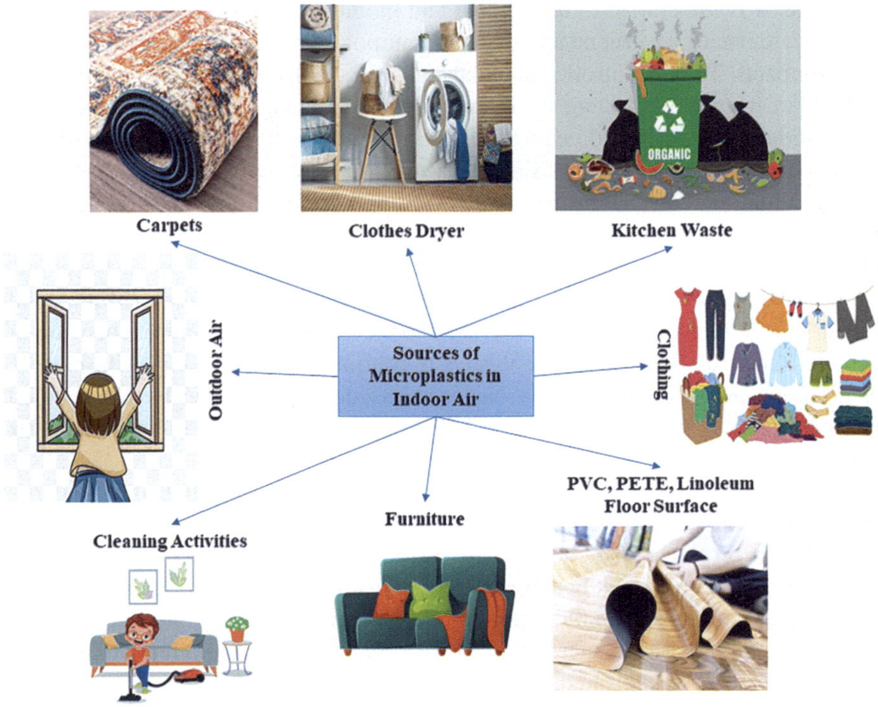

Fig. 10.1 Sources of microplastics in indoor air

Carpets were identified as the main source of microplastics, particularly polyethylene, polyester, polystyrene, and polyacrylic. The authors also proposed that polyvinyl chloride floor varnishes and linoleum surfaces could contribute to the presence of polyvinyl chloride fibers. Sheraz et al. (2023) emphasized that kitchen waste within the indoor environment is a notable contributor to indoor microplastics. Another significant origin of microplastics indoors is the breakdown of plastics found in packaging materials. It's essential to note that not all plastics degrade indoors, but some are more susceptible than others. The propensity for a specific type of plastic to degrade indoors, leading to the release of airborne microplastics, depends on factors like the type of plastic, environmental conditions, and exposure duration (Ageel et al. 2022; Kacprzak and Tijing 2022). Another method by which microplastics can enter indoor environments is through the application of heat to plastic materials. This process can lead to the release of both microplastics and potentially harmful chemical compounds into the indoor air. For instance, heating plastic containers holding food can cause the degradation of plastic structures, resulting in the emission of microplastics into the indoor environment (Bhat et al. 2023; Kacprzak and Tijing 2022). Microplastics can pass through indoor environments from outdoor sources through processes such as wind dispersal and adherence to clothing (Zhang et al. 2020). Specific lighting, like LED bulbs, may hasten the breakdown of plastic materials, resulting in the emission of microplastics in indoor air. Additionally, regular cleaning practices, such as sweeping and vacuuming, can disturb settled dust containing microplastics, leading to their dispersal in indoor air (Hale et al. 2022; O'Brien et al. 2023; Zhai et al. 2023).

10.2.2 Occurrence of Microplastics in Indoor Air

Microplastics have been studied in various compartments (sediment, water, food, cellular tissue, etc.) all over the world. However, the volume of studies investigating MPs in indoor air is limited, although this compartment is a potential platform for MP deposition that demands prominent human health concerns (Bhat 2023). All studies conducted globally have detected measurable levels of microplastics in indoor environments, indicating a widespread presence of these particles in this setting. Table 10.1 shows a global overview of the occurrence and abundance of microplastics, along with the sampling methods of those studies. The abundance of microplastics denotes the concentration/presence of MPs per unit sample. In indoor environments, the shape of microplastics can exhibit significant variation, influenced by their sources and types. Nonetheless, findings from numerous studies on atmospheric microplastics suggest that indoor airborne microplastics primarily manifest as fibers and fragments, indicative of their origin in synthetic textiles. It's worth noting that these microplastic fibers and fragments may possess diverse chemical compositions (Kek et al. 2024). On the contrary, microplastics in marine environments often display diverse forms, including fragments, fibers, and granules. These microplastics derive from various sources, such as plastic debris and paint particles, as elucidated in the study (De-la-Torre et al. 2023). Size refers to the length (sometimes including width, which depends on the shape of MPs) of the MPs present in the samples. Different studies classify the size range in different manners, as shown in Table 10.1. Studies have documented a range of sizes for airborne MPs in indoor settings, with the measurements varying based on factors like the sampling location, type of indoor environment, and the specific sampling technique utilized (Dris et al. 2017). The smaller size distribution and enrichment of fragments observed in outdoor airborne MPs suggest more intricate source dynamic.

This phenomenon may be attributed to particulate matter or minerals containing elements such as Al, Ca, Na, Cl, and others in ambient particle samples that could accelerate microplastic degradation (Yao et al. 2022). Another important property of microplastics is the polymer type that indicates the origin of the microplastics. Indoor MPs, typically originating from synthetic textiles, exhibit less diversity of polymer types than their outdoor counterparts (Kek et al. 2024). As Table 10.1 shows, polyethylene was found in most of the studies, and some of the studies found it as the highest percentage (Choi et al. 2022; Field et al. 2022; Jenner et al. 2021; Kashfi et al. 2022; Perera et al. 2023; Saygin et al. 2023; Yao et al. 2022). Synthetic fiber components, including polyester (PES), polyethylene terephthalate (PET), and cellulose, are often associated with products, including furniture, clothes, and household items (Choi et al. 2022; Zhang et al. 2020). Numerous industries use polyethylene for a variety of purposes. Some of these sectors include the production

Table 10.1 Overview of the occurrences of microplastics in indoor air around the world

Country	References	Sampling information	Abundance	Shape information	Size information	Color Information	Polymer Types
India	Nandi et al. (2024)	24 h dust samples were collected using a respirable dust sampler in post-monsoon and winter	– 8.3×10^4 particles/m^2/day during post-monsoon – 1.03×10^5 particles/m^2/day during winter	– Fibers were the most dominant, and fragments were the least	– Fibers ranged 3.6–6.9 μm in winter	– White/ transparent fibers were dominant	– Polyvinyl chloride was e dominant
Pakistan	Aslam et al. (2022)	40 house dust samples from 2 cities	– Ranged from 159.05–234.05 MP/m^2	– Fiber was the dominant type (96.94% and 98.12%)	– No information	– White/ transparent color was the most dominant	– Polyester, polyethylene, copolymers of polypropylene, and polyurethane
China	Zhai et al. (2023)	Passive sampling from four points in four areas (dormitories, offices, dining halls, laboratories)	Average microplastic abundance: – Dormitory(highest): (14088.05 pcs/$m^{3)}$ – Dining hall: 4308.26 pcs/m^3 – Office: 13097.13 pcs/m^3 – Laboratory: 7512.55 pcs/m^3	– Mostly fibrous, blocky, granular, and irregular	– Average particle size —66.15 μm	– Dark MPs were dominant.	– Polyamide, polyethylene, petroleum resin, polyoxymethylene, cis-butadiene rubber, polyvinyl chloride, ethylene-vinyl acetate copolymer, polyvinyl butyral, and other rare microplastic material

Japan	Field et al. (2022)	Air sampler was placed around an operating theater (OT) and nearby anesthetic room (1 week sampling period, 12 h intervals)	– OT average: 1924 ± 3105 MP/m2/day – Anesthetic room average: 541 ± 969 MP/m2/day	– Fragment MPs were dominant (78%)	– 21–30 μm length and 10–250 μm width were the dominant size MPs	– Transparent MPs were dominant (84%)	– Polyethylene terephthalate (37%), polypropylene (25%), polyethylene (7%) and nylon (13%)
South Korea	Choi et al. (2022)	Samples were collected from five points at different months for 48 h long using an air sampler	– Ranged from 0.45 to 6.64 particles/m^3	– Non-fiber MPs were dominant	– 50–100 μm MPs were dominant (48–96%)	– No information	– Dominant components were polyethylene and polypropylene (68%)
Malaysia	Dewika et al. (2023)	Air samples were collected over 6 weeks (8 h per day) in classrooms and offices using both active and passive air samplers	– Mean MP counts in office156 ± 48 MPs/m^3; $n = 6$ – Mean MP counts in the classroom-104 ± 13 MPs/m^3; $n = 6$	– Fibers in office —76% – Fibers in classroom—66%	– Size in passive sampler—73% accounted for >500 μm – Size in active sampler—43% accounted for 100–500 μm	– Transparent MP was dominant	– Rayon was in the highest percentage in the classroom (79.3%) – Polyvinyl alcohol (PVA) was in the highest percentage in the office (39.5%)

(continued)

Table 10.1 (continued)

Country	References	Sampling information	Abundance	Shape information	Size information	Color Information	Polymer Types
Spain	Torres-Agullo et al. (2022)	An air sampling pump was engaged to collect samples from three urban indoor environments houses, public transport, and working places)	– Average MP in buses (highest): 17.3 ± 2.4 MPs/m^3 – Subway: 5.8 ± 1.9 MPs/m^3 – Houses: 4.8 ± 1.6 MPs/m^3 – Workplaces: 4.2 ± 1.6 MPs/m^3	– Fiber MPs were dominant	– Fibers <100 μm represented the most abundant size range	– Black was the dominant color among the MPs	– Polyamide (51%), polyester (48%), and PP (1%)
Turkey	Saygin et al. (2023)	Dust samples from 33 house floors in 4 urban districts	– Size-dependent abundance (refer to size information)	– Fragment/shapeless was the most dominant	– <5 mm MP was dominant (ranged from 1.16×10^4 to 4.90×10^4 mg/kg)	– Gray was the most dominant color (21.6 ± 2.1%), followed by white (18.9 ± 3.2%) and brown (17.6 ± 3.2%)	– Low-density polyethylene (LD-PE) was in the highest percentage (21.54 ± 2.05%)
UK	Jenner et al. (2021)	Passive samples were taken over 6-months from 20 houses	– The average concentration was 1414 ± 1022 MP/m^2/day	– Fibrous particles were the most abundant (90%)	– 5–250 μm was the most dominant (90%)	– No information	– Polyethylene terephthalate presents in 90% of samples

Australia	Perera et al. (2023)	Indoor air samples from seven locations for 6 h using a stainless-steel air sampler	– The highest indoor MP at a child care site- 2.25 ± 0.38 particles/m^3; – The lowest indoor MP inside a vehicle- 0.20 ± 0.14 particles/m^3	– Fibers—98% – Fragments—2%	– Fibers ranged from 71 to 4950 μm in length	– Black MP was dominant – Red MP was the lowest	Polyethylene terephthalate was the most dominant polymer
USA	Yao et al. (2022)	Passive sampling in four locations (one office, one hallway, one classroom, and one household)	– Average (1.96 ± 1.09) × 10^4 fibers/m^2/day	– Fibers dominant	– 35 to 1000 μm size range was dominant	– Blue and gray	– Polyethylene dominant
Iran	Kashfi et al. (2022)	Dust samples from 30 buildings	– Ranged from 27 to 112 MP/mg of dust	– Fiber MP was the dominant type	– Most abundant detected MPs were related to the size of >1000 μm	– White/ transparent was dominant	– Polyethylene was the most common polymer type
Kuwait	Uddin et al. (2022)	Air samples from four types of buildings using a six-stage cascade impactor	– Ranged from 3.2 to 27.1 particles/m^3	– Dominantly fibers (91%), with few fragments in lower-size fractions (9%)	– 0.45–2800 μm	– Red dominant	– Polyester, nylon, polyamide
29 countries	Soltani et al. (2022)	Dust samples from 108 homes in 29 countries over a 1-month period	– Ranged from 1257 ± 224 to 3518 ± 490 fibers/m^2/day	– Fiber was dominant and accounted for 98.6%	– 200–400 μm fiber length was the most prevalent	– Transparent fibers were dominant (28–48%)	– Natural cellulose fibers were dominant (31 to 41%)

of boxes, pipes, cleaning bottles, dispensing bottles, plastic bags for computer components, and laboratory equipment. Polyethylene is a widely used plastic polymer that is manufactured in large quantities worldwide (Sheraz et al. 2023). Because of their large molecular weight and hydrophobic character (caused by the -CH groups), these particles are particularly noteworthy because they make it easier for environmental contaminants to adsorb onto them. Because polyethylene (PE) is prone to photooxidation, UV stabilizers, and other additives must be included in its composition (Castro et al. 2020).

10.3 Microplastics in Outdoor Air

10.3.1 Sources of Microplastics in Outdoor Air

Given the diverse and extensive characteristics of plastic-based materials used in outdoor environments, particularly in urban areas, they can include any materials that contain plastic (Jahandari 2023). Widespread microplastic contamination both below and beyond the planetary barrier level highlights densely populated metropolitan areas as major sources of microplastic pollution (González-Pleiter et al. 2021). Microplastic in outdoor air can be deduced from agricultural activities, industrial activities, construction activities, painting, sea spray, plastic littering, waste incineration/landfi,ll, road dust, road paints, vehicle tires, and drying of clothes, bed sheets, pillows, etc., in the outdoor environment (Fig. 10.2).

Roblin et al. (2020) hypothesized that either the long-range movement of microplastics during dry deposition or marine sources, such as sea spray, might be the cause of the preponderance of small-sized MPs at the coastal location with less fluctuation. Another notable observation made by Wang et al. (2020) is that microplastics from the Philippines were found over the South China Sea, suggesting that microplastics may travel from continental sources into the marine atmosphere. On the other hand, the fact that the microplastic degradation pattern in lakes and marine beaches is comparable indicates that air fallout is the main source of microplastics in water bodies (Wang et al. 2023). Dris et al. (2016) concluded that microplastics in atmospheric fallout could originate from various sources, including synthetic fibers from clothing and household items, the degradation of larger plastic objects, and activities such as landfill disposal or waste incineration. When textiles like clothes, bedding, pillows, and blankets are left to dry in exposed areas like balconies, UV radiation degradation occurs, releasing microplastics into the atmosphere (Truong et al. 2021). Moreover, packaging and textile products can serve as sources of polypropylene, a type of microplastic polymer detected in the samples (X. Wang et al. 2020).

Klein and Fischer 2019 noted that as secondary sources, MPs could originate from tires, road dust, and road paints. On the other hand, Yukioka et al. 2020 summarized that the outdoor MPs might have come from rubber products like tires, hoses, electrical insulation, plastic bags (polyethylene and polypropylene), bands, shoes, and soles. Bahrami et al. (2022) estimation suggests that around 84% of the

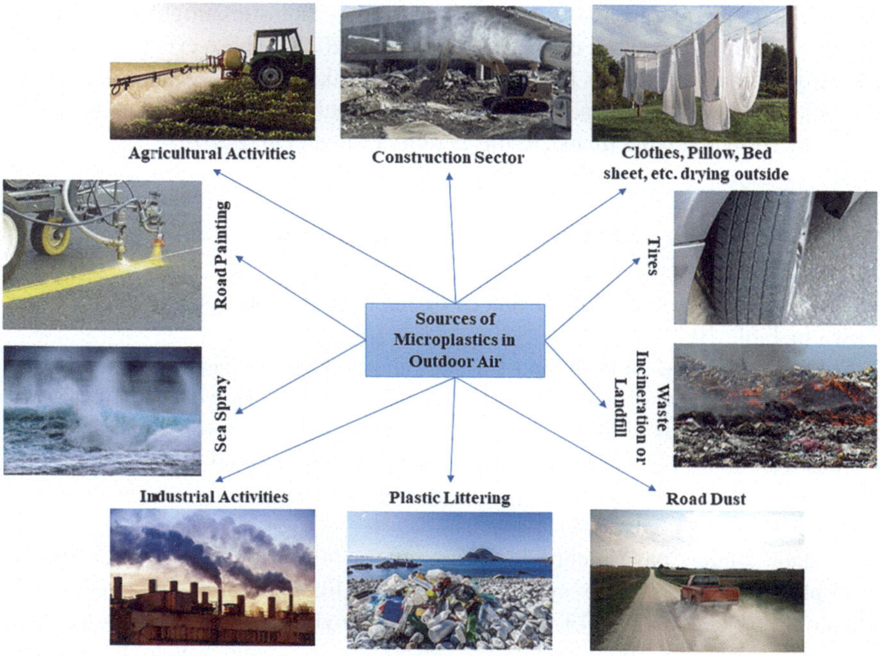

Fig. 10.2 Sources of microplastics in outdoor air

microplastics discovered in the western regions of the USA come from road debris, with the remaining 16% originating from soil dust from farms and coastal environments. As cars drive on the roadways, microscopic plastic particles are released into the air by the tires, brakes, and road surfaces (such as asphalt and road paint). Although road surfaces and rubber tires have different compositions, the majority of rubber tires are composed of synthetic and natural polymers such as carbon black, natural rubber, Styrene-butadiene rubber, and additives (Järlskog et al. 2021). Y. Ding et al. (2021) suggested that phenoxy resin may be traced back to the physical and mechanical tearing of paints and additives used in buildings. In contrast, polyethylene, polyethylene terephthalate, and polypropylene are probably industry products.

Agricultural activities are a significant source of MPs in the outdoor air, as reported by Y. Huang et al. 2020. Mulching (Batool et al. 2022), sewage sludge and compost fertilizers (Brahney et al. 2020), greenhouses and seeding trays (Bank and Hansson 2022), drip irrigation, fertilizer bags, pesticide containers, silage bags (Lwanga et al. 2022), livestock feeders, agricultural films, and crop protection covers (Jahandari 2023) constitute the primary sources of microplastics in farmlands. Marine microplastics can contribute to atmospheric microplastic episodes in urban areas, particularly those situated along coastlines. These microplastics, originating from terrestrial environments, have low density and can accumulate at the surface of the ocean's mixed layer (Brahney et al. 2020). Consequently, wave action and bubble-burst ejection are two ways that marine microplastics might be dispersed into the atmosphere (Batool et al. 2022; Jahandari 2023). These events include the expulsion of plastic particles from the ocean by breaking waves, which causes bubbles holding trapped air to ascend to the top and rupture. The particles are now able to be carried by the wind thanks to this activity (Allen et al. 2020).

10.3.2 Occurrence of Microplastic in Outdoor Air

An overview of the global occurrences of microplastics in outdoor air has been shown in Table 10.2.

Studies conducted on both land surfaces and bodies of water have consistently documented the deposition of atmospheric microplastics. However, there is variability in the methods used to collect samples among different research endeavors. Various techniques have been employed, such as gathering road dust, actively sampling suspended air particles, and passively collecting atmospheric deposition over a set period (Wang et al. 2023).

To date, studies have identified various types of microplastics present in the atmosphere. These include synthetic polymers like polyurethane (PUR), PE, polytetrafluoroethylene (PTFE or Teflon), polyvinyl acetate (PVA), epoxy resin (EP), PET, rayon (RY), polyacrylonitrile (PAN), ethylene vinyl acetate (EVA), PES, poly(N-methyl acrylamide) (PAA), and alkyd resin (ALK). Additionally, natural materials like cotton and wool have also been detected as microplastics in the atmosphere (Enyoh et al. 2019). As depicted in Table 10.1, different polymer types were predominated, most likely as a result of the many sources of microplastics found in the outside environment. However, polyvinyl chloride (PVC) and PE turned out to be the most common varieties. Microplastics are found in various forms or shapes, including fragments, foam, films, granules, fibers, and microbeads, while studies from around the world primarily observed a dominance of fragment-shaped microplastics in outdoor air (Table 10.2). The presence of such a diverse range of microplastic types and forms suggests the wide array of sources contributing to atmospheric microplastic pollution. Furthermore, there was variation in color dominance, which is indicative of the many sources of microplastics present in the outside environment. All investigations have consistently revealed the existence of microplastics in the environment, highlighting their extensive diffusion despite variations in technique.

10.4 Indoor Versus Outdoor Airborne Microplastics

The indoor environment exhibits a notably higher level of microplastic pollution than the outdoor environment, compared to their average microplastic counts. Since individuals predominantly inhabit indoor spaces, they face an elevated risk of ingesting and inhaling microplastics indoors, potentially resulting in health hazards. Both indoor and outdoor surroundings are dominated by fibers, which are found in

Table 10.2 Overview of the occurrences of microplastics in outdoor air around the world

Country	Reference	Sampling Information	Abundance	Shape information	Size Information	Color Information	Polymer Types
India	Patchaiyappan et al. (2021)	Street dust samples from 16 locations using a passive sampler	– Average: 227.94 ± 89.82 particles/100 g samples	– Fragments MPs were dominant (92.6%)	– No information	– No information	– Polyvinyl chloride (PVC) was the dominant polymer (24%).
Pakistan	Sharaf Din et al. (2024)	Gravimetric sampler was installed on the rooftop, where samples were collected for 2 weeks	– Average: 0.93 ± 0.32 items/m^3	– <50 μm: Fragments dominant – 50–1000 μm: Fibers dominant	– <50 μm MPs were the most abundant	– Transparent/white MPs were dominant	– Polyethylene terephthalate (PET) was dominant over other polymers (42.2%). Polyethylene (PE) was next dominant (30%)
China	Liao et al. (2021)	The LB-120 F intelligent medium flow total suspended particle sampler was used to gather 63 outdoor air samples	– Average: 1583 ± 1180 n/m^3	– Fragments MPs were predominant (>90%)	– 5–30 μm size ranged airborne MPs dominated (98.9%)	– No information	– Polyethylene (PE, 26.8%). – Polystyrene (PS, 17.8%) – Polyester (17.2%)
Japan	Yukioka et al. (2020)	12 surface road dust samples for 5 days	– Average: 2.0 ± 1.6 pieces/m^2	– No information	– 1400 ± 947 μm on average	– No information.	– Composed of 13 polymer types, where polyethylene was dominant in all size ranges, except in 100–5000 μm, where PET was the dominant MPs

(continued)

Table 10.2 (continued)

Country	Reference	Sampling Information	Abundance	Shape information	Size Information	Color Information	Polymer Types
Vietnam	Yukioka et al. (2020)	12 surface road dust samples for 7 days	– 19.7 ± 13.7 pieces/m^2	– Fragments MPs were predominant (68–81%)	– 791 ± 530 μm on average	– Transparent/ white dominant (27%)	– Composed of 14 polymer types, where polyethylene was dominant in all size ranges
Nepal	Yukioka et al. (2020)	13 surface road dust samples for 4 days	– 12.5 ± 10.1 pieces/m^2	– Fragments MPs were predominant (73–76%)	– 907 ± 675 μm on average	– Green dominant (26%)	– Composed of 15 polymer types, where polyethylene was dominant in all size ranges
Philippine		12 h dust sampling in 17 locations using a respirable dust sampler (2 weeks sampling)	– Average: 9.11 ± 5.46 SAMPs (suspended atmospheric microplastics)	– Fibrous MPs were the dominant type (88%)	– 1000–5000 μm size range was the dominant group (57%)	– Black MPs were the most dominant color (39%)	– Polyester was the most dominant polymer (74%)
Germany	Klein and Fischer (2019)	Biweekly atmospheric deposition samples (108 samples) were taken at six different sites for 12 weeks using nine bulk samplers	– A median abundance between 136.5 and 512.0 MPs/m^2/d	– Fragments (95%) were markedly dominating compared to fibers (5%)	– <63 μm size was most common	– No information	– Polyethylenes/ ethylvinyl acetate copolymers were dominating significantly (48.8% and 22.0%)

UK	Wright et al. (2020)	Nine-story rooftop atmospheric deposition samples were collected for 2 weeks (twice a week) using an aluminum rain gauge	– Ranged from 575 to 1008 MPs/m^2/d	– Fibrous MPs were dominant, accounting for 92%	– 400–500 μm size range was dominant	– No information	– 15 different petrochemical-based polymers were identified, where polyacrylonitrile (PAN) was the most abundant (67%)
USA	Yao et al. (2022)	The total atmospheric deposition sample was collected using four stainless steel funnels on the rooftop	– 3.93×10^4 particles/m^2/day around	– Fibers were the most dominant	– No information	– Blue fibers were the dominant color	– Polyvinyl chloride (PVC) was the dominant polymer (67%)
Iran	Shahraki et al. (2022)	Dust samples from 15 sampling points in 5 area	– Average: 9.8 pcs/100 g	– Fragment MPs were predominant (64%)	– Size range of 0.9–3.8 mm was dominant	– Transparent dominant (46%)	– Polyethylene was the dominant polymer (49%)

far greater abundance indoors than outside. Although a wide variety of polymers have been found in outdoor contexts, the polymer diversity is limited in interior environments, perhaps because there are fewer sources of microplastics there. Outdoor microplastic sources are numerous and may be local or distant, making them difficult to avoid, whereas indoor sources are more limited and traceable. More research is necessary to fully understand how weather factors like wind and precipitation affect the amount of microplastics in the environment. Factors influencing microplastic abundance indoors are well-defined across studies, with human activities being a common contributor. Small-sized microplastics are more prevalent in indoor air compared to outdoor air, as evidenced by studies focusing on the most abundant size range.

10.5 Health Hazard for Airborne Microplastics

There is currently little knowledge about the toxicity of microplastics and their long-term health effects, which necessitates quick action. Microplastics have the ability to penetrate the distal lung as well as the central airway once they are inhaled. They can have a number of detrimental consequences inside the respiratory system, including oxidative stress, inflammation, cellular damage, DNA damage, and cytokine release (Levermore et al. 2020; Vethaak and Legler 2021). Microplastics within the size range of 0.1–10 μm have the potential to traverse both the blood–brain barrier and the placenta. Additionally, particles smaller than 150 μm or 2.5 μm may pass through the gastrointestinal epithelium or enter the systemic circulation through endocytosis (Campanale et al. 2020). Fig. 10.3 shows the impacts of microplastics on human health when they enter the human body through inhalation of airborne microplastics.

Airborne microplastics present a potential health hazard, particularly for the respiratory system. Although the long-term risks remain uncertain, continuous research is essential to comprehend their impact fully. Airborne MPs are a relatively new area of research, but there's growing concern about their potential health hazards. Table 10.3 shows a specific list of health hazards caused by microplastics.

10.5.1 Statistical Approaches for Health Risk Assessment of Airborne Microplastics

Several equations have been used so far to assess the human health risks associated with airborne microplastics. Assessing human health risks associated with airborne microplastics involves various factors and considerations. While there is ongoing research in this field, specific equations for quantitatively assessing these risks may not yet be fully established.

Calculating the exposure to airborne microplastics is a common method to assess human health risk. Eqs. 10.1 and 10.2 are often used to determine the estimated

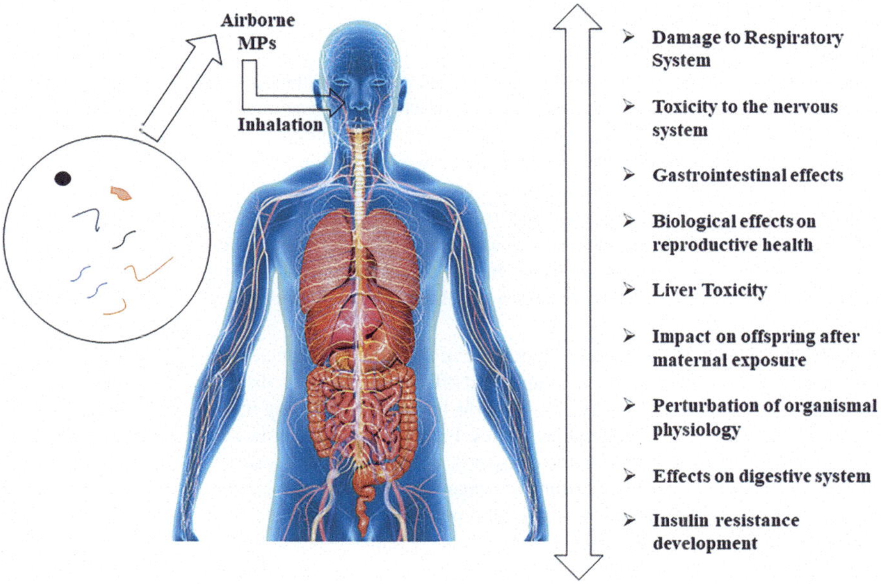

Fig. 10.3 Health hazard of airborne microplastics

daily intake (EDI) for ingestion and inhalation of microplastics (Saygin et al. 2023; Zhu et al. 2022). The higher the EDI value, the greater the risks.

$$EDI_{ing} = \frac{Cp \times f \times IR}{W} \tag{10.1}$$

$$EDI_{inh} = \frac{Cp \times f \times IR \times \frac{1}{PEF}}{W} \tag{10.2}$$

where,

EDI_{ing} = Estimated daily intake of MPs through ingestion

EDI_{inh} = Estimated daily intake of MPs through inhalation

Cp = MP concentrations

f = The total fraction of exposure in the indoor environment (i.e., hours spent by individuals in the indoor environment on a particular day); values vary from adult to child and can be found in Johnson-Restrepo and Kannan 2009

IR = Indoor dust intake rate for humans, and that is for infants 0.04, toddlers 0.03, children 0.02, and adults 0.02 g/day (Saygin et al. 2023)

W = Body weight of infants, toddlers, children, and adults

PEF = Particulate emission factor

On the other hand, Perera et al. 2023 determined the scenario-specific daily human exposure to airborne MPs by following Eq. 10.3.

$$E_{iw} = \sum_{jl}^{JL} C_j \times V_{il} \times T_{ijl} \tag{10.3}$$

Table 10.3 Microplastics-induced health effects

Health effects	Specifications	References
Neurotoxicity	Toxicity to the nervous system and potential behavioral changes. Chemicals like phthalates and flame retardants in MPs may have neurotoxic effects and impact cognitive function and neurological development, particularly in children	González-Pleiter et al. (2021), Pang et al. (2021)
Gastrointestinal effects	Disturb the gastrointestinal system through inflammation, disruption of gut microbiota (irritating the gut lining, triggering inflammation, and increasing intestinal permeability), cell damage, and toxicity	Chen et al. (2023), Hirt and Body-Malapel (2020), Oleksiuk et al. (2022)
Effects on reproductive health	Microplastics accumulate in the human placenta, affecting pregnancies and offspring. They also disrupt hormones by either interfering with or imitating the body's natural hormones. In animals, microplastics can harm reproductive organs such as ovaries and testes, potentially leading to issues with sperm quality, egg production, and overall reproductive function	Jassim (2023), Tanveer et al. (2023), Zurub et al. (2024)
Liver toxicity	Exposure to microplastics can result in adverse effects on the liver, potentially leading to damage and disease. Research on animals exposed to microplastics has revealed indications of liver damage, including inflammation, oxidative stress, and fatty liver changes. Although microplastics have been detected in human liver tissue, the health implications of this finding necessitate further investigation	Cheng et al. (2022), Haldar et al. (2023), Yee et al. (2021)
Effects on maternity and offspring	Taking into account the developmental origins of the health and disease (DOHaD) framework, exposure to microplastics in utero can influence the lifelong health of the offspring through programming effects. A recent finding has uncovered the presence of microplastics in human placental tissue, prompting concerns regarding the possibility of transfer to the developing fetus and its potential health implications. MPs can also mimic or interfere with the body's natural hormones, disrupting essential processes crucial for a healthy pregnancy, such as implantation and fetal development	Ragusa et al. (2021), Tanveer et al. (2023), Zurub et al. (2024)
Developing insulin resistance	Research conducted on mice indicated that ingestion of smaller-sized microplastic beads (0.5 μm) at higher doses (1 μg/mL) resulted in elevated levels of fasting plasma insulin and increased homeostatic model assessment of insulin resistance (HOMA-IR) in the mice. Certain studies indicated that microplastics could directly disrupt metabolic processes associated with insulin sensitivity, although the precise mechanisms involved are still being investigated	Haldar et al. (2023), Zhao et al. (2022)

(continued)

Table 10.3 (continued)

Health effects	Specifications	References
Effects on digestive system	Exposure to microplastics can lead to health concerns related to the digestive system. The gut microbiome, consisting of a diverse array of bacteria essential for digestion, immunity, and overall health, may be particularly affected. Microplastics have the potential to disrupt this delicate balance by physically harming gut bacteria, adsorbing essential nutrients that bacteria depend on, and interfering with bacterial communication	Bunea and Dinischiotu (2023), Siri et al. (2021), Yan et al. (2022)

where,

E_{iw} = Person-specific time-weighted integrated exposure i: female or male within different age groups (2–4/5–17/18–64/65 or above) over w period of time (w: working day/weekend)

C_j = Concentration of MPs in microenvironment j

V_{il} = Rate of inhalation for person i at l level of activity (l: sleep or nap and passive/light/moderate/high)

T_{ijl} = Time that person i spends in microenvironment j doing activity l

J = Total microenvironments number that person i move through during the certain time period w

L = Levels of different activity that performs throughout the time period w

Saygin et al. (2023) took into account the plastic debris' mass concentration as well as its polymer type in order to evaluate the possible dangers associated with it in settled home dust. An integrated pollution load index (PLI) and polymer hazard index may be used to assess the level of plastic debris pollution in settled house dust (PHI). This holistic approach enables more informed decision-making and targeted interventions to mitigate the risks associated with plastic pollution in indoor environments.

The PLI provides a comprehensive measure of pollution by considering multiple pollutants or parameters. In the context of plastic debris in settled house dust, the PLI can be calculated by comparing the mass concentrations of plastic debris in dust samples from different locations or over time. This index helps quantify the overall level of plastic pollution in the environment. Additionally, the PHI focuses specifically on the hazards posed by different polymer types of plastic debris. Since not all plastics have the same toxicity or environmental impact, the PHI allows for a more nuanced assessment by considering the polymer composition of the plastic debris found in house dust. By assigning hazard values to different polymer types based on their known environmental effects, the PHI provides insight into the potential risks associated with specific types of plastic debris. The PLI and PHI can be calculated using Eqs. 10.4, 10.5, and 10.6.

$$CF_i = \frac{C_i}{C_{oi}} \tag{10.4}$$

$$LI = \sqrt{CF_i} \tag{10.5}$$

$$PHI = \sum P_n \times S_n \tag{10.6}$$

where,

CF_i = MPs quotient

PLI = Pollution load index

PHI = Polymer hazard index

C_{oi} = The background value that is the lowest concentration of MPs in collected samples

C_i = Concentration of MPs

P_n = Percentage of a given polymer type

S_n = Polymer type's hazard score in the PHI computation

The PLI and PHI can be categorized according to the calculated values. For example, Saygin et al. (2023) categorized PLI by extreme danger > danger > high > minor (PLI higher to lower) and PHI by high > extreme danger > medium > danger > minor (PHI higher to lower).

Zuri et al. 2023 summarized assessing human health risks by determining the total daily intake (TDI) values using Eqs. 10.7, 10.8, 10.9, and 10.10.

$$ID = \frac{[MP] \times V_{air} \times EF}{BW} \tag{10.7}$$

$$DI = \frac{[MP] \times DIR}{BW} \tag{10.8}$$

$$EDI = \frac{\sum [MP]_{dust}}{BW} \tag{10.9}$$

$$TDI = ID + DI + EDI \tag{10.10}$$

where,

ID = Daily amount of MP potentially inhaled with air

DI = Daily amount of MP ingested via dust intake

EDI = Estimated daily intake of MP

TDI = Total daily intake

[MP] = Concentration of MP in one m^3 of indoor and outdoor air

V_{air} = Daily volume of air inhaled

EF = Exposure factor and an average

bw = Average weight

DIR = Dust ingestion rate

$[MP]_{dust}$ = Average concentration of MPs in dust

10.6 Conclusion

Both indoor and outdoor settings contain microplastics. The related dangers to human health can be expected to be higher inside due to their higher abundance and the maximal exposure of people in indoor environments. While research in this area is still in its early stages, advancements are being made in quantifying microplastics in various matrices, enabling the estimation of total daily intake (TDI) by considering dietary intake (DI), inhalation dose (ID), and environmental dose (EDI). However, data on dermal exposure is currently lacking. Although it is unknown how these factors relate to the abundance of microplastics, the outside environment is home to a variety of sources and factors that affect microplastic abundance. On the other hand, indoor air has been shown to have a few sources, including carpets and fabrics, in addition to variables influencing the quantity of microplastics. While the full extent of the health effects of airborne microplastics remains to be fully elucidated, emerging evidence suggests that chronic exposure to these ubiquitous pollutants may have adverse implications for respiratory, cardiovascular, immune, neurological, and overall systemic health. Further research is needed to understand better the mechanisms of action, dose–response relationships, and long-term health impacts of airborne microplastics, as well as to develop effective strategies for mitigating exposure and protecting public health.

References

Afrin S, Uddin MK, Rahman MM (2020) Microplastics contamination in the soil from urban landfill site, Dhaka, Bangladesh. Heliyon 6(11):e05572

Ageel HK, Harrad S, Abdalflah MA-E (2022) Occurrence{,} human exposure{,} and risk of microplastics in the indoor environment. Environ Sci: Processes Impacts 24(1):17–31. https://doi.org/10.1039/D1EM00301A

Alam FC, Sembiring E, Muntalif BS, Suendo V (2019) Microplastic distribution in surface water and Sediment River around slum and industrial area (case study: Ciwalengke River, Majalaya District, Indonesia). Chemosphere 224:637–645

Allen S et al (2020) Examination of the ocean as a source for atmospheric microplastics. PLoS One 15(5):1–14. https://doi.org/10.1371/journal.pone.0232746

Aslam I, Qadir A, Ahmad SR (2022) A preliminary assessment of microplastics in indoor dust of a developing country in South Asia. Environ Monit Assess 194(5):340. https://doi.org/10.1007/s10661-022-09928-3

Bahrami A, Haghighat F, Bahloul A (2022) Vacuum cleaner as a source of abiotic and biological air pollution in buildings: a review. Adv Build Energy Res 16(3):414–425. https://doi.org/10.1080/17512549.2020.1863859

Bank, Michael S, Hansson SV (2022) The microplastic cycle: an introduction to a complex issue. In: Michael S Bank (ed) Microplastic in the environment: pattern and process. Springer International Publishing, Cham, pp 1–16. https://doi.org/10.1007/978-3-030-78627-4_1

Batool I, Qadir A, Levermore JM, Kelly FJ (2022) Dynamics of airborne microplastics, appraisal and distributional behaviour in atmosphere; a review. Sci Total Environ 806:150745. https://www.sciencedirect.com/science/article/pii/S004896972105823X

Bhat MA (2023) Indoor microplastics: a comprehensive review and bibliometric analysis. Environ Sci Pollut Res Int 30(58):121269–121291. https://doi.org/10.1007/s11356-023-30902-0

Bhat MA, Gedik K, Gaga EO (2023) Atmospheric micro (Nano) plastics: future growing concerns for human health. Air Qual Atmos Health 16(2):233–262. https://doi.org/10.1007/s11869-022-01272-2

Brahney J et al (2020) Plastic rain in protected areas of the United States. Science 368(6496):1257–1260. https://www.science.org/doi/abs/10.1126/science.aaz5819

Bunea A-C, Dinischiotu A (2023) Human exposure to micro- and Nanoplastic: biological effects and health consequence. Reviews in Biological and Biomedical Sciences 5(1):1–13. https://doi.org/10.31178/rbbs.2022-2023.5.1.1

Campanale C et al (2020) A detailed review study on potential effects of microplastics and additives of concern on human health. Int J Environ Res Public Health 17(4):1212. https://www.mdpi.com/1660-4601/17/4/1212

Castro GB, Bernegossi AC, Felipe MC, Corbi JJ (2020) Is the development of daphnia magna neonates affected by short-term exposure to polyethylene microplastics? J Environ Sci Health A 55(8):935–946. https://doi.org/10.1080/10934529.2020.1756656

Chen Y et al (2023) Biological effects of polystyrene micro- and Nano-plastics on human intestinal organoid-derived epithelial tissue models without and with M cells. Nanomedicine 50:102680. https://www.sciencedirect.com/science/article/pii/S154996342300031X

Cheng W et al (2022) Polystyrene microplastics induce hepatotoxicity and disrupt lipid metabolism in the liver organoids. Sci Total Environ 806:150328. https://www.sciencedirect.com/science/article/pii/S004896972105405X

Choi H et al (2022) Comparison of microplastic characteristics in the indoor and outdoor air of urban areas of South Korea. Water Air Soil Pollut 233(5):169. https://doi.org/10.1007/s11270-022-05650-5

De-la-Torre GE et al (2023) Plastic and paint debris in marine protected areas of Peru. Sci Total Environ 901:165788. https://www.sciencedirect.com/science/article/pii/S004896972304411X

Dewika M et al (2023) Assessing microplastic distribution in indoor environments: a case study of a University Facility in Kuala Lumpur, pp 1–24. https://ssrn.com/abstract=4611351

Di Dong C et al (2020) polystyrene microplastic particles: in vitro pulmonary toxicity assessment. J Hazard Mater 385(October 2019):121575. https://doi.org/10.1016/j.jhazmat.2019.121575

Ding L et al (2019) Microplastics in surface waters and sediments of the Wei River, in the northwest of China. Sci Total Environ 667:427–434

Ding Y et al (2021) The abundance and characteristics of atmospheric microplastic deposition in the northwestern South China Sea in the fall. Atmos Environ 253:118389. https://www.sciencedirect.com/science/article/pii/S1352231021002089

Dris R et al (2016) Synthetic fibers in atmospheric fallout: a source of microplastics in the environment? Mar Pollut Bull 104(1):290–293. https://www.sciencedirect.com/science/article/pii/S0025326X16300066

Dris R et al (2017) A first overview of textile fibers, including microplastics, in indoor and outdoor environments. Environ Pollut 221:453–458

Enyoh CE et al (2019) Airborne microplastics: a review study on method for analysis, occurrence, movement and risks. Environ Monit Assess 191(11):668

Field DT et al (2022) Microplastics in the surgical environment. Environ Int 170(August):107630. https://doi.org/10.1016/j.envint.2022.107630

Gasperi J et al (2018) Microplastics in air: are we breathing it in? Curr Opin Environ Sci Health 1:1–5. https://doi.org/10.1016/j.coesh.2017.10.002

González-Pleiter M et al (2021) Occurrence and transport of microplastics sampled within and above the planetary boundary layer. Sci Total Environ 761:143213. https://www.sciencedirect.com/science/article/pii/S0048969720367449

Haldar S et al (2023) Nanoplastics toxicity specific to liver in inducing metabolic dysfunction—a comprehensive review. *Genes* 14(3)):590. https://www.mdpi.com/2073-4425/14/3/590

Hale RC et al (2022) Durable plastic goods: a source of microplastics and chemical additives in the built and natural environments. Environ Sci Technol Lett 9(10):798–807. https://doi.org/10.1021/acs.estlett.2c00417

Haque MR et al (2023) Assessment of microplastics pollution in aquatic species (fish, crab, and snail), water, and sediment from the Buriganga River, Bangladesh: an ecological risk appraisals. Sci Total Environ 857:159344. https://doi.org/10.1016/J.SCITOTENV.2022.159344

Hirt N, Body-Malapel M (2020) Immunotoxicity and intestinal effects of Nano- and Microplastics: a review of the literature. Part Fibre Toxicol 17(1):57. https://doi.org/10.1186/s12989-020-00387-7

Hossain MS et al (2019) Microplastics in fishes from the Northern Bay of Bengal. Sci Total Environ 690:821–830

Huang Y et al (2020) Agricultural plastic mulching as a source of microplastics in the terrestrial environment. Environ Pollut 260:114096. https://www.sciencedirect.com/science/article/pii/S0269749119362475

Huang J et al (2021) Microplastic pollution in soils and groundwater: characteristics, analytical methods and impacts. Chem Eng J 425:131870

Jahandari A (2023) Microplastics in the urban atmosphere: sources, occurrences, distribution, and potential health implications. J Hazard Mater Adv 12(April):100346. https://doi.org/10.1016/j.hazadv.2023.100346

Järlskog I et al (2021) Traffic-related microplastic particles, metals, and organic pollutants in an urban area under reconstruction. Sci Total Environ 774:145503. https://www.sciencedirect.com/science/article/pii/S0048969721005714

Jassim AK (2023) Effect of microplastic on the human health. In: In *advances and challenges in microplastics*, ed. IntechOpen, El-Sayed Salama. Rijeka. https://doi.org/10.5772/intechopen.107149

Jenner LC, Sadofsky LR, Danopoulos E, Rotchell JM (2021) Household indoor microplastics within the Humber region (United Kingdom): quantification and chemical characterisation of particles present. Atmos Environ 259:118512. https://www.sciencedirect.com/science/article/pii/S1352231021003332

Jeon MS et al (2022) Polystyrene microplastic particles induce Autophagic cell death in BEAS-2B human bronchial epithelial cells. Environ Toxicol 38(2):359–367. https://doi.org/10.1002/tox.23705

Johnson-Restrepo B, Kannan K (2009) An assessment of sources and pathways of human exposure to Polybrominated diphenyl ethers in the United States. Chemosphere 76(4):542–548. https://www.sciencedirect.com/science/article/pii/S004565350900294X

Kacprzak S, Tijing LD (2022) Microplastics in indoor environment: sources, mitigation and fate. J Environ Chem Eng 10(2):107359. https://doi.org/10.1016/j.jece.2022.107359

Kashfi FS et al (2022) Occurrence and exposure assessment of microplastics in indoor dusts of buildings with different applications in Bushehr and shiraz cities, Iran. Sci Total Environ 829:154651. https://doi.org/10.1016/j.scitotenv.2022.154651

Kek HY et al (2024) Critical review on airborne microplastics: an indoor air contaminant of emerging concern. Environ Res 245(December 2023):118055. https://doi.org/10.1016/j.envres.2023.118055

Khan HM, Setu S (2022) Microplastic ingestion by fishes from Jamuna River, Bangladesh. Environ Nat Res J 20(2):157–167

Kirstein IV, Gomiero A, Vollertsen J (2021) Microplastic pollution in drinking water. Curr Opin Toxicol 28:70–75

Klein M, Fischer EK (2019) Microplastic abundance in atmospheric deposition within the metropolitan area of Hamburg, Germany. Sci Total Environ 685:96–103. https://doi.org/10.1016/j.scitotenv.2019.05.405

Leslie HA et al (2022) Discovery and quantification of plastic particle pollution in human blood. Environ Int 163:107199

Levermore JM, Smith TEL, Kelly FJ, Wright SL (2020) Detection of microplastics in ambient particulate matter using Raman spectral imaging and Chemometric analysis. Anal Chem 92(13):8732–8740. https://doi.org/10.1021/acs.analchem.9b05445

Liao Z et al (2021) Airborne microplastics in indoor and outdoor environments of a Coastal City in eastern China. J Hazard Mater 417(May):126007. https://doi.org/10.1016/j.jhazmat.2021.126007

Lwanga EH et al (2022) Review of microplastic sources, transport pathways and correlations with other soil stressors: a journey from agricultural sites into the environment. Chem Biol Technol Agric 9(1):20. https://doi.org/10.1186/s40538-021-00278-9

Mercy, Fariha Tahsin, A K M Rashidul Alam, and Md Ahedul Akbor. 2022. Abundance and characteristics of microplastics in major urban wetlands of Dhaka, Bangladesh.

Molina E, Benedé S (2022) Is there evidence of health risks from exposure to micro- and Nanoplastics in foods? Front Nutr 9:910094. https://doi.org/10.3389/fnut.2022.910094

Moorchilot VS, Aravind UK, Aravindakumar CT (2023) Occurrence of azo-dyes, plasticizers, and PAH-bound microplastics: an emerging source and sink for hazardous compounds in indoor environments? Air Qual Atmos Health 17(2):425–438. https://doi.org/10.1007/s11869-023-01455-5

Murray CJL (2020) Global burden of 87 risk factors in 204 countries and territories, 1990–2019: a systematic analysis for the global burden of disease study 2019. Lancet 396(10258):1223–1249. https://doi.org/10.1016/S0140-6736(20)30752-2

Nandi S, Kumar RN, Dhandapani A, Iqbal J (2024) Characterization of microplastics in outdoor and indoor air in Ranchi, Jharkhand, India: first insights from the region. Environ Pollut 346(November 2023):123543. https://doi.org/10.1016/j.envpol.2024.123543

O'Brien S et al (2020) Airborne emissions of microplastic Fibres from domestic laundry dryers. Sci Total Environ 747:141175. https://www.sciencedirect.com/science/article/pii/S0048969720347045

O'Brien S et al (2023) There's something in the air: a review of sources, prevalence and behaviour of microplastics in the atmosphere. Sci Total Environ 874:162193

OECD (2022) Global plastics outlook. OECD. https://www.oecd-ilibrary.org/environment/global-plastics-outlook_aa1edf33-en

Oleksiuk K et al (2022) Health risks of environmental exposure to microplastics. J Educ Health Sport 13(1):79–84

Oliveri Conti G et al (2020) Micro- and Nano-plastics in edible fruit and vegetables. The first diet risks assessment for the general population. Environ Res 187:109677

Osorio ED, Tanchuling MA, Diola MB (2021) Microplastics occurrence in surface waters and sediments in five river mouths of Manila Bay. Front Environ Sci 9:364

Pang LY et al (2021) Microplastics - an emerging silent menace to public health. Life Sci Med Biomed 5(1) https://biomescientia.com/index.php/lsmb/article/view/72

Panno SV et al (2019) Microplastic contamination in karst groundwater systems. Ground Water 57(2):189–196

Patchaiyappan A et al (2021) Prevalence and characteristics of microplastics present in the street dust collected from Chennai Metropolitan City, India. Chemosphere 269:128757. https://doi.org/10.1016/J.CHEMOSPHERE.2020.128757

Perera K, Ziajahromi S, Nash SB, Leusch FDL (2023) Microplastics in Australian indoor air: abundance, characteristics, and implications for human exposure. Sci Total Environ 889(May):164292. https://doi.org/10.1016/j.scitotenv.2023.164292

Ragusa A et al (2021) Plasticenta: first evidence of microplastics in human placenta. Environ Int 146:106274. https://www.sciencedirect.com/science/article/pii/S0160412020322297

Rana M, Mishail M, Haque R, Tasnim SS, Mostafizur Rahman M (2023) The potential contribution of microplastic pollution by organic fertilizers in agricultural soils of Bangladesh: quantification, characterization, and risk appraisals. Front Environ Sci 11:1205603

Rayhan MRI et al (2024) Exposure of polychlorinated biphenyls via indoor dust particles and their health risks in Dhaka City, Bangladesh. J Hazard Mater Adv 14:100421. https://www.sciencedirect.com/science/article/pii/S2772416624000226

Roblin B, Ryan M, Vreugdenhil A, Aherne J (2020) Ambient atmospheric deposition of anthropogenic microfibers and microplastics on the Western periphery of Europe (Ireland). Environ Sci Technol 54(18):11100–11108. https://doi.org/10.1021/acs.est.0c04000

Rochman CM (2018) Microplastics research—from sink to source. Science 360(6384):28–29. https://www.science.org/doi/abs/10.1126/science.aar7734

Saygin H, Baysal A, Zora ST, Tilkili B (2023) A characterization and an exposure risk assessment of microplastics in settled house floor dust in Istanbul, Turkey. Environ Sci Pollut Res Int 30(57):121030–121049. https://doi.org/10.1007/s11356-023-30543-3

Schwabl P et al (2019) Detection of various microplastics in human stool: a prospective case series. Ann Intern Med 171(7):453–457

Shahraki M et al (2022) Microplastics in atmospheric dust samples of Sistan: sources and distribution. J Environ Health Sci Eng 20(2):931–936

Sharaf Din K et al (2024) Exploration of microplastic concentration in indoor and outdoor air samples: morphological, polymeric, and elemental analysis. Sci Total Environ 908(November 2023):168398. https://doi.org/10.1016/j.scitotenv.2023.168398

Sheraz M, Kim J, Kim J (2023) Nano/microplastics in indoor air: a critical review of synthesis routes for toxicity testing and preventative measure strategies. Process Saf Environ Prot 180(June):274–304. https://doi.org/10.1016/j.psep.2023.10.002

Siri C et al (2021) Adsorption of progesterone onto microplastics and its desorption in simulated gastric and intestinal fluids. Environ Sci: Processes Impacts 23(10):1566–1577. https://doi.org/10.1039/D1EM00226K

Soltani NS, Taylor MP, Wilson SP (2021) Quantification and exposure assessment of microplastics in Australian indoor house dust. Environ Pollut 283:117064. https://www.sciencedirect.com/science/article/pii/S0269749121006461

Soltani NS, Taylor MP, Wilson SP (2022) International quantification of microplastics in indoor dust: prevalence, exposure and risk assessment. Environ Pollut 312:119957. https://www.sciencedirect.com/science/article/pii/S026974912201171X

Su L, Nan B, Craig NJ, Pettigrove V (2020) Temporal and spatial variations of microplastics in roadside dust from rural and urban Victoria, Australia: implications for diffuse pollution. Chemosphere 252:126567

Tanveer M et al (2023) Microplastics: unraveling the signaling pathways involved in reproductive health. Environ Sci Pollut Res 30(42):95077–95085. https://doi.org/10.1007/s11356-023-29273-3

Torres-Agullo A, Karanasiou A, Moreno T, Lacorte S (2022) Airborne microplastic particle concentrations and characterization in indoor urban microenvironments. Environ Pollut 308(July):119707. https://doi.org/10.1016/j.envpol.2022.119707

Truong T-N-S et al (2021) Microplastic in atmospheric fallouts of a developing southeast Asian megacity under tropical climate. Chemosphere 272:129874. https://www.sciencedirect.com/science/article/pii/S004565352100343X

Uddin S et al (2022) A preliminary assessment of size-fractionated microplastics in indoor aerosol—Kuwait's baseline. Toxics 10(2):2–17. https://doi.org/10.3390/toxics10020071

Van Cauwenberghe L et al (2015) Microplastics in sediments: a review of techniques, occurrence and effects. Mar Environ Res 111:5–17

Vethaak AD, Legler J (2021) Microplastics and human health. Science 371(6530):672–674. https://www.science.org/doi/abs/10.1126/science.abe5041

Wang X et al (2020) Atmospheric microplastic over the South China Sea and East Indian Ocean: abundance, distribution and source. J Hazard Mater 389:121846. https://www.sciencedirect.com/science/article/pii/S030438941931800X

Wang C, Babel S, Lichtfouse E (2023) Microplastic occurrence, fate, impact, and remediation, pp 211–236. https://doi.org/10.1007/978-3-031-36351-1_9

Welden NA, Lusher A (2020) Chapter 9 - microplastics: from origin to impacts. In: Letcher TM (ed) Plastic waste and recycling. Academic Press, pp 223–249. https://www.sciencedirect.com/science/article/pii/B9780128178805000098.

Wright SL et al (2020) Atmospheric microplastic deposition in an urban environment and an evaluation of transport. Environ Int 136(September 2019):105411. https://doi.org/10.1016/j.envint.2019.105411

Yan Z et al (2022) Analysis of microplastics in human feces reveals a correlation between fecal microplastics and inflammatory bowel disease status. Environ Sci Technol 56(1):414–421
Yao Y, Glamoclija M, Murphy A, Gao Y (2022) Characterization of microplastics in indoor and ambient air in northern New Jersey. Environ Res 207(May 2021):112142. https://doi.org/10.1016/j.envres.2021.112142
Yee MS-L et al (2021) Impact of microplastics and Nanoplastics on human health. Nanomaterials 11(2):496. https://www.mdpi.com/2079-4991/11/2/496
Yuan Z, Nag R, Cummins E (2022) Human health concerns regarding microplastics in the aquatic environment - from marine to food systems. Sci Total Environ 823:153730. https://www.sciencedirect.com/science/article/pii/S0048969722008221
Yukioka S et al (2020) Occurrence and characteristics of microplastics in surface road dust in Kusatsu (Japan), Da Nang (Vietnam), and Kathmandu (Nepal). Environ Pollut 256:113447. https://doi.org/10.1016/j.envpol.2019.113447
Zhai X et al (2023) Characterization and quantification of microplastics in indoor environments. Heliyon 9(5):e15901. https://doi.org/10.1016/j.heliyon.2023.e15901
Zhang Q et al (2020) Microplastic fallout in different indoor environments. Environ Sci Technol 54(11):6530–6539
Zhao J et al (2022) Polystyrene bead ingestion promotes adiposity and Cardiometabolic disease in mice. Ecotoxicol Environ Saf 232:113239. https://www.sciencedirect.com/science/article/pii/S0147651322000793
Zhu J et al (2022) Microplastics in dust from different indoor environments. Sci Total Environ 833(April):155256. https://doi.org/10.1016/j.scitotenv.2022.155256
Zuri G, Karanasiou A, Lacorte S (2023) Microplastics: human exposure assessment through air, water, and food. Environ Int 179(May):108150. https://doi.org/10.1016/j.envint.2023.108150
Zurub RE, Cariaco Y, Wade MG, Bainbridge SA (2024) Microplastics exposure: implications for human fertility, pregnancy and child health. Front Endocrinol 14:1330396. https://doi.org/10.3389/fendo.2023.1330396

Microplastics and Microbe Interaction on Human and Animal Health

11

Jaya Verma, Sapna Devi, and Monika Yadav

Abstract

Owing to their widespread occurrence in many ecosystems and possible effects on both biotic and abiotic components, microplastics have become an urgent environmental concern. An overview of microplastic–bacteria interactions, potential human and animal health impacts, and wider environmental implications of microplastic presence are given in this chapter. Microplastics could be easily found in various domains such as freshwater and oceans, soil, and even our atmosphere. Along with affecting microbial communities, the interaction between microplastic and bacteria also has an impact on the transfer of pollutants, ecosystem functioning, and nutrient cycling. This raises questions about the safety of food and the exposure of people to dangerous substances, calling for more investigation into the long-term human health effects of microplastics. Fish, seabirds, and marine mammals are among the aquatic creatures that may confuse microplastics for food, which could result in ingesting them, internal damage, and nutrient dilution. Furthermore, absorbed contaminants and additives seeping from microplastics can cause toxicity, irregularities in reproduction, and behavioral changes in impacted species. These issues reveal that microplastics can linger in the nature for longer duration, divide into smaller particles called nanoplastics, and infiltrate different trophic levels. The compila-

J. Verma
Department of Microbiology, Faculty of Science, Guru Nanak Dev University, Amritsar, Punjab, India

S. Devi
Department of Food Science and Technology, Khalsa College of Amritsar, Amritsar, Punjab, India

M. Yadav (✉)
Department of Applied Sciences and Humanities (Environmental Science), Faculty of Engineering and Technology, Rama University, Kanpur, Uttar Pradesh, India

A. Kumar, V. Singh (eds.), *Microplastics Pollution and its Remediation*,
https://doi.org/10.1007/978-981-97-4068-0_11

tion, origins, characterization, interactions with microorganisms, and detrimental effects of microplastics are included in the current study.

Keywords

Microorganisms · Biodegradation · Neurodegenerative Disease · Nuclear Magnetic Resonance · Oxidative Stress

11.1 Introduction

We presently exist in the plastic age, with an annual production and manufacturing volume of over 370 million tonnes of plastic (Brussels, Belgium 2020). Polyethylene terephthalate (PET or polyester), polyacrylonitrile (PAN or acrylic), polyethylene (PE), polyamide (nylon), polyvinyl chloride (PVC), polystyrene (PS), styrene butadiene rubber (used in car tires, for example) and polypropylene (PP) are only a few of the several components found in plastics. These plastic polymers are all sourced from carbonaceous materials like natural gas and petroleum (Boucher and Friot 2017; Bilal et al. 2021; Hidalgo-Ruz et al. 2012; Thompson et al. 2009). In the past few years, a sharp rise in the demand for plastics has been noticed. Plastic products are safe and sustainable alternatives since they assist individuals, communities, and the food sector in particular. Plastic trash has unavoidably resulted from this widespread use of plastic. Approxemetly 34.6% of 30 million tonnes of plastic garbage that European nations will gather by 2020 will have undergone recycling. By 2050, it is anticipated that the continued use of recent production and management of waste techniques will result in plastic waste of 12 billion tonnes in the environment (Jambeck et al. 2015). Microplastics (MPs) are plastic pollutants that are emitted in minute pieces—typically smaller than 5 mm—into the environment. These pollutants are frequently made in an industrial setting. Pollutants referred to as primary MPs are created intentionally when plastic debris of this particular size is produced (Browne et al. 2011). Primary microplastics are meant to be used in cleaning products, cosmetic abrasives, personal hygiene products, medication vectors, and other applications, according to Auta et al. (2017). Secondary MPs are the tiny particles produced by applying degradation techniques, such as mechanical segregation or thermo-oxidation on big plastic fragments (Andrady et al. 2011). When big plastics break down owing to a variety of intricate circumstances, secondary microplastics are produced (for example, UV radiation, waves, wind, heat, physical distortion, etc.) (Rocha-Santos and Duarte (2015) and Cole et al. (2011)). Microplastics readily absorb and solidify other environmental contaminants like heavy metals and organic contaminants, because of their vast surface area and high water density (Brennecke et al. 2016; Koelmans et al. 2016). Microplastics can carry pollutants both within and across different areas of the environment because of their surface contact. In addition to the obvious environmental problems associated with microplastic pollution, there is a chance that global plastic pollution will contaminate the food chain and have detrimental effects on human health (Bradney et al. 2019).

11.2 Sources of Microplastics

Smaller plastic particles, about 5 mm in size, are known as microplastics. These can either be tiny particles that enter into the environment or pieces of bigger, more massive plastics that are present there (sustainable investment institute). Depending on their origin, microplastics are divided into two main categories: primary and secondary. Primary microplastics are those that are discharged into the environment in their most basic form. Nevertheless, laundry plastics deteriorate and wear down into smaller particles in the environment, which is how secondary microplastics are created (Boucher and Friot 2017). Microplastics are frequently found in synthetic textiles, car tires, road markings, cosmetics, and other products.

- *Synthetic textiles:* When synthetic clothing is washed, microplastics may fall out of it.
- *Vehicle tires:* As tires deteriorate from use, they release microplastics into the air.
- *Road markings:* Microplastics may be present in the paint used for these markings.
- *Cosmetics and personal care items:* Microbeads in other cosmetics and exfoliating items might add to the microplastic pollution problem.
- *Plastic pellets*: They are raw materials utilized in the manufacturing of plastic products.
- *Marine coatings:* Paints for the sea might discharge microplastics into the ocean.
- *City dust:* Urban dust may include microplastics.
- *Agriculture use:* Applications for polymer-based products including microplastics in agriculture are numerous. Their main use is in nutrient prills, which are composed of nutrient mixes coated with polymers to promote soil nutrient diffusion (Kershaw and Rochman 2015). Fertilizer prills are thought to make up around 8000 tons of polymers in the EU. Although the exact amount of microplastics in these polymers is unknown, if it were quantified, they might rank among the main sources of microplastic pollution.
- *Additional sources:* Oil and gas industry, pharmaceutical and healthcare industries, plastic bio-beads, packaging, textile printing, automobile molding, and more.

11.3 Characterization of Microplastics

Microplastics are microscopic plastic particles that are found in environmental samples. To characterize them, many approaches are used to detect, quantify, and analyze these particles (Huang et al. 2023). Spectroscopy and microscopy are common approaches used for characterizing objects as shown in Fig. 11.1.

(a) *Microscopy:* Dissection, polarized light, scanning electron, fluorescence, and atomic force microscopy are among the microscopic methods used to identify and quantify microplastics.

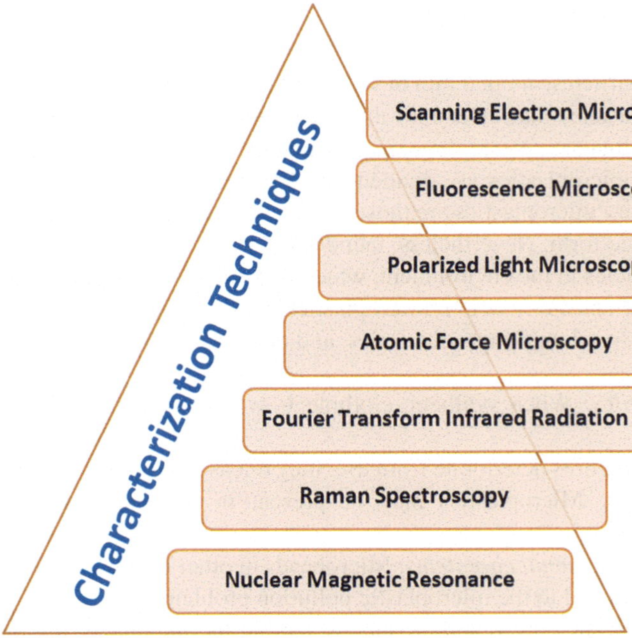

Fig. 11.1 Various characterization techniques of microplastics

- *Dissecting microscopy:* This technique uses a dissecting microscope to detect and measure microplastics in samples visually. This is also known as Stereoscopic Microscopy.
- *Polarized light microscopy:* It is a technique used to detect and measure microplastics based on their optical characteristics when exposed to polarized light.
- *Fluorescence microscopy:* This technique uses the fluorescence of specific types of plastics to detect their fluorescent qualities in order to identify and quantify microplastics.
- *Scanning electron microscopy (SEM):* High magnification surface morphology can be seen using scanning electron microscopy (SEM), which enables precise identification and quantification.
- *Atomic force microscopy (AFM):* It is a high-resolution microscopy method that can be used to recognize and measure microplastics by analyzing their surface topography and physical characteristics.

(b) *Spectroscopy:* A range of methods, such as NIR spectroscopy, Raman spectroscopy, FTIR, etc., can be used to describe distinct types of microplastics. Depending on the chemical and molecular structure of the various forms of microplastics, these approaches offer distinct spectrum information that may be utilized to identify and distinguish between them.

- *Fourier-transform infrared spectroscopy (FTIR):* By examining how the material's molecular bonds adsorb infrared light, FTIR spectroscopy can be utilized for identification of variety of microplastics. This makes it possible to identify particular functional groups and chemical compositions that the microplastics contain.

- *Raman spectroscopy:* It gives details on the molecular bonds' vibrational modes in microplastics, enabling the distinction of various polymer kinds based on their distinct Raman Spectra.
- *Near-infrared spectroscopy (NIR):* Using near-infrared spectroscopy (NIR), which analyzes the adsorption of near-infrared light to reveal details about the molecular makeup and structure of the microplastics, it is possible to identify the microplastics.
- *Nuclear magnetic resonance (NMR):* Based on their distinct spectrum properties, nuclear magnetic resonance (NMR) and photoluminescence spectroscopy are also utilized in this form of spectroscopy to identify and differentiate microplastics.

These spectroscopic techniques are crucial for precisely identifying and quantifying various microplastic kinds. The precise identification and measurement of distinct microplastic types in a range of environmental samples are made possible by these spectroscopic techniques, which also offer crucial insights into the origins and patterns of microplastic pollution.

11.4 Interaction of Microplastics with Microbes

Microplastics accumulate in different environments and there is a growing interest in comprehending the ways in which microbial communities interact with them (Khalid et al. 2021a). Microbes, which include fungi, bacteria, and archaea, are essential to the balance of the environment and overall health of ecosystems (Saccá et al. 2017). This section explores the complex mechanisms of interaction between microplastics and microorganisms. The wide distribution of microplastics in the environment has given rise to widespread concern about their substantial impact on ecosystems and human health (Ghosh et al. 2023). Numerous processes, such as the breakdown of large plastic objects, the addition of microbeads to products for personal hygiene, and the elimination of synthetic fibers from clothes, result in the production of these microscopic plastic particles (Miraj et al. 2021). Microplastics' chemical and physical effects on living things have been extensively researched, but little has been discovered regarding how they interact with populations of microbial organisms (Mammo et al. 2020; Nath et al. 2023). Understanding the various forms of microplastics and their extensive dissemination is crucial before exploring the relationship with microorganisms. Microplastics have been categorized into two main types: primary microplastics, particles that are purposefully made, and secondary microplastics, which are left over after bigger plastic objects break down (Zobkov and Esiukova 2018). Numerous ecosystems, including freshwater and marine environments, soils, and even the atmosphere, have been indicated to contain microplastics. However, their sources and pathways into these ecosystems vary significantly (Akdogan and Guven 2019).

11.4.1 Mechanisms of Microplastic–Microbe Interaction

The complex and multifaceted mechanisms of microplastic–microbe interaction include adhesion, biofilm formation, biodegradation, and ecological consequences (Crichton 2023). While microbial activity holds promise for decreasing the existence of microplastics in environment, it also expresses concerns over possible ecological disruptions and the transfer of contaminants through ecosystems (Jain et al. 2023). Microplastics and microbes engage in a multitude of interactions through diverse physical, chemical, and biological mechanisms (Qiu et al. 2022). Because of their small size and increased surface area, microplastics are excellent substrates for microbial colonization (Rogers et al. 2020, b). Biofilm formation is a common microbial strategy that involves microorganisms attaching to microplastics' surface, leading to the generation of complex communities of microbes (He et al. 2022). Chemical interactions between microplastics and microbes involve the exchange of substances, such as organic pollutants, between the plastic surface and the microbial cells (Mei et al. 2020). Microorganisms release enzymes, which are capable of decomposing specific types of plastics, aiding in the degradation of microplastics into tinier fragments (Othman et al. 2021).

11.4.2 Adhesion and Biofilm Formation

A key process facilitating interactions between microorganisms and microplastics involves the attachment of microbes on surface of microplastic particles (Rogers et al. 2020, b). Microbes have different structures including pili, fimbriae, and extracellular polymeric substances (EPS) that help them attach to surfaces (Muhammad et al. 2020). The polymer composition of many microplastics creates a hydrophobic surface that attracts microbial colonization (Wang et al. 2021). Microorganisms can attach themselves to microplastics and produce EPS, which creates a protective matrix. This matrix facilitates the synthesis of biofilms on the microplastic surface (Sooriyakumar et al. 2022). The presence of biofilms can influence the physical and chemical properties of microplastics, potentially changing how they interact and persist in the surrounding ecosystem (Rummel et al. 2017). The matrix of EPS formed by biofilms can increase the buoyancy of microplastics, affecting their vertical distribution in water column. Biofilms can also perform as hotspots for the sorption of other pollutants, causing accumulation of toxic compounds on the surface of microplastics.

11.4.3 Biodegradation of Microplastic

One promising approach to mitigate the environmental effects of microplastics is the application of microbial biodegradation. This process breaks down materials into simpler compounds through microorganisms (Miri et al. 2022). Biodegradable plastics and natural biodegradation processes offer hope for reducing the accumulation of microplastics and their harmful effects (Arpia et al. 2021). Research indicates the important role that microbial activity plays in the degradation of microplastics, which are recognized as persistent pollutants (Mishra et al. 2022; Jain et al. 2023) (Fig. 11.2).

Certain bacteria, such as members of the genera *Pseudomonas*, *Bacillus*, and *Ideonella*, have been identified for their ability to disintegrate plastic polymers (Atanasova et al. 2021). Enzymes like lipases and esterases, produced by microorganisms, are capable of initiating the depolymerization of plastic polymers, thereby

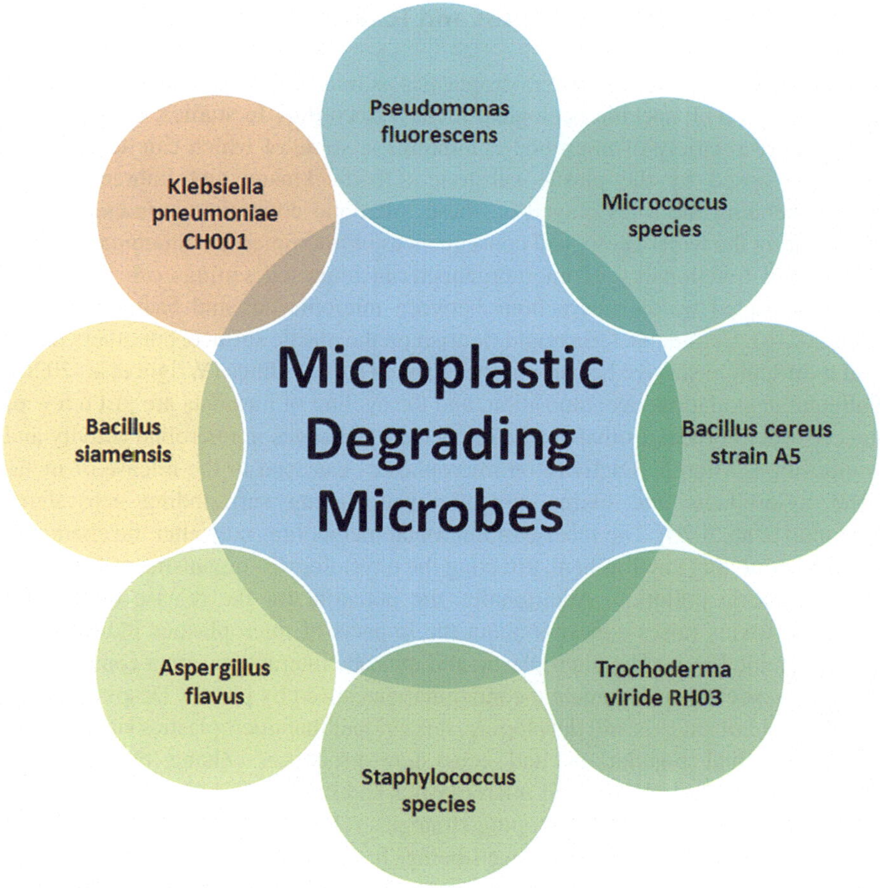

Fig. 11.2 Various microorganisms used for degradation of microplastics

resulting in the development of tiny molecules (Mohanan et al. 2020). Researchers have identified bacterial strains capable of enzymatically degrading common plastics like polypropylene (PP) and polyethylene (PE) (Wróbel et al. 2023). Scientists aim to develop biotechnological solutions for degrading microplastics in natural environments and waste management facilities by harnessing microbial capabilities. This process is a promising avenue for mitigating the environmental impact of microplastics. The efficiency of microbial biodegradation can differ depending on factors including the kind of polymer, the composition of microbial communities, and the environmental conditions (Yuan et al. 2020). While microplastic particle size can be decreased by biodegradation, it can additionally lead to the formation of nanoplastics, which can create new environmental concerns and challenges. It is essential to comprehend the working of microbial plastic degradation to develop strategies that can help mitigate plastic pollution.

11.4.4 Microplastics as Microbial Habitats and Release of Microbial Metabolites and Toxins

Microbial communities can use microplastics as microhabitats to create distinctive environments with specific ecological niches. According to studies, microplastics can support a variety of microbial communities, some of which can adapt to the problems posed by the plastic substrate (Oberbeckmann and Labrenz 2020). Comprehending the roles played by these microbial communities is essential to appreciating the wider ecological consequences of microplastic contamination. The variety and functionality of microorganisms in impacted settings can be significantly impacted by the interactions between microplastics and bacteria (Khalid et al. 2021b). Certain bacteria might flourish on the plastic surface, but others might find it difficult to survive because of the changed surroundings (Wright et al. 2020). Pollutant degradation, decomposition, and the cycling of nutrients are just a few of the ecosystem processes that can be impacted by changes in microbial activity and composition. Microbial activity on microplastics can lead to the release of metabolic by-products and toxins, potentially impacting surrounding ecosystems (Mammo et al. 2020). The release of microbial metabolites may alter the chemistry of the surrounding environment, affecting the physiology of organisms in proximity to microplastic pollution. Additionally, the potential for the release of harmful microbial toxins raises concerns about the impacts of microplastics indirectly on higher trophic levels. The microplastic and microbe interactions are a complex and dynamic aspect of environmental contamination caused by plastic. Despite the fact that this field of study is still developing, it is evident that microplastics significantly affect microbial populations and ecosystem processes (Zhang et al. 2021). Understanding the link between microplastics and microbes is vital for mitigating plastic pollution's effects. Developing strategies relies on this insight to safeguard both the environment and human well-being. It is essential to continue interdisciplinary research and collaboration to unravel the complexities of this interaction and pave the way for sustainable solutions.

11.5 Adverse Effects of Microplastic

11.5.1 Effect on Humans

1. *Gastrointestinal Problems:* Microplastics exposure has been associated with various health concerns, including digestive issues. According to research, consuming microplastic particles through contaminated water or food may cause a number of gastrointestinal problems (Zhao et al. 2023). These issues could involve changes in intestinal permeability, modification of gut microbiota, constipation, irritable bowel syndrome, and digestive system inflammation (Zhao et al. 2023; Qiao et al. 2019). The symbiotic association between natural gut microbiome population and hosts is significantly impacted by microplastic exposure, which can result in dysbiosis, a disruption. The immune system of the host may be negatively impacted by dysbiosis, which may result in chronic illnesses, heightened susceptibility to pathogenic infections, and alterations in genetic expression and capacity of the gut microbiota (Fackelmann and Sommer 2019; Deng et al. 2020). In an investigation, the microplastic impact on the microbiota of human gut and associated alterations in the digestive tract were shown (Tamargo et al. 2022), which also revealed the first probable breakdown of polymers during human digestion. Researchers replicated the movement of microplastics composed of PET through the digestive tract by combining static and dynamic models. The outcomes exhibited that digestive tract, particularly the colon, experienced structural changes related to PET microplastics. Furthermore, microplastics altered the gut microbial ecology, which may have aided in the formation of biofilms. These results imply that the digestive system may be harmed by microplastics. Research has shown that people with inflammatory bowel disease (IBD) tend to have higher amounts of microplastics in their feces compared to healthy individuals. This finding suggests there might be a link between the presence of microplastics and the onset or progression of IBD (Yan et al. 2021).
2. *Inflammation:* The human A549 lung cells were shown to be impacted by inflammation in in vitro experiments using varying-sized polystyrene particles (535 nm and 202 nm). The lung cells treated with the larger particles expressed more IL-8 than the same cells subjected to 64 nm particles (Brown et al. 2001). Furthermore, in response to unaltered or carboxylated polystyrene nanoparticles leukemia, human gastric adenocarcinoma, and cells of histiocytic lymphoma significantly enhances the expression of IL-8 genes and IL-6. This implies that rather than the particle's charge, the enhanced inflammatory reactions to polystyrene particles are probably driven by the particle's composition or by the fact of existence of particles (Forte et al. 2016; Prietl et al. 2014).
3. *Oxidative Stress and Cytotoxicity:* MPs can induce oxidative stress by oxidizing molecules that stick to their surface and producing reactive oxygen radicals that their host releases during inflammation (Kelly and Fussell 2012). Nanoparticles were utilized by Forte et al. (2016) to induce pro-inflammatory responses in human stomach cancer cells and A549 lung cells. Greater polyethylene particles

(0.3–10 m) have been demonstrated by Green et al. (1998) to increase the production of cytokines, which are agents that cause inflammation. These agents include IL-6, IL-1b, and TNF-a. It was discovered that MPs were cytotoxic as a result of inflammation and oxidative damage. Research on animals and in vitro experiments have demonstrated that MPs could be used by specific cells, like macrophages. MPs constituted a risk since they interacted with intracellular organelles easily and were not membrane-bound (Geiser et al. 2005). + Microplastics at concentrations of 0.05–10 mg/L developed substantial amounts of reactive oxygen species, which an experimental investigation indicated contributed to cellular damage in epithelial cells and human brain. Furukuma and Fujii's (2016) in vitro research indicates that microplastics particles have been exhibiting cytotoxic nature (Schirinzi et al. 2017). Additionally, when lung epithelial cells and macrophages were exposed to nanoplastics in vitro, their generation of reactive oxygen radicals increased. This, in turn, caused cytolysis and the endoplasmic reticulum to aggregate unfolded protein particles (Chiu et al. 2015). Adenomatous cells in the small intestine were cytotoxically affected by polystyrene particles. People who are exposed to microplastics may therefore result in oxidative damage and cytotoxicity. Furthermore, even at high concentrations, no significant cytotoxic effects were discovered in several in vitro studies (Magrì et al. 2018).

4. *Metabolic Homeostasis:* Recent research has shown that microplastics and nanoplastics can affect and alter metabolism of cells in both in vivo and in vitro settings, in addition to inducing inflammation and fatalities. Because of interactions between the nanoparticle and the cytoplasmic membrane, nanoparticles based on polystyrene may affect the signaling systems of epithelial cells of lungs. Human lung cells' basolateral K+ ion channels have been shown to be activated by negative charge exposure to 20 nm carboxylated polystyrene nanoparticles. The nanoplastic particles induced a concentration-dependent and long-lasting increase in short-circuit currents via ion channels being active and promoting Cl − and HCO3− ion outflow (McCarthy et al. 2011). Moreover, it was shown that the endocytic pathway causes human cancer cell lines HepG-2, HCT116, and A549 and macrophages to form large vesicle-like structures in response to 30 nm polystyrene nanoparticles. Thus, by preventing the dispersion of proteins involved in cytokinesis and vesicle transport, the growth of binucleated cells is encouraged (Xia et al. 2016). Furthermore, a severe oral exposure to polystyrene nanoparticles that is positively charged may interfere with intestinal iron transit and cellular uptake (Mahler et al. 2012).
5. *Immune System Dysfunction:* Based on observations regarding their dispersion and human sensitivity, MPs may induce systemic or local immune responses when exposed. However, ambient exposure to MPs was sufficient to weaken immune systems in physiologically vulnerable people, following immunosuppression or autoimmune disorders (Prata 2018; Prata et al. 2020). MP initiated autoimmune illnesses may be caused by immune modulators produced, inappropriate immune cell stimulation, or chronic cell damage. These occurrences would cause the body to produce antibodies directed against self-antigens

(Farhat et al. 2011). Furthermore, systemic lupus erythematosus and autoimmune rheumatic illnesses have been connected to MP exposure (Fernandes et al. 2015; Bernatsky et al. 2016).

6. *Transfer of Cells to Other Tissues:* After exposure, MPs may use the circulatory system to reach distal regions. MP internalization resulted in vascular edema, blockages, elevated blood pressure in the respiratory system, response to inflammation, as well as cytotoxicity to blood cells in the cardiovascular system (Wright and Kelly 2017; Campanale et al. 2020). Experiments conducted in vitro have indicated that endothelial wall adherence and red blood cell coagulation can be induced by nanoparticle exposure (Barshtein et al. 2016). MPs hastened the process of hemolysis and aided in the generation of the pro-inflammatory chemical histamine (Hwang et al. 2019). Increased epithelial membrane porosity brought on by inflammation is the most important means of MP transfer. On the contrary, renal active function was severely compromised by the particle buildup. When MPs are transported to distal tissues, they may result in prolonged inflammation, impaired organ performance, and a higher chance of neoplasia (Prata et al. 2020). Because MPs activate osteoclasts when they get to the bone, they might be a factor in bone loss. There are discrepancies in the data, such as the fact that MPs are present in different physiological tissues and that the mass mismatch between MP tissue load and actual consumption is not taken into consideration. Therefore, while assessing MPs' absorption, dispersion, and impacts, it is important to take into account their toxicokinetic characteristics.
7. *Carcinogenicity:* Human interaction with plastic products has long been linked to cancerous growths. However, as of yet, no solid evidence has been found. By damaging DNA, irritation and prolonged inflammation brought on by MPs ingestion may promote cancer (Prata 2018). According to Chang (2010), pro-inflammatory substances that activated the vasculature and caused oxidative stress and prolonged irritation were linked to the development of malignancies.

11.5.2 Effect on Animals

1. *Disruption of Immunological Function and Inflammation:* Blue mussels (Mytilus edulis L.) exposed to PE experience severe inflammatory reactions and notable histological alterations (von Moos et al. 2012). Larger PET particles, up to 100 μm in diameter, are found extracellularly in the tissue, while smaller ones, ranging in size from 0.5 μm to 20 μm, are kept in the cytoplasm of histiocytes inside the joint capsule (Wright and Kelly 2017). Additionally, autoimmune rheumatic illness and systemic lupus erythematosus have been linked to particulate matter exposure (Fernandes et al. 2015; Bernatsky et al. 2016). Researchers found that the lysosomal compartment, peroxisomal proliferation, and immune response were among the cellular effects in mussels given PE and PS treatments to absorb pyrene (Avio et al. 2015). Different sized plastic particles accumulated and induced swelling in the livers, gills, and intestines of zebrafish in another study (Lu et al. 2016). For example, PS MPs enhanced the gut's production of

interferon, IL-1β, and IL-1α in adult male zebrafish, suggesting swelling and microplastic dysbiosis (Jin et al. 2018). Animals given doses of 10–150 μm PS MPs showed signs of minor gastrointestinal distress (Li et al. 2020).

2. *Oxidative Stress and Cytotoxicity:* Oxidative stress occurred due to enhancing in free radicals and their by-products, a decrease in the antioxidant enzymes that prevent oxidative damage, or malfunction; cytotoxicity, on the other hand, is determined by cell viability and is mediated by molecular release related to cytotoxic granules (Dusinska et al. 2017). Following exposure to MPs, oxidative stress was observed in zebrafish (Danio rerio) and mice (Deng et al. 2020; Lu et al. 2016). These results demonstrate that organisms exposed to MPs experience oxidative stress and microplastics initiated toxicity (Rochman et al. 2013). The hazardous chemicals are quickly broken down and removed by organisms to prevent addition in the body. The cytochrome P450 system makes it easier for molecules to absorb oxygen atoms during initial metabolism. In order to create more hydrophilic compounds, foreign chemicals such as glutathione, glucuronic acid, or sulfates are associated with cytochrome P450 system molecules in the second phase (Capó et al. 2021). However, an in vitro study discovered that MPs were harmful for health of humans. Data showed that exposure to 0.001–10 ng/L doses of polystyrene nanoplastics (PS-NPs) did not lead to cytotoxicity, but it did modify the lethal impact of human medications in cell lines of marine fish (Almeida et al. 2019), and exposure to PVC microplastics was found to alter antioxidant activities (Xia et al. 2020). Thus, only the most potentially concentrated cytotoxic PS-NPs were exposed. When organophosphorus flame retardants (OPERs) were combined with PE and PS beads (0.5–1.0 μm) in mice (Mus musculus), the resultant oxidative stress caused by OPFR was greater than when OFPR was used alone (Deng et al. 2020). The inward and outward flow of body fluids may be disrupted when NPs reach the circulation, even though grain development and the protein-induced coalescence of NPs were caused by similar blood proteins. Furthermore, the peppery furrow shell clam *Scrobicularia plana* showed antioxidant (glutathione peroxidase, superoxide dismutase, and catalase) and biotransformation (glutathione S-transferases) enzyme expression following exposure to low-density polyethylene (LDPE) microplastics (11–13 μm) (O'Donovan et al. 2018). Furthermore, Yang et al. (2020) report that the digestive tracts of goldfish (Carassius auratus) larvae collected substantial concentrations of 70 nm and 50 μm MPs., potentially leading to oxidative stress. The overexpression of reactive oxygen species (ROS) and mitogen-activated protein kinase (MAPK)-hypoxia-inducible factor 1 (HIF-1)/nuclear factor kappaB (NFκB) was observed when Daphnia pulex is exposed to nanoplastics of polystyrene. Recent studies have shown that after 90 days, prolonged microplastics exposure to Sparus aurata resulted in oxidative damage in the liver (Capó et al. 2021). Therefore, in marine and aquatic species, cytotoxicity may be a major mechanism of MPs and NPs poisoning responses, and exposure to MPs and NPs may result in histological changes and an induction of oxidative stress.

3. *Translocation and Circulation:* Investigations in the lab using mussels and crabs (Browne et al. 2008; Watts et al. 2016) have documented the transfer of micro-

plastics into the gastrointestinal system (GIs). NPs may translocate across the gastrointestinal barrier and then enter the circulatory system, based on their surface-associated charges and size (Revel et al. 2018). The exposure of juvenile planktivorous fish (Acanthochromis polyacanthus), an abundant and common coral reefs species in the Indo-Pacific region, to microplastics was evaluated. After a fish was exposed to plastic for 1 week under five various treatment of plastic concentration, the stomach of the fish contained a maximum of 2102 (300 μm) particles of plastic. The plastics had a mean diameter of 2 mm, which was of similar size as the particles of natural food. When the size of the plastic particles was decreased to about 25% of the size of the meal particles, the amount of polymers in the GIs increased considerably (Critchell and Hoogenboom 2018). Diets heavy rich in high-fructose carbs and saturated fats, as well as starvation, may cause the gastrointestinal mucosa to become more permeable (West-Eberhard, 2019). Moreover, particles less than 5 μm may be able to cross the digestive system's wall, which could lead to bioaccumulation if release is exceeded by absorption or if the particles are broken down in organs or tissues (Roch et al. 2020). Because of their capacity to translocate, MPs and NPs can disclose tissue that is farther away. It has been shown that MPs/NPs can go from the mouth to the liver in rats, as well as into the circulatory system and other organs that are distant (Jani et al. 1990). Browne et al. (2008) explained how this debris was consumed, moved, and accumulated by mussels (Mytilus edulis). In this initial trial, microplastic collected in gastrointestinal tract after consumption. In addition, the mussels were processed with microplastic and seawater (Browne et al. 2008). In experimental settings, researchers have discovered that microplastics can travel throughout the gastrointestinal tracts of both mussels and crabs. After 30 days of exposure, fish exposed to PVC microplastics exhibited significant changes. Their livers showed histological abnormalities, and the level of the enzyme GSTa initially increased but then decreased as the concentration of microplastics rose. Additionally, microplastics altered the expression level of the gene CYP1A (Xia et al. 2020). Translocation occurs more frequently during inflammation because epithelial barriers become more porous (Prata et al. 2020).

4. *Neurotoxicity and Neurodegenerative Diseases:* Prüst et al. (2020) claim that oxidative stress causes neuro-inflammation and cellular damage in the brain, which raises the risk of neurodevelopmental and neurodegenerative diseases.
5. *Deoxyribonucleic Acid Damage:* Adult Daphnia were shown to display modification of gene expression related to stress (e.g., HSP70, HSP60, and GST) and various genes related to body structure and function (e.g., SERCA) 48 h after exposure. Gonzalez-Soto et al. (2019) investigated mussels determining the effects of nutritional consumption of polystyrene microplastics of 0.5 and 4.5 μm solely with absorption of benzo[a]pyrene (BaP), in order to comprehend the impacts of MP size and the amount of absorbed BaP on the organism. They discovered that when it came to DNA damage caused by deoxyribonucleic acid and the cell makeup of digestive tubules (DTs), smaller MPs were more hazardous than larger ones. It has been observed that being exposed to LDPE (low-density

polyethylene) microplastics (11–13 μm) without absorbing pollutants BaP and PFOS cause genotoxicity in peppery furrow shell clams (*Scrobicularia plana)*, resulting in single- and double-strand DNA breaks (O'Donovan et al. 2018). According to Xu et al. (2019), exposure to polystyrene nanoplastics changed cell variability, caused arrest of S phase in cell cycle, elevated the inflammatory gene transcription, and changed the level of cellular cycle proteins. On the contrary, in the marine water species Chlamys farrier, exposure to plastics haven't exhibited rise in BDE-209 DNA damage to hemoglobin (Xia et al. 2020). Furthermore, larvae exposed to PVC microplastics had changed gene expression associated to antioxidants in their livers. The dispersion and lethality of polystyrene nanoplastics in zebrafish, along with the generation of antioxidant genes and hypoactivity, are influenced by the exposure pathway in aquatic environments (Zhang et al. 2020).

11.6 Conclusions

In conclusion, because of their interactions with microorganisms and the possible impacts on humans and animals, microplastics present a serious threat to our environment. Studies have indicated that microplastics have tendency to perform as carriers of bacterial colonization, resulting in the development of biofilms that have the capacity to alter ecosystems and provide health hazards. Several studies have demonstrated that microplastics have the possibility to both absorb and transfer dangerous substances and microorganisms, hence intensifying the possible risks linked to their consumption or contact. This has sparked worries about the spread of diseases and pollutants from microplastics to higher-order creatures in the food chain, such as humans. The health and survival of these species may be negatively impacted by the ingestion of microplastics due to physical injury, nutrient dilution, and the release of hazardous compounds. Although investigation on the long-term consequences of microplastics on health of human are still in progress, mounting data points to possible dangers, including oxidative stress, inflammation, and biological process disruption brought on by the buildup of these particles in different tissues. These results highlight the critical need for more research, preventative mitigation strategies, and public education on the origins, effects, and possible dangers of microplastics.

References

Andrady AL (2011) Microplastics in the marine environment. Marine pollution bulletin, 62(8), 1596–1605.

Akdogan Z, Guven B (2019) Microplastics in the environment: a critical review of current understanding and identification of future research needs. Environ Pollut 254:113011

Almeida M, Martins MA, Soares AM, Cuesta A, Oliveira M (2019) Polystyrene nanoplastics alter the cytotoxicity of human pharmaceuticals on marine fish cell lines. Environmental Toxicology and Pharmacology, 69, 57–65.

Arpia AA, Chen WH, Ubando AT, Naqvi SR, Culaba AB (2021) Microplastic degradation as a sustainable concurrent approach for producing biofuel and obliterating hazardous environmental effects: a state-of-the-art review. J Hazard Mater 418:126381
Atanasova N, Stoitsova S, Paunova-Krasteva T, Kambourova M (2021) Plastic degradation by extremophilic bacteria. Int J Mol Sci 22(11):5610
Auta HS, Emenike CU, Fauziah SH (2017) Distribution and importance of microplastics in the marine environment: a review of the sources, fate, effects, and potential solutions. Environ Int 102:165–176
Avio CG, Gorbi S, Milan M, Benedetti M, Fattorini D, d'Errico G, Pauletto M, Bargelloni L, Regoli F (2015) Pollutants bioavailability and toxicological risk from microplastics to marine mussels. Environ Pollut 198:211–222
Barshtein G, Livshits L, Shvartsman LD, Shlomai NO, Yedgar S, Arbell D (2016) Polystyrene nanoparticles activate erythrocyte aggregation andAdhesion to endothelial cells. Cell BiochemBiophys 74(1):19–27. https://doi.org/10.1007/s12013-015-0705-6
Bernatsky S, Smargiassi A, Barnabe C, Svenson LW, Brand A, Martin RV et al (2016) Fine particulate air Pollution and systemic AutoimmuneRheumatic disease in two Canadian provinces. Environ Res 146:85–91
Bilal M, Ali H, Ullah R, Hussain I, Khan BA (2021) Overview of microplastics threat in aquatic animals since 1960 to 2020 Int. J Res Anal Rev 8:950–964
Boucher J, Friot D (2017) Primary microplastics in the oceans: a global evaluation of sources, vol 10. IUCN, Gland, Switzerland
Bradney L, Wijesekara H, Palansooriya KN, Obadamudalige N, Bolan NS, Ok YS, Rinklebe J, Kim KH, Kirkham MB (2019) Particulate plastics as a vector for toxic trace-element uptake by aquatic and terrestrial organisms and human health risk. Environ Int 131:104937
Brennecke D, Duarte B, Paiva F, Caçador I, Canning-Clode J (2016) Microplastics as vector for heavy metal contamination from the marine environment. Estuar Coast Shelf Sci 178:189–195
Brown DM, Wilson MR, MacNee W, Stone V, Donaldson K (2001) Size-dependent Proinflammatory effects of UltrafinePolystyrene particles: a role for surface area and oxidative stress in the enhanced activity of Ultrafines. Toxicol ApplPharmacol 175:191–199
Browne MA, Crump P, Niven SJ, Teuten E, Tonkin A, Galloway T, Thompson R (2011) Accumulation of microplastic on shorelines Woldwide: sources and sinks. Environ Sci Technol 45(21):9175–9179
Browne MA, Dissanayake A, Galloway TS, Lowe DM, Thompson RC (2008) Ingested microscopic plastic translocates to the circulatory system of the mussel,Mytilus edulis (L.). Environ Sci Technol 42:5026–5031
Campanale C, Massarelli C, Savino I, Locaputo V, Uricchio VF (2020) A Detailed Review Study on Potential Effects of Microplastics and Additives ofConcern on Human HealthIjerph 17(4):1212
Capó X, Company JJ, Alomar C, Compa M, Sureda A, Grau A, Hansjosten B, López-Vázquez J, Quintana JB, Rodil R, Deudero S (2021) Long-term exposureto virgin and seawater exposed microplastic enriched-diet causes liver oxidativestress and inflammation in gilthead seabream Sparus aurata, linnaeus 1758. SciTotal Environ 767:144976
Chang C (2010) The immune effects of naturally occurring and synthetic nanoparticles. J Autoimmun 34(3):J234–J246
Chiu H-W, Xia T, Lee Y-H, Chen C-W, Tsai J-C, Wang Y-J (2015) Cationic polystyrene Nanospheres induce Autophagic cell Deaththrough the induction of endoplasmic reticulum stress. Nanoscale 7(2):736–746. https://doi.org/10.1039/c4nr05509h
Cole M, Lindeque P, Halsband C, Galloway TS (2011) Microplastics as contaminants in the marine environment: a review. Mar Pollut Bull 62(12):2588–2597
Crichton CRMH (2023) Plastic degradation in natural marine sediment and the inbuilt adaptation of Mātauranga Māori. (Doctoral dissertation,. ResearchSpace@ Auckland
Critchell K, Hoogenboom MO (2018) Effects of microplastic exposure on the bodycondition and behaviour of planktivorous reef fish (Acanthochromispolyacanthus). PLoS One 13:e019330

Dusinska M, Runden-Pran E, Schnekenburger J, Kanno J (2017) Chapter 3—ToxicityTests: in vitro and in vivo. In: Fadeel B, Pietroiusti A, Shvedova AA (eds) Adverse effects of engineered nanomaterials, 2nd edn. Academic Press, pp 51–82

Farhat SCL, Silva CA, Orione MAM, Campos LMA, Sallum AME, Braga ALF (2011) Air Pollution in autoimmune rheumatic diseases: a review. Autoimmun Rev 11(1):14–21

Fernandes EC, Silva CA, Braga ALF, Sallum AME, Campos LMA, Farhat SCL (2015) Exposure to air pollutants and disease activity inJuvenile-onset systemic lupus erythematosus patients. Arthritis Care Res 67(11):1609–1614

Forte M, Iachetta G, Tussellino M, Carotenuto R, Prisco M, De Falco M et al (2016) Polystyrene nanoparticles internalization in human GastricAdenocarcinoma cells. Toxicol Vitro 31:126–136

Furukuma S, Fujii N (2016) Vitro cytotoxicity evaluation of Plasticmarine debris by colony-forming assay. Jpn J Environ Toxicol 19(2):71–81

Fackelmann G, Sommer S (2019) Microplastics and the gut microbiome: how chronically exposed species may suffer from gut dysbiosis. Mar Pollut Bull 143:193–203

Geiser M, Rothen-Rutishauser B, Kapp N, Schürch S, Kreyling W, Schulz H et al (2005) Ultrafine particles cross cellular membranes by nonphagocytic mechanisms in lungs and in cultured cells. Environ Health Perspect 113(11):1555–1560. https://doi.org/10.1289/ehp.800

Ghosh S, Sinha JK, Ghosh S, Vashisth K, Han S, Bhaskar R (2023) Microplastics as an emerging threat to the global environment and human health. Sustain For 15(14):10821

Gonzalez-Soto N, Hatfield J, Katsumiti A, Duroudier N, Lacave JM, Bilbao E, Orbea A, Navarro E, Cajaraville MP (2019) Impacts of dietary exposure todifferent sized polystyrene microplastics alone and with sorbed benzo[a]pyrene onbiomarkers and whole organism responses in mussels Mytilus galloprovincialis. Sci.Total Environ. 684:548–566

Green T, Fisher J, Stone M, Wroblewski BM, Ingham E (1998) Polyethylene particles of a 'critical size' are necessary for the induction ofCytokines by macrophages in vitro. Biomaterials 19(24):2297–2302. https://doi.org/10.1016/s0142-9612(98)00140-9

He S, Jia M, Xiang Y, Song B, Xiong W, Cao J, Zeng G (2022) Biofilm on microplastics in aqueous environment: physicochemical properties and environmental implications. J Hazard Mater 424:127286

Hidalgo-Ruz V, Gutow L, Thompson RC, Thiel M (2012) Microplastics in the marine environment: a review of the methods used for identification and quantification. Environ Sci Technol 46:3060–3075

Huang Z, Hu B, Wang H (2023) Analytical methods for microplastics in the environment: a review. Environ Chem Lett 21(1):383–401

Hwang J, Choi D, Han S, Choi J, Hong J (2019) An assessment of theToxicity of polypropylene MPs in human derived cells. Sci Total Environ 684:657–669. https://doi.org/10.1016/j.scitotenv.2019.05.071

Jain R, Gaur A, Suravajhala R, Chauhan U, Pant M, Tripathi V, Pant G (2023) Microplastic pollution: understanding microbial degradation and strategies for pollutant reduction. Sci Total Environ 905:167098

Jani P, Halbert GW, Langridge J, Florence AT (1990) Nanoparticle uptake by the rat gastrointestinal mucosa: quantitation and particle size dependency. Journal of pharmacy and pharmacology, 42(12), 821–826.

Jambeck JR, Geyer R, Wilcox C, Siegler TR, Perryman M, Andrady A, Narayan R, Law KLM, Pollution. (2015) Plastic waste inputs from land into the ocean. Science 347(6223):768–771

Jin Y, Xia J, Pan Z, Yang J, Wang W, Fu Z (2018) Polystyrene microplastics inducemicrobiota dysbiosis and inflammation in the gut of adult zebrafish. Environ Pollut 235:322–329

Kelly FJ, Fussell JC (2012) Size, source and chemical composition asDeterminants of toxicity attributable to ambient particulate matter. AtmosEnviron 60:504–526. https://doi.org/10.1016/j.atmosenv.2012.06.039

Kershaw PJ, Rochman CM (2015) Sources, fate and effects of microplastics in the marine environment: part two of a global assessment. GESAMP working group 40 2nd phase. International Maritime Organization

Khalid N, Aqeel M, Noman A, Hashem M, Mostafa YS, Alhaithloul HAS, Alghanem SM (2021a) Linking effects of microplastics to ecological impacts in marine environments. Chemosphere 264:128541

Khalid N, Aqeel M, Noman A, Khan SM, Akhter N (2021b) Interactions and effects of microplastics with heavy metals in aquatic and terrestrial environments. Environ Pollut 290:118104

Koelmans AA, Bakir A, Burton GA, Janssen CR (2016) Microplastic as a vector for Chemicals in the Aquatic Environment: critical review and model-supported reinterpretation of empirical studies. Environ Sci Technol 50(7):3315–3326

Li B, Ding Y, Cheng X, Sheng D, Xu Z, Rong Q, Wu Y, Zhao H, Ji Z, Zhang Y (2020) Polyethylene microplastics affect the distribution of gut microbiota andinflammation development in mice. Chemosphere 244:125492

Lu Y, Zhang Y, Deng Y, Jiang W, Zhao Y, Geng J, Ding L, Ren H (2016) Uptakeand accumulation of polystyrene microplastics in zebrafish (Danio rerio) and toxiceffects in liver. Environ Sci Technol 50:4054–4406

Magrì D, Sánchez-Moreno P, Caputo G, Gatto F, Veronesi M, Bardi G et al (2018) Laser ablation as a versatile tool to mimic polyethylene TerephthalateNanoplastic pollutants: characterization and toxicology assessment. ACS Nano 12(8):7690–7700. https://doi.org/10.1021/acsnano.8b01331

Mahler GJ, Esch MB, Tako E, Southard TL, Archer SD, Glahn RP, Shuler ML (2012) Oral exposure to polystyrene nanoparticlesaffects iron absorption. Nat Nanotechnol 7:264–271

Mammo FK, Amoah ID, Gani KM, Pillay L, Ratha SK, Bux F, Kumari S (2020) Microplastics in the environment: interactions with microbes and chemical contaminants. Sci Total Environ 743:140518

McCarthy J, Gong X, Nahirney D, Duszyk M, Radomski M (2011) Polystyrene nanoparticles activate ion transport in humanairway epithelial cells. Int J Nanomedicine 6:1343–1356

Mei W, Chen G, Bao J, Song M, Li Y, Luo C (2020) Interactions between microplastics and organic compounds in aquatic environments: a mini review. Sci Total Environ 736:139472

Miraj SS, Parveen N, Zedan HS (2021) Plastic microbeads: small yet mighty concerning. Int J Environ Health Res 31(7):788–804

Miri S, Saini R, Davoodi SM, Pulicharla R, Brar SK, Magdouli S (2022) Biodegradation of microplastics: better late than never. Chemosphere 286:131670

Mishra S, Swain S, Sahoo M, Mishra S, Das AP (2022) Microbial colonization and degradation of microplastics in aquatic ecosystem: a review. Geomicrobiol J 39(3–5):259–269

Mohanan N, Montazer Z, Sharma PK, Levin DB (2020) Microbial and enzymatic degradation of synthetic plastics. Front Microbiol 11:580709

Muhammad MH, Idris AL, Fan X, Guo Y, Yu Y, Jin X, Huang T (2020) Beyond risk: bacterial biofilms and their regulating approaches. Front Microbiol 11:928

Nath J, De J, Sur S, Banerjee P (2023) Interaction of microbes with microplastics and Nanoplastics in the agroecosystems-impact on antimicrobial resistance. Pathogens 12(7):888

Oberbeckmann S, Labrenz M (2020) Marine microbial assemblages on microplastics: diversity, adaptation, and role in degradation. Annu Rev Mar Sci 12:209–232

O'Donovan S, Mestre NC, Abel S, Fonseca TG, Carteny CC, Cormier B, Keiter SH, Bebianno MJ (2018) Ecotoxicological effects of chemical contaminants adsorbed to microplastics in the clam Scrobicularia plana. Frontiers in marine science. 5, 143.

Othman AR, Hasan HA, Muhamad MH, Ismail NI, Abdullah SRS (2021) Microbial degradation of microplastics by enzymatic processes: a review. Environ Chem Lett 19:3057–3073

Plastics Europe (2020) Plastics—the facts: an analysis of European plastics production demand and waste data. PlasticsEurope, Brussels, Belgium

Prata JC (2018) Airborne MPs: consequences to human health? Environ.Pollut. 234:115–126

Prata JC, da Costa JP, Lopes I, Duarte AC, Rocha-Santos T (2020) Environmental exposure to MPs: an overview on possible human health effects. Sci Total Environ 702:134455. https://doi.org/10.1016/j.scitotenv.2019.134455

Prietl B, Meindl C, Roblegg E, Pieber TR, Lanzer G, Fröhlich E (2014) Nano-sized and micro-sized polystyrene particles affectphagocyte function. Cell Biol Toxicol 30:1–16

Prüst M, Meijer J, Westerink RHS (2020) The plastic brain: neurotoxicity of micro-and nanoplastics. Part Fibre Toxicol 17:24

Qiu X, Qi Z, Ouyang Z, Liu P, Guo X (2022) Interactions between microplastics and microorganisms in the environment: modes of action and influencing factors. Gondwana Res 108:102–119

Qiao R, Deng Y, Zhang S, Wolosker MB, Zhu Q, Ren H, Zhang Y (2019) Accumulation of different shapes of microplastics initiates intestinal injury and gutmicrobiota dysbiosis in the gut of zebrafish. Chemosphere 236:124334

Revel M, Châtel A, Mouneyrac C (2018) Micro (nano) plastics: a threat to human health?. Current Opinion in Environmental Science & Health, 1, 17–23.

Roch S, Friedrich C, Brinker A (2020) Uptake routes of microplastics in fishes:practical and theoretical approaches to test existing theories. Sci Rep 10:3896

Rocha-Santos T, Duarte AC (2015) A critical overview of the analytical approaches to the occurrence, the fate and the behavior of microplastics in the environment. Trends Anal Chem 65:47–53

Rochman CM, Hoh E, Kurobe T, Teh SJ (2013) Ingested plastic transfers hazardouschemicals to fish and induces hepatic stress. Sci Rep 3:326

Rogers KL, Carreres-Calabuig JA, Gorokhova E, Posth NR (2020) Micro-by-micro interactions: how microorganisms influence the fate of marine microplastics. Limnology and Oceanography Letters 5(1):18–36

Rummel CD, Jahnke A, Gorokhova E, Kühnel D, Schmitt-Jansen M (2017) Impacts of biofilm formation on the fate and potential effects of microplastic in the aquatic environment. Environ Sci Technol Lett 4(7):258–267

Saccá ML, Barra Caracciolo A, Di Lenola M, Grenni P (2017) Ecosystem services provided by soil microorganisms in soil biological communities and ecosystem resilience. Springer International Publishing, pp 9–24

Schirinzi GF, Pérez-Pomeda I, Sanchís J, Rossini C, Farré M, Barceló D (2017) Cytotoxic effects of commonly used nanomaterials and MPs onCerebral and epithelial human cells. Environ Res 159:579–587. https://doi.org/10.1016/j.envres.2017.08.043

Sooriyakumar P, Bolan N, Kumar M, Singh L, Yu Y, Li Y, Siddique KH (2022) Biofilm formation and its implications on the properties and fate of microplastics in aquatic environments: a review. J Hazard Mater Adv 6:100077

Tamargo N, Molinero JJ, Reinosa V, Alcolea-Rodriguez R, Portela MA, Banares JFF, Moreno-Arribas MV (2022) PET microplastics affect humangut microbiota communitiesduring simulated gastrointestinal digestion, first evidence of plausible polymer biodegradation duringhuman digestion. Sci Rep 12(1):52

Thompson RC, Swan SH, Moore CJ, Vom Saal FS (2009) Our plastic age. Philos Trans R Soc B Biol Sci 364(1526):1973–1976

Moos V, Burkhardt-Holm N, Kohler P (2012) Uptake and effects of microplasticson cells and tissue of the blue mussel Mytilus edulis L.after an experimentalexposure. Environ Sci Technol 46:11327–11335

Wang J, Guo X, Xue J (2021) Biofilm-developed microplastics as vectors of pollutants in aquatic environments. Environ Sci Technol 55(19):12780–12790

Watts AJR, Urbina MA, Goodhead R, Moger J, Lewis C, Galloway TS (2016) Effect of microplastic on the gills of the shore crab Carcinusmaenas. Environ. Sci.Technol. 50:5364–5369

West-Eberhard MJ. Nutrition, the visceral immune system, and the evolutionary origins of pathogenic obesity. Proc Natl Acad Sci U S A. 2019;116(3):723–731. https://doi.org/10.1073/pnas.1809046116. Epub 2018 Dec 31. PMID: 30598443; PMCID: PMC6338860

Wright RJ, Erni-Cassola G, Zadjelovic V, Latva M, Christie-Oleza JA (2020) Marine plastic debris: a new surface for microbial colonization. Environ Sci Technol 54(19):11657–11672

Wright SL, Kelly FJ (2017) Plastic and human health: a micro issue? Environ SciTechnol 51:6634–6647

Wróbel M, Szymańska S, Kowalkowski T, Hrynkiewicz K (2023) Selection of microorganisms capable of polyethylene (PE) and polypropylene (PP) degradation. Microbiol Res 267:127251

Xia B, Zhang J, Zhao X, Feng J, Teng Y, Chen B, Sun X, Zhu L, Sun X, Qu K (2020) Polystyrene microplastics increase uptake, elimination and cytotoxicity ofdecabromodiphenyl ether (BDE-209) in the marine scallop Chlamys farreri. EnvironPollut 258:113657

Xia L, Gu W, Zhang M, Chang Y-N, Chen K, Bai X, Yu L, Li J, Li S, Xing G (2016) Endocytosed nanoparticles hold endosomes and stimulate binucleated cells formation. Part Fibre Toxicol 13:63

Xu M, Halimu G, Zhang Q, Song Y, Fu X, Li Y, Li Y, Zhang H (2019) Internalization and toxicity: a preliminary study of effects of nanoplastic particles onhuman lung epithelial cell. Sci Total Environ 694:13379

Deng Y, Yan Z, Shen R, Wang M, Huang Y, Ren H, Zhang Y, Lemos B (2020) Microplastics release phthalate esters and cause aggravated adverse effects in themouse gut. Environ Int 143:105916

Zhao Y, Liu S, Xu H (2023) Effects of microplastic and engineered nanomaterials on inflammatory bowel disease: a review. Chemosphere 326:13848

Yang H, Xiong H, Mi K, Xue W, Wei W, Zhang Y (2020) Toxicity comparison ofnano- sized and micron-sized microplastics to goldfish Carassius auratus larvae.J. HazardMater 388:122058

Yuan J, Ma J, Sun Y, Zhou T, Zhao Y, Yu F (2020) Microbial degradation and other environmental aspects of microplastics/plastics. Sci Total Environ 715:136968

Yan Z, Liu Y, Zhang T, Zhang F, Ren H, Zhang Y (2021) Analysis of microplastics in human feces reveals a correlation between fecal microplastics and inflammatory bowel disease status. Environmental Science Technology 56(1):414–421

Zhang R, Silic MR, Schaber A, Wasel O, Freeman JL, Sepúlveda MS (2020) Exposure route affects the distribution and toxicity of polystyrene nanoplastics in zebrafish. Sci Total Environ 724:13806

Zhang X, Li Y, Ouyang D, Lei J, Tan Q, Xie L, Yan W (2021) Systematical review of interactions between microplastics and microorganisms in the soil environment. J Hazard Mater 418:126288

Zobkov MB, Esiukova EE (2018) Microplastics in a marine environment: review of methods for sampling, processing, and analyzing microplastics in water, bottom sediments, and coastal deposits. Oceanology 58:137–143

Microplastics in Human Health

12

D. Rajnandini, Vanshika Sahani, Mantasa Praveen, Debolina, Reet Arora, Jagriti Saraswat, Kirandeep Kaur, Ashok Saini, and Indra Mani

Abstract

In the early twenty-first century, the discovery of microplastics—tiny plastic particles less than 5 mm—initiated a surge in research. Microplastics, characterized by their origin, shape, and monomeric units, have been detected across freshwater and marine environments, raising concerns about their impact on marine biota. Recent studies have uncovered microplastics in diverse sources, including food products and human consumables, yet the physiological effects on human health remain poorly understood. A considerable amount of plastic waste generated decades ago still resides in landfills. Its degradation, along with the production of plastic products, has led to the accumulation of microplastics (size <5 mm) and nanoplastics (size <1 μm) in our surroundings, including both the terrestrial as well as the aquatic environments. The atmosphere we reside in, the food and water that we consume, and the personal care products that we use have been categorized as the most common sources of exposure to microplastics by humans. These tiny contaminants come in variable shapes, sizes, and types and affect human health considerably. Despite the presence of biological barriers,

D. Rajnandini and Vanshika Sahani contributed equally with all other contributors.

D. Rajnandini · V. Sahani · M. Praveen · Debolina · R. Arora · J. Saraswat · K. Kaur · A. Saini (✉)
Department of Microbiology, Institute of Home Economics, University of Delhi,
New Delhi, India
e-mail: ashok.saini@ihe.du.ac.in

I. Mani
Department of Microbiology, Gargi College, University of Delh, New Delhi, India
e-mail: indra.mani@gargi.du.ac.in

A. Kumar, V. Singh (eds.), *Microplastics Pollution and its Remediation*,
https://doi.org/10.1007/978-981-97-4068-0_12

these particles can navigate their way through the cells to enter the body and pose a threat at chemical, cellular, and organ levels, leading to severe conditions like alteration of gene expression, inflammation in the GI tract, and several other metabolic effects. Another fate of these microplastics could be their biodegradation inside the body, potentially via enzymatic digestion by the residing gut microflora and the intracellular mitochondrial enzymes. The chapter thoroughly explores the prevalence of microplastics in the surroundings, their types, their journey to translocate from the source to inside the human body, and the consequences of the same.

Keywords

Plastic · Microplastics · Nanoplastics · Pollution · Polymers · Human health

12.1 Introduction

In 1907, work on synthetic polymers, such as plastic, started. Since then, plastic production and consumption have increased dramatically over the next few decades. The mid-1900s saw the beginning of concerns about plastic pollution, which have grown ever since. Microplastics (MPs) are small plastic particles that are smaller than 5 mm in size. They were identified in the early twenty-first century (Thompson et al. 2004). According to Gigault et al. (2018), particles smaller than 1 μm are referred to as nanoplastics, or NPs. Then, in the decades that followed, there was a noticeable and quick increase in the amount of studies conducted on microplastics, revealing their hazy existence.

These tiny plastic particles have been found in both freshwater and marine environments. Major concerns regarding the marine biota are also raised by the discovery of MPs in wastewater discharge worldwide (Eriksen et al. 2018; Andrady 2011; Cole et al. 2011). It has been discovered that marine life is seriously harmed by plastic waste. When floating plastics are mistaken for food by marine creatures like albatrosses, fulmars, shearwaters, and petrels, they enter the food chain. Besides 26 species of cetaceans, sea turtles, etc., it is known that 44% of all seabird species consume plastic material including bags, fishing line etc. (Moore 2008).

The fragmentation of larger particles caused by ongoing weather changes and surface water effluents found in water bodies results in microplastic particulates, which are isolated and identified as microbeads from personal care products (Eriksen et al. 2018; Andrady 2011; Cole et al. 2011). According to a study done by Cox et al. in 2019, it has been estimated that humans between 39,000 and 52,000 microplastic particles annually from a variety of sources, with the amount varying depending on age and gender (Cox et al. 2019).

The objectives of this chapter are to discuss (Abbasi et al. 2021) the types of microplastics on the basis of their shapes, origin, and monomeric unit (Acharya et al. 2021), various sources of microplastics by which they enter the human body, and (Akindoyo et al. 2016) the impact of microplastics on human health and well-being.

12.2 Types of Microplastics

Understanding the different types of microplastics is essential to conduct additional research on them. Microplastics can be classified based on their origin (Rochman et al. 2019), forms (Frias et al. 2018), and monomeric units.

12.2.1 Classification based on shape

The first way to categorize microplastics is by shape; they mostly consist of films, filaments, ropes, pieces, pellets, rubber, microbeads, foams, and sponges (Frias et al. 2018); some of these forms are covered in more detail in the section that follows (Rochman et al. 2019).

12.2.1.1 Films

They are thin and pliable, in addition to being flat. They do not break apart with ease; instead, they wrinkle, crease, or fold over. They can either be found in a plethora of colors or be partially or completely transparent.

12.2.1.2 Fragments

Fragments are irregularly shaped rigid structures that possess non-uniform thickness throughout.

12.2.1.3 Pellets/Needles

Pellets are usually large (3-5 mm). They are either cylindrical or spherical in shape and are found in numerous colors.

12.2.1.4 Foams

They are cloud-like, soft, and compressible. They are mostly white or opaque, but they can also be found in other colors.

12.2.1.5 Fibers

They are the microplastics that are generated from textile, due to abrasion, wear, and tear, etc. They are one of the leading causes of microplastic pollution (Acharya et al. 2021). Along with being rigid, they are unbranched too (Lehmann et al. 2021).

12.2.2 Classification based on source of origin

Additionally, we can divide microplastics into two groups (Fig. 12.1) according to where they came from:

12.2.2.1 Primary Microplastics

Plastic particles smaller than five millimeters are purposefully produced. They are made either directly, as microbeads in personal care and cosmetic products (PCCPs)

TYPES	PRIMARY MICROPLASTICS	SECONDARY MICROPLASTICS
DEFINITION	They are intentionally manufactured for specific purposes.	They are a result of degradation and fermentation processes.
SIZE	Their size is <5mm.	They are found in variable sizes.
EXAMPLE	Pre-production pellets for industrial abrasives.	Fibers obtained from discarded fish gear, textile.

Fig. 12.1 Differences between primary and secondary microplastics

and as abrasives for industrial uses, or indirectly, as preproduction pellets (Rochman et al. 2019).

12.2.2.2 Secondary Microplastics

These are particles of microplastics that are left over after the breakdown of primary microplastic particles by environmental and other processes (Rochman et al. 2019). These processes can be chemical, physical, or biological (Bermúdez and Swarzenski 2021). Additionally, they are created while using a product—for instance, when washing garments and removing microfibers.

Most primary microplastics used in toxicology research and investigations nowadays are homogeneous.

Microplastics can be separated from plastic drinking bottles, tap water, food packaging containers, papers, and more. They can be composed of a range of materials, including polypropylene (PP), polystyrene (PS), polyethylene terephthalate (PET), polyvinyl chloride (PVC), polyethylene (PE), high density polyethylene (HDPE), low density polyethylene (LDPE), and more.

12.2.3 Classification based on the monomeric units

Depending on the kind of monomeric unit they are made of, microplastics differ. Here is a detailed discussion of few of them:

12.2.3.1 Polyethylene Terephthalate (PET)

Monomeric Unit: Ethylene terephthalate with alternating ($C_{10}H_8O_4$) units (Sin and Tueen 2023).

Polycondensation of ethylene glycol and terephthalic acid is the main process involved in PET production (Nisticò 2020). However, it shows semicrystalline properties when it is cooled down slowly. PET is a transparent (in case of amorphous) or translucent (if semicrystalline) lightweight thermoplastic (Sin and Tueen 2023). This property of PET is widely exploited for producing transparent water bottles. It is robust and highly inert in comparison to other commercially available plastics (Nisticò 2020).

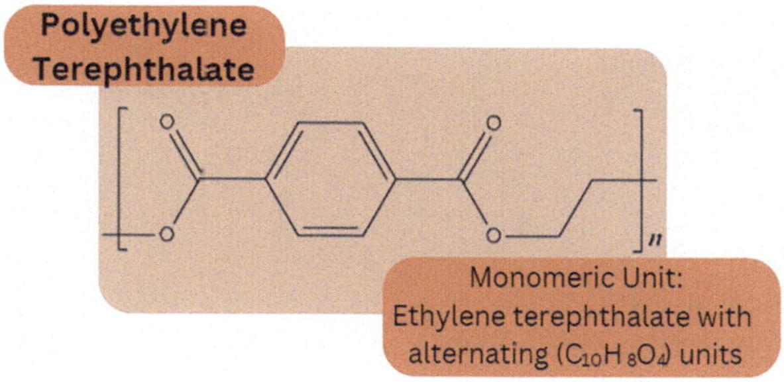

12.2.3.2 Polypropylene (PP)

Monomeric Unit: Propylene (Maddah 2016).

Propylene gas is produced through high-temperature cracking of propane and petroleum hydrocarbons (Koerner and Koerner 2018). The addition polymerization reaction, which is the mechanism used to produce polymers, comes next. From a structural standpoint, polypropylene is a vinyl polymer where each carbon molecule has a methyl group connected to it. Because it is inexpensive and has qualities like low density, high melting temperature, and chemical inertness, polypropylene is the best material for long-term use. Furthermore, it has a great deal of versatility. Blending polypropylene with other polymers yields polymers with superior properties (Maddah 2016).

The following varieties of polypropylene can be distinguished (Maddah 2016)

(a) *Homo-polymer polypropylene, or HPP*—This is a solid that is semicrystalline and made entirely of propylene monomers.
(b) *Random copolymer polypropylene (RCP)*: In this kind of polypropylene, ethylene is present in the PP chains as a comonomer at concentrations ranging from 1% to 8%.
(c) *Impact copolymer polypropylene (ICP):* Basically, it's a blend of RCP and HPP phases. It has between 45% and 65% ethylene concentration.

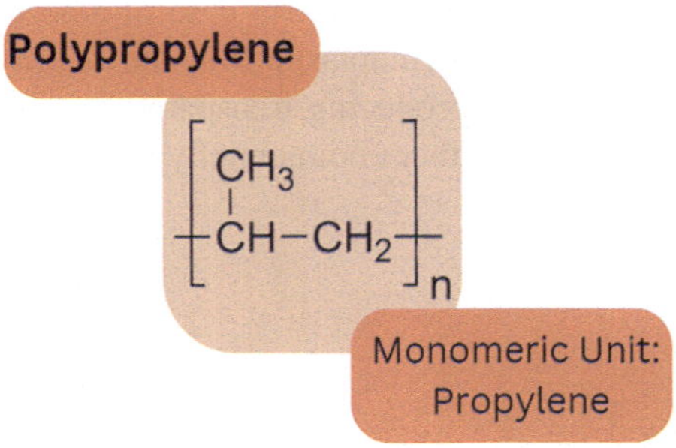

12.2.3.3 Polystyrene (PS)

Monomeric Unit: Styrene (Ho et al. 2017).

Polystyrene (PS) can be obtained by the polymerization of styrene monomer. The polymerization could be of free-radical, ionic, or coordination type (Wünsch 2000). Polystyrene is a colorless and transparent thermoplastic. As it is an inexpensive, rigid material, it is used in construction materials, packaging foams, disposable cups, etc. Although styrene is a liquid, polystyrene can be foamed or solid. Polystyrene is generally found in four types, namely: general purpose polystyrene (GPPS) which is also referred to as aligned polystyrene (PS), high impact polystyrene (HIPS), Polystyrene foam, and expanded polystyrene (EPS) foam (Ho et al. 2017).

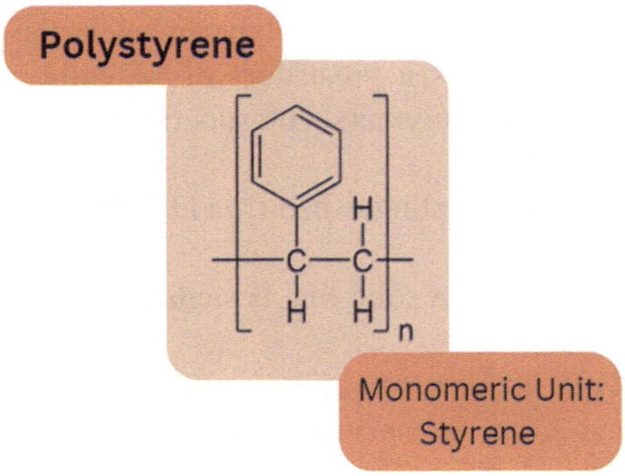

12.2.3.4 Polyethylene (PE)

Monomeric Unit: Ethylene (Paxton et al. 2019).

Ethylene is the monomeric unit of polyethylene (PE) (Paxton et al. 2019). Ethane polymerization results in the production of polyethylene, a long-chain polymer. Because the side chains obtained vary depending on the production process, many kinds of polyethylene can be obtained. High-density polyethylene (HD-PE) and low-density polyethylene (LD-PE) are the two types of polyethylene that are produced (Danso et al. 2019). According to Bouwmeester et al. (2015), it is the most produced synthetic polymer. Typically, polyethylene is made as microbeads, which have been found in toothpaste and other personal hygiene products (Bouwmeester et al. 2015).

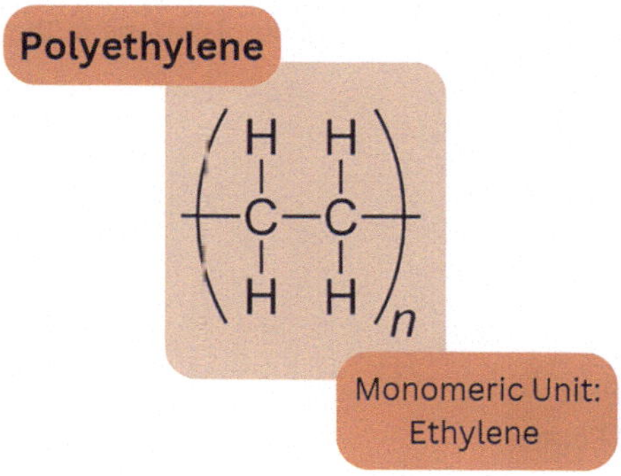

12.2.3.5 Polyvinyl Chloride (PVC)

Monomeric Unit: Vinyl Chloride (Comăniţă et al. 2015).

PVC is a thermoplastic that is obtained by polymerization of vinyl chloride units to obtain macromolecular chains (Lithner et al. 2011). These vinyl chloride units are obtained by thermal cracking of ethylene dichloride. Three methods, namely—mass, emulsion, and suspension can be used to carry out polymerization. PVC is one of the most frequently used polymer compounds in the world because it is inexpensive, lightweight, inert, and easy to maintain. It is particularly useful in the production of tires, floors, tubes and plumbing, cables and wires, and other materials (Akindoyo et al. 2016).

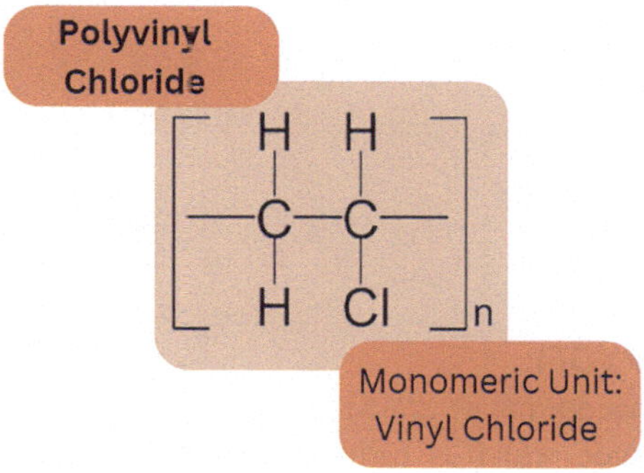

12.2.3.6 Polyurethane (PU)

Monomeric Unit: Isocyanate and polyol (Das and Mahanwar 2020).

When a substance bears more than one functional group, it is known as a polyol. The isocyanate group bears the formula R-N=C-O and could be either aliphatic or aromatic. Their reaction forms urethane linkage repeats in the presence of a chain extender and other additives, hence forming the polymer. Both these units determine the properties of PUR like softness, hardness, flexibility, etc., and hence their alteration could lead to significant changes in the polymer. PURs are classified as condensation polymers (Simón et al. 2018). PUR is a lightweight, highly resistant, and flexible polymer (Kemona and Piotrowska 2020) with wide applications in the automotive, textile, and medical industries.

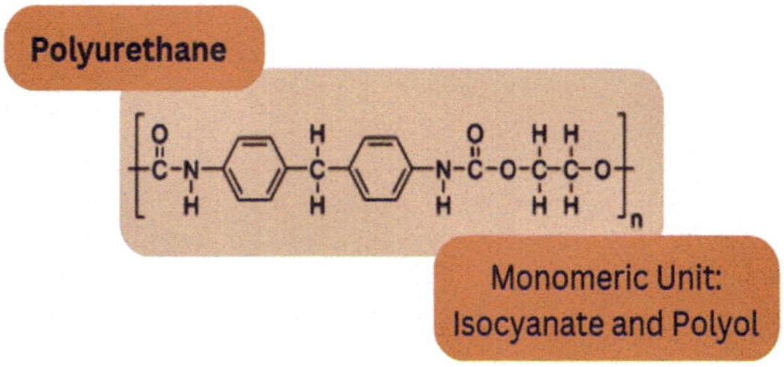

12.2.3.7 Polycarbonate

Monomeric Unit: Bisphenol A (BPA) (Carwile et al. 2009).

A reaction involving condensation combining phenol and acetone in an acidic atmosphere yields bisphenol A. Polycarbonate is produced when carbonyl chloride and BPA combine. The process is interfacial. Thermally resistant upto 135 °C, polycarbonate is a thermoplastic that is strong, stiff, and durable. It finds widespread application in the automotive, electronic, and lighting sectors.

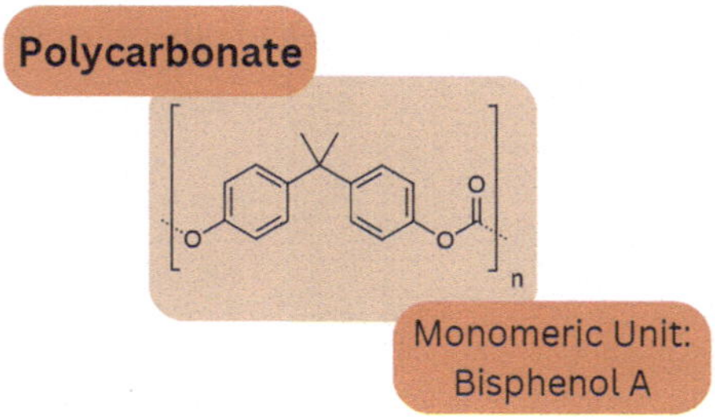

12.2.3.8 Polyvinylidene Fluoride (PVDF)

Monomeric Unit: Vinylidene fluoride (Alaaeddin et al. 2018).

PVDF is a semicrystalline fluoropolymer, which is formed through a radical-initiated polymerization process. This polymer possesses good chemical and aging resistance (Kang and Cao 2014). In addition to these properties, PVDF has surprisingly promising pyroelectric and piezoelectric properties. All of these advantages enable the use of this polymer in multiple industries like those of insulations, sensors, biomaterials, and photonic devices, commonly in the form of hybrid membranes and sheets (Alaaeddin et al. 2018).

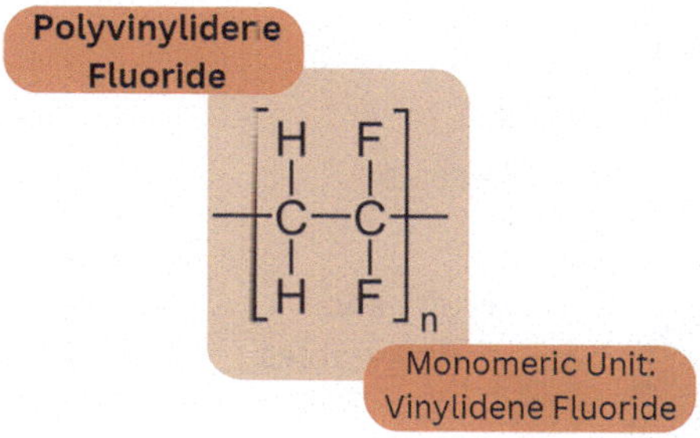

12.3 Sources of Microplastic

Microplastics are released into the environment as a result of the diversity of plastic goods that humans use on a daily basis and then trash into the environment. Moreover, waste treatment systems handle millions of tons of plastic trash, and occasionally, incorrect disposal of this material causes the release of microplastics into soil or aquatic habitats (Filho et al. 2019; Lebreton et al. 2017). Then, based on the quantity of plastic particles in the surroundings, individuals are exposed to the microplastics either directly or indirectly. As many food items and other edible elements are collected from watersheds and various water bodies are act as primary sources of microplastics particles. Microplastics are mostly identified and detected in extensively consumed dietary products like sea foods and table salts (Santillo et al. 2017).

According to more recent researches, microplastics have been found in soil (Zhang et al. 2018), fish (Park et al. 2020; Barboza et al. 2020), the tea bags (Hernandez et al. 2019, b), containerized food (Fadare et al. 2020), drinking water and beverages (Kosuth et al. 2018), garments, skin care products, food containing plastic sheets (Mason et al. 2018), shorelines (Eo et al. 2019, Scheurer and Bigalke, 2018), and more (Fig. 12.2).

12.3.1 Consuming Liquids and Water

Nano and microplastics (NMPs) in surface water and groundwater have raised concerns. Surprisingly, packaging, rather than water sources, emerges as a significant contributor to MPs in bottled water. Abrasion, mechanical stress, and the opening and shutting of bottles are some of the reasons why plastic bottles—including glass ones—can leak significant volumes of MPs (Winkler et al. 2019). Microplastics can be found in the packaging of a variety of goods, such as baby bottles and electric kettles. MPs have been found in drinks other than water, such as wine, beer, and snacks (Kankanige and Babel, 2020). It is possible that these drinks were made with filtered municipal tap water. Beyond direct drinking water, drinking water is used in food processing, which could increase the amount of nanoplastics in processed foods. The ramifications highlight the necessity of acknowledging bottled water

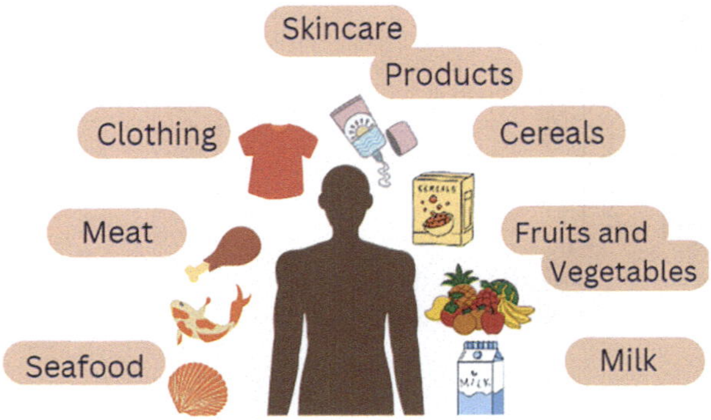

Fig. 12.2 Different sources of microplastics which affect humans

packaging as a substantial source of microplastics. It is imperative to address plastic contamination in drinking water supplies, and packaging is crucial to safeguard water quality and food safety.

12.3.2 Personal Care Products

In order to achieve a variety of effects in personal care products (PCPs), including exfoliation, abrasion, glitter, emulsification, viscosity management, and skin conditioning, MPs are purposefully included (Yurtsever, 2019). Direct MP translocation into human tissue is a possible concern associated with toothpaste and cosmetics applied to the mucosa. After utilizing these products, partial rinses or swallowing can introduce MPs into the gastrointestinal (GI) tract. MPs are specifically added to PCPs to accomplish a range of actions such as cleaning, abrasion, sparkle emulsifying viscous leadership, and skin conditioning (Yurtsever, 2019). NMPs may be released into the air when contact lenses are worn. Hydrogel polymers are commonly utilized to make contact lenses. After adhering to the lenses, these particles could pierce the corneal epithelial cells and cause damage by continuing to come into contact with them (Burgener and Bhamla, 2021).

12.3.3 Atmosphere

There are several broad sources connected to the existence of airborne microplastics in the environment. Synthetic fabrics, the deterioration of rubber from synthetic automobile tires, and pollution due to urbanization are the main sources of these microplastics. According to Boucher and Friot (2017), wind-induced transport is thought to be responsible for 7% of ocean contamination. Moreover, plastic pieces from apparel and home furnishings (materials inside buildings, garbage incineration, landfills) are sources of airborne microplastics (Dris et al. 2016, b). Synthetic fibers on exterior surfaces, such as automobile seats, have been successfully recognized by forensic tests (Grieve and Biermann, 1997). Notably, Dris et al. (2016) found that synthetic clothing plays a significant role in airborne microplastic pollution, with the kind and amount of fibers varying according to seasonal and fashion trends.

Microplastics are released into the outside environment by a number of sources, including garbage disposal, road dust, sports grounds, building sites, and tire abrasion (mostly from Styrene-Butadiene Rubber, or SBR).

12.3.3.1 Abrasion

The mechanical wear of tissue components brought on by rubbing against another surface is known as abrasion. This wear results in the release of fibers and particles, or fragments, which frequently collect in household dust. It's thought that a large portion of the microplastics found in household dust are caused by mechanical stress on surfaces, especially furniture and floors.

12.3.3.2 Sports Arenas

Artificial turf and sand/fiber combinations are two important sources of microplastic production in the overall setting of sports arenas. A bonded substrate (base layer), a synthetic elastic layer (generally Polypropylene, PP), fibers (Polyethylene, PP, Polyamide, PA), and filler granules (Styrene-Butadiene rubber products, SBR, Polyurethane, PUR, occasionally in combination with sand) are the standard components of modern artificial turf.

12.3.4 Food

The primary way humans come into contact with microplastics is through food consumption. The different food sources are as follows.

12.3.4.1 Aquatic

Seafood is a vital protein source for 19% of the global population. Mussels, as filter feeders, unintentionally ingest nanoplastics, potentially making them vectors for human exposure (Gündoğdu et al. 2020; Nalbone et al. 2021; Sparks et al. 2021; Kumar et al. 2021). Various marine species, including prawns, squids, and sardines, exhibit significant variability in polymer types (Ribeiro et al. 2020). Commercially grown bivalves, such as *Crassostrea gigas* and *Mytilus edulis*, show contamination in soft tissues, posing potential risks for consumers (Van Cauwenberghe and Janssen, 2014). European shellfish consumers may face exposure to 11,000 microplastics annually. Contamination sources include storm discharges, landfills, untreated sewage, tourist activities, coastal recreation, inadequate industrial control, windblown debris, and riverine transport (Barnes et al. 2009; Fendall and Sewell, 2009).

12.3.4.2 Poultry and dairy products

According to studies, packaging has been found to be a major source of the microplastics that are linked to chicken tissues (Kedzierski et al. 2020). Nevertheless, MPs were not found in the flesh that is normally consumed by humans.

Milk and Milk Products: The presence of MPs in milk has not been thoroughly studied. For instance, Pac et al. (2021) discovered 88 MPs/L in unpasteurized milk and 694 MPs/L in processed milk; Kutralam-Muniasamy et al. (2020) documented 3–11 MPs/L in Mexican branded milk; and Diaz-Basantes et al. (2020) observed a typical of 40 MPs/L in Ecuadorian milk.

12.3.4.3 Cereals

When humans consume cereals containing MPs directly, then cereals acts as a direct vector; while ingestion of animal products that contain NMPs act as an indirect vector. There is mounting evidence of pollution of the terrestrial environment, with particular emphasis on agricultural soils used for food production (Harms et al. 2021; Rillig et al. 2017; Steinmetz et al. 2016; Wang et al. 2021). According to Domenech and Marcos (2021), sewage sludge fertilizer (Corradini et al. 2019; Lian et al. 2021; Weithmann et al. 2018), contaminated irrigation water, and fertilizers can all lead to NMP contamination in cereals. Industrial processing and packing may also be the source of NMP cereal contamination (Dessð et al. 2021).

12.3.4.4 Fruits and Vegetables

Fruits and vegetables also add MPs to the total amount of MPs found in food along with cereals. There is a dearth of information regarding NMP contamination in commercial fruits and vegetables. The MPs in a number of Italian fruits and vegetables supplied for commercial consumption were quantitatively assessed by Conti et al. (2020), showing fruits being more polluted than vegetables. NMP accumulation in plants produced in a lab has been found in the edible tissues of radish (Tympa et al. 2021) and cucumbers (Li et al. 2020). To find out how much microplastic is in fruits and vegetables, a recent study was carried out in Italy (Oliveri et al. 2020).

12.3.4.5 Raw and Processed Foods:

The presence of NMPs in other types of processed foods is not much investigated. Li et al. (2020) discovered MPs in edible microalgae, such as Nori seaweed, which serves as a highly effective vector for the transmission of non-pathogenic protoplasts to people (Yang et al. 2021). Hernandez et al. (2019) found that at brewing temperature, a single plastic tea bag can emit around 11.6 billion microplastics. Additionally, several investigations have found MPs and other fibers in salt (Fadare et al. 2021; Fischer et al. 2019; Gündoğdu, 2018), and honey (Diaz-Basantes et al. 2020). These findings all point to the possibility that eating raw or processed food can expose people to NMPs.

12.4 Microplastic Exposure Routes for Humans

Humans are exposed to various sources of microplastics as discussed above. The exposure of microplastic to humans can be direct or indirect. These exposures are monitored carefully to trace their impact on human health and their associated risk.

12.4.1 Direct Exposure Conditions

Humans are directly exposed to microplastics when they eat food kept in plastic containers, such as throwaway cups, containers, and water bottles that are often constructed of PET. Food packaging material is a primary source of direct exposure. The majority of food consumed by people is typically wrapped in thin plastic sheets, which constantly come into touch with food and discharge MPs into it. A study shows that soaking a nylon tea bag in water at 95 °C released 1.13 ± 0.07 mg of nylon-6 (Cella et al. 2022).

12.4.2 Indirect Exposure Conditions

MPs emitted by numerous everyday utility products, including wet wipes (Lee et al. 2021), polyester fabrics (Galvío et al. 2020; Hernández et al. 2017), bulky items, and soil conditions, are indirectly exposed to humans. Humans can inhale MP fibers, which are typically discharged into the atmosphere (O'Brien et al. 2023; Soltani et al. 2021).

MPs are also released into the atmosphere by tire wear particles (Rasmussen et al. 2023), landfills (Kabir et al. 2023; Shi et al. 2023), and human movement while wearing synthetic garments (Dris et al. 2016, b).

12.5 Routes of Exposure of Microplastics

There are three major routes of exposure of microplastics to humans including oral, dermal, and inhalation.

12.5.1 Oral Intake

MPs can be directly ingested with the food or water stored in plastic containers or packaging material (Li et al. 2020; Cella et al. 2022).

12.5.2 Dermal Exposure

Although microplastics often do not cross the skin barrier, the danger of exposure rises as they accumulate on the skin. Thereby, it functions as an alternative exposure pathway. The usage of creams for the face and other items containing microplastics, as well as contact with the skin with wet tissues (Lee et al. 2021) are potential routes of exposure (Li et al. 2020). Plastic additives such as triclosan (TCS), bisphenols (BPs), phthalates, and bromine-containing flame retardants (BFRs) may also be absorbed by the skin during this type of exposure (Wu et al. 2022).

12.5.3 Inhalation

The atmosphere is filled with MPs from dumps (Kabir et al. 2023; Shi et al. 2023), tire wear particulates (Rasmussen et al. 2023), and MPs created by human activity and products used by humans such as clothing textiles madeup of synthetic materials (Dris et al. 2016). In addition to these, MPs released from bulk materials, such as those found in soil and agricultural settings, also become airborne and may be breathed by people (O'Brien et al. 2023; Soltani et al. 2021). According to Kumar et al. (2022), the MPs that are present in the atmosphere typically have sizes ranges from 10–8000 μm.

12.6 Factors Affecting the Release of MPs in Environment

The discharge of MPs from the source materials is influenced by a number of environmental parameters, including temperature, UV exposure, the qualities of food simulants (oil and water mixtures, acidic/basic), and exposure duration.

12.6.1 Temperature

It quickens the plastic's disintegration into MPs from throwaway containers (Zhou et al. 2023). By slicing the polymer chains, a temperature gradient facilitates the breakdown of PE films and the fragmentation of polymers into MPs (Beachell and Nemphos 1959).

12.6.2 pH

Under basic and acidic circumstances (pH 3 and 9, respectively), a PS container produced a significant amount of PS-MPs, which increased with exposure time and the temperatures of the liquid simulants (Wang et al. 2021). The processes of oxidation and hydrolysis controlled the release mechanism.

12.6.3 UV

In addition to breaking down plastics into MPs, photoaging speeds up aging. Sunlight's ultraviolet (UV) rays, specifically UV-A and UV-B, accelerate oxidation and cause microcracks on the MP surfaces, which is how they largely deteriorate (Hüffer et al. 2018). Regular use of plastic products contributes to significant but unidentified releases of MPs under different usage scenarios (Sait et al. 2021; Mohajerani et al. 2020; Almroth et al. 2017; Kalpana et al. 2019).

12.7 Impact of Microplastics on Human Health

Concerns regarding the possible effects of microplastics on human health are raised by their pervasive and contaminating presence in our food, personal care items, water bodies, and atmosphere. PCPs contain microplastics that can enter the body through the skin, enter the respiratory system by inhalation, and be eaten by sources of food and water. Once within the body, their small size enables them to pass through strong biological barriers. This section addresses the effects of microplastics' translocation in human tissues as well as their potential target tissues.

12.7.1 Relocation inside Human Tissues

This section discusses the possible translocation mechanism of plastic particles into the skin, which is different from the pathways within the lung and GIT.

12.7.1.1 Human Skin

NMPs can migrate into the deeper layers of the skin tissue when applied onto the skin's surface. A potential pathway for migration through the skin barrier involves the movement of particles across hair follicles, sebaceous glands (responsible for the production of sebum), and sweat glands (Desai et al. 2010; Schneider et al. 2009).

12.7.1.2 Gastrointestinal Tract

When NMPs enter the body through eating, the mucus layer serves as their first line of defense. It covers the inside of the digestive tract, which is crucial for preserving intestinal homeostasis. It also acts as a permeable barrier to stop pathogens from getting to the gut epithelial cells (Vancamelbeke and Vermeire 2017; Rackaityte and Lynch 2020).

When the gut microbiota is out of equilibrium, many illnesses can arise. Tamargo et al. (2022) used the globally certified Dynamical Gastroenterology Simulator simgi® model (which simulates the key functional parts of the digestive tract) and feces from healthy donors to study the effects of MP digestion on the human gut flora. The microbiota changed in response to microplastic exposure, and biofilm development was seen (Tamargo et al. 2022).

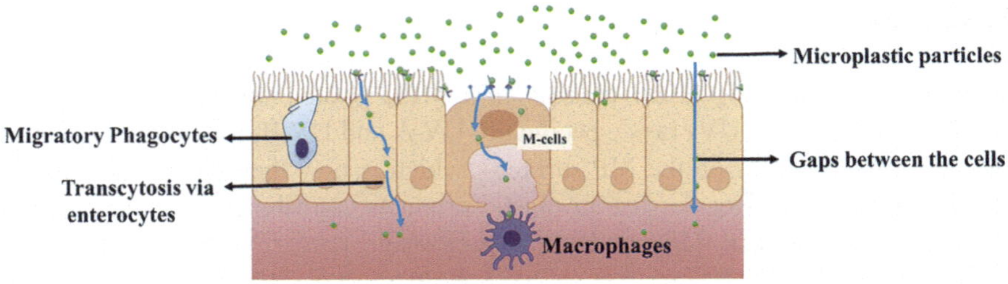

Fig. 12.3 Schematic illustration of the internalization of microplastics (MPs) in the GI tract: MPs enter the GI through different ways, such as phagocytes, enterocytes and gaps between cells

After entering the intestine, the microplastics can cross the mucus layer by various ways (Fig. 12.3):

1. Transcytosis by enterocytes, i.e., transported in bound vesicles and secreted across the layer.
2. Internalization by M cells.
3. Through gaps between the cells.
4. Internalization by macrophages and migratory phagocytes.

12.7.2 Lungs

NMPs can be inhaled and thereby enter the body. In the nasopharyngeal region, hair and mucus often capture PM10, while PM2.5 can enter the bronchial tubes and alveoli. According to Cooper and Loxham (2019) and Schraufnagel (2020), PM0.1 has the ability to translocate directly across the alveolar epithelium. Nonetheless, the epithelium layer has a mucus layer for capturing inhaled particles, just like the GIT. Cilia can assist the movement of the microplastics particles inside the mucus towards the mouth so they can be evacuated (Schraufnagel 2020).

Microplastics once reaching the alveoli can further migrate inside the alveoli by various possible mechanisms (Fig. 12.4):

1. Transcytosis by type I pneumocytes.
2. Through intracellular spaces.
3. Through alveolar macrophages.
4. Through lymphatic drainage.
5. By type II pneumocytes.

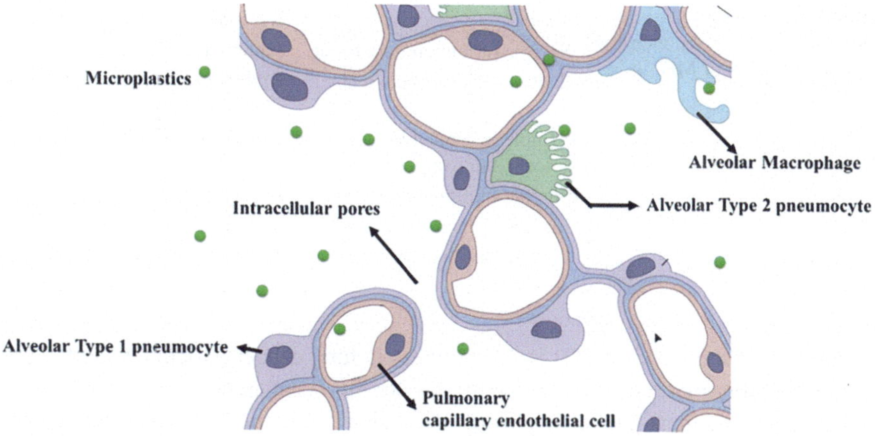

Fig. 12.4 Mechanisms of absorption of microplastics in the alveolar region of the lungs: Internalization of microplastics occurs in the lungs through various ways, such as intracellular pores, alveolar macrophages, lymphatic drainage and Transcytosis by type I and II pneumocytes

12.7.3 Transport of NMP Via the Lung's and the GIT's Biological Barriers

Numerous immune cells react to NMPs once they have penetrated the epithelial barrier of the skin, gastrointestinal system, or lung. These cells include mast cells, eosinophils, lymphocytes of the types T and B, dendritic cells, and macrophages. Any germs that have made it through are consumed and eliminated (Bain and Schridde 2018). According to Bain and Schridde (2018) and Grainger et al. (2017), macrophages can cause the inflammatory reactions mentioned, but in a healthy organism, they typically respond to the invader silently.

While in vitro experiments suggest that macrophages may internalize NMPs even in the absence of direct evidence linking the transportation of NMPs across biological barriers to inflammatory responses (Ramsperger et al. 2022; Stock et al. 2021), NMPs have been shown to trigger inflammatory reactions in macrophage (Völkl et al. 2022) as well as epithelial cells upon contact (Wu et al. 2019). To find out if tissue and blood specimens contained MPs, several investigations were conducted on them. A summary of some of them can be seen in Table 12.1.

12.8 Fate of MPs Within the Body

According to several studies (Prata 2018; Hurley et al. 2016; Liepins and Pearce 1976; Dehghani et al. 2017; Dong et al. 2020; Rezaei et al. 2019), microplastics carried by the wind may cause respiratory discomfort, cytotoxic and inflammatory effects, and autoimmune illnesses in men. It has been demonstrated that 50 nm polystyrene particles have cytotoxic and genotoxic effects on macrophages and lung

Table 12.1 Research demonstrating the existence and identification of MPs in several tissues

Tissues	References
Sample of human lung tissue	Jenner et al. (2022)
Sample of human placenta	Braun et al. (2021), Ragusa et al. (2021)
Sample of human kidney, spleen, and liver	Horvatits et al. (2022)
Sample of human blood	Leslie et al. (2022)
Sample of heart tissue	Yang et al. (2023)
Coroner autopsy sample	Amato-Lourenço et al. (2021)

epithelial cells (Paget et al. 2015). Microfibers in lung cancer patients degrade very little, indicating that these artificial particles are bio-permanent.

Despite being inert, the microplastics' size, shape, chemical makeup, and hydrophobicity can harm tissues and cells (Wright's and Kelly 2017). Because of their hydrophobicity and high surface area to volume ratio, various other contaminants such as heavy metals, antibiotics, and a wide range of hydrophobic persistent organic contaminants became attached to MPs. The affinity between these chemicals and microplastics has the potential to result in their entry into the human body.

12.8.1 Microplastic Tissue Buildup in Mice

Research has shown that MPs were deposited in the stomach, kidney, and liver of mice. The MP particle size had a significant impact on the distribution pattern and tissue-accumulation kinetics. Additional analysis revealed that MP exposure caused oxidative damage, disruption of lipid, and energy metabolism.

12.8.2 Effects of Exposure to Microplastics on the Body's Levels of Structural Organization

Early research on uptake and impacts on lungs (Song et al. 2022) and intestinal (Hou et al. 2022) organoid structures revealed impaired lung organoid development without detrimental effects on the mature organoids. It suggests that exposure to airborne NMP may hinder the development of tissue in the lungs in young infants. According to Burkhart et al. (1999), interstitial lung illnesses and plastic product exposure were connected in workers. This suggests that NMP transport and the ensuing inflammation response are generally conceivable in people.

12.8.2.1 Chemical Level

Exposure to microplastics may change the expression of some genes, including those involved in immune response, development, energy metabolism, cellular stress response, and immunological functions (Liu et al. 2019a, b).

12.8.2.2 Cellular Level

Extremely small particles have been shown to be able to pass via a variety of channels, including protein-mediated uptake, endocytosis, passive diffusion, sticky contact, the blood–brain barrier (BBB) and placenta (Vethaak and Leslie 2016; Krug and Wick 2011; Monti et al. 2015).

12.8.2.3 Organ Level

It is established that microplastics, upon entering the tissues, can spread to other tissues or organs. Studies have shown renal impairment in mice due to the accumulation of microplastics (Meng et al. 2022) and behavioral disorders in fish due to the crossing of BBB by nanoparticles (Mattsson et al. 2017).

12.8.3 Effect of Microplastics on various Systems

The following highlights the possible effects of microplastics on the human digestive, respiratory, endocrine, reproductive, and immunological systems:

12.8.3.1 Digestive System

It has been established that ingesting microplastics can irritate the GIT and result in inflammation (Bouwmeester et al. 2015). According to Jin et al. (2019), microplastics have been found to impact the intestinal microbiome, which can result in gastrointestinal issues like bloating and pain. This is due to alterations in the composition of both beneficial and detrimental bacteria in the gut. It is also known that microplastics can emit substances that are hazardous to the gastrointestinal system, such as heavy metals and polycyclic aromatic hydrocarbons, which can cause nausea and vomiting when swallowed orally (Abbasi et al. 2021).

Research has indicated that PS microplastics induce zebrafish gut microbiota imbalance and inflammation (Jin et al. 2018), whereas exposure to PS particles in mice's gut results in a problem related to lipid metabolism (Lu et al. 2018). Exposure to microplastics can alter the gut flora, which can pose a number of health problems.

12.8.3.2 Respiratory System

Because microplastics induce oxidative stress in the airways upon inhalation, they have been associated with wheezing and inflammation that results in shortness of breath, weakness, and dizziness due to a decreased blood oxygen concentration (Wright and Kelly 2017). Research has been conducted on the correlation between nanoplastics and mitochondrial damage in human respiratory system cells (Lin et al. 2022). When exposed to high amounts, environmental toxins like PS have the ability to harm human lung cells. Moreover, this raises the risk of developing persistent obstructive pulmonary disease. In this case, microplastics may serve as transporters of these poisons (Dong et al. 2020). Additionally, Dong et al. (2020) discovered a link between pulmonary illnesses in humans and microplastics.

12.8.3.3 Endocrine System

Microplastics contain endocrine-disrupting chemicals or EDCs, which alter the working of the endocrine system. A notable example of EDCs is BPA which affects estrogen activity in humans significantly (Hengstler et al. 2011).

12.8.3.4 Immune System

Human immune cells are capable of inducing inflammation when exposed to microplastics. The control of immunological and stress-related processes was impacted by microplastic exposure, which led to altered homeostasis and increased energy expenditure in mussels (Détrée and Gallardo-Escárate 2018). PS microplastics cause autophagic cell division within epithelial cells of human lungs by activating complement-related genes in the larval stages of zebrafish (Veneman et al. 2017) and when they are ingested by macrophages (Chiu et al. 2015).

12.8.4 Metabolic Effects

Microplastics frequently result in a range of metabolic and endocrine problems. They may also result in miscarriage and infertility. Due to their interference with the hormone system's ability to operate, they cause disruption in the endocrine system (Vandenberg et al. 2017). Microplastics act as transporters for harmful compounds such as bisphenol A, which enter the body and cause disorders related to reproductive and endocrine functions (Campanale et al. 2020).

Recent research using Raman micro-spectroscopy also found microplastics in the placentas of six pregnant women (Ragusa et al. 2021). Investigations of the possible harm that microplastics could do to the human body's immune system are necessary. According to animal studies, repeated exposure to microplastics might cause abnormalities in homeostasis and chronic inflammation (Détrée and Gallardo-Escárate 2018). Using human lung cells as a subject, it was discovered that by controlling the expression of proteins and genes connected to the immune response, microplastics can trigger innate immunity (Chiu et al. 2015).

12.8.5 Other Effects

Microplastics have detrimental health effects mainly through inflammation, oxidative stress (increased formation of reactive oxygen species), disturbances in lipid metabolism, abnormalities in the gut microbiota, and neurotoxicity, according to findings from research on human cells and mice. Human gastric cancer cells' production of the essential inflammatory chemicals IL-6 and IL-8 is markedly increased with exposure to 44 nm PS nanoparticles (Forte et al. 2016; Lee et al. 2023). According to documented animal studies, mice exposed to PS microplastics experience disruptions in their liver's lipid metabolism, increased oxidative stress, and increased acetylcholine esterase activity (Deng et al. 2017). They may also experience an imbalance in the microbiota within their gastrointestinal tracts (Liang et al. 2018).

Shan and colleagues (2022) orally exposed mice to PS nanoparticles for 7 days. They discovered accumulation of nanoparticles in the central nervous system, leading to microglia activation and neuron damage (Shan et al. 2022). Additional data reveals that exposure to PS microplastics induced cognitive dysfunction in mice (Estrela et al. 2021), alongside alterations in locomotor function and anticholinesterase activity (Liu et al. 2022).

12.9 Biodegradation of Nonbiodegradable Polymers inside the Body

The disintegration of polymers caused by a particular biological activity is known as biodegradation. All polymers can decompose under the right circumstances (Ali et al. 1994). In a physiological setting, polymers can degrade by hydrolysis, oxidation, or decarboxylation (Bischoff 1972). The type of polymer, prior weathering, and surface area (which is connected to size) can all affect how much physiological breakdown occurs in microplastics. Plastics biodegrade slowly, yet it can nevertheless disrupt regular metabolic processes by, for example, competing for, or inactivating enzymes (Kulkarni 1965). Microplastics have the potential to absorb organic molecules found within the body when they get inside. Their degradation may be influenced by these chemical molecules.

12.9.1 Digestion in Digestive System

Both the resident microflora and enzymes may be the cause of degradation. Significant structural breakdown of the polyethylene terephthalate microplastics was observed by the local microflora (Tamargo et al. 2022). According to studies by Schwabl et al (2019) and Zhang et al. (2021), microplastics found in human feces may degrade in the lumen.

12.9.1.1 Digestion in Hydrolysis

Nylon, polyurethane, and polyester are examples of hydrophilic polymers, which are more prone to hydrolysis and are thus more quickly broken down in the human body. Hydrophilic polymers also have hydrolyzable links and heterochains, particularly those that contain oxygen and nitrogen (Bischoff 1972).

$$Polyurethane + H_2O \rightarrow Polyol + Isocyanate$$

12.9.1.2 Digestion By Oxidation

Homochain polymers that are hydrolytically stable are primarily broken down by oxidation, such as polyethylene and polypropylene (Ali et al. 1994). Biological activity can create highly oxidizing surroundings, such as those generated by active phagocytes, though the human system lacks many of the conditions commonly linked to polymer oxidation (such as high temperature and exposure to electromagnetic radiation) (Ali et al. 1994).

$$Polyethylene + O_2 \rightarrow Oxidized\ Polyethylene$$

12.9.2 Degradation Via the Monooxygenase System and Intracellular Mitochondrial Enzymes

After 2 days at 37 °C, the intracellular enzymes produced by mitochondria and the monooxygenase system were in charge of the oxidation of polyethylene in bacteria (*Bacillus brevis* and *Pseudomonas putida*) and liver (mice, rat, and cats) homogenates (Wasserbauer et al. 1990). In particular, cytochrome P-450 was detected in rat liver homogenates (Beranová et al. 1990). As a matter of fact, these enzymes can transfer hydrogen molecules from the polymer, resulting in free radicals that are unstable and able to spread the reaction along the chain (Kulkarni 1965).

12.10 Conclusion and Future Perspectives

The prevalence of plastic debris has escalated since the inception of synthetic polymers in 1907, with significant environmental ramifications. Plastic pollution spans terrestrial, marine, and even remote areas, with mega- and macro-plastics accumulating predominantly in the Northern Hemisphere. Despite constituting a small portion of discarded waste, plastics constitute a considerable proportion of shoreline debris, persisting for hundreds to thousands of years. Plastic pollution exerts detrimental effects on marine fauna, with various species ingesting plastics and suffering adverse consequences. The physiological ramifications of microplastics upon entering the human body is still largely unknown. There have been studieson the cells of human body in vitro, which displayed changes in the gut microbiota upon exposure to microplastics but the consequences of it in vivo are still unknown. The extensive use of microplastics in the production of packaging materials, clothing, and several other materials of everyday use, in addition to their improper disposal has increased the risk of microplastic exposure by several folds in recent years. Despite being inert, microplastics have been shown to cause damage to cells and tissues due to their hydrophobic properties. Studies conducted on mice have shown that MPs accumulation in their organs has led to oxidative stress and affected lipid metabolism. In order to minimize human exposure to microplastics, it is crucial to curb microplastic pollution and learn more about its alternatives and proper disposal techniques. Several other studies need to be conducted in addition to the implementation of strict measures by the authorities wherever needed.

Conflict of Interest

None.

References

Abbasi S, Moore F, Keshavarzi B (2021) PET-microplastics as a vector for polycyclic aromatic hydrocarbons in a simulated plant rhizosphere zone. Environ Technol Innov 21:101370

Acharya S, Rumi SS, Hu Y, Abidi N (2021) Microfibers from synthetic textiles as a major source of microplastics in the environment: a review. Text Res J 91(17–18):2136–2156. https://doi.org/10.1177/0040517521991244

Akindoyo JO, Beg MDH, Ghazali S, Islam MR, Jeyaratnam N, Yuvaraj AR (2016) Polyurethane types, synthesis and applications—a review. RSC Adv 6(115):114453–114482. https://doi.org/10.1039/c6ra14525f

Alaaeddin MH, Sapuan S, Zuhri M, Zainudin E, Al-Oqla FM (2018) Properties and common industrial applications of polyvinyl fluoride (PVF) and Polyvinylidene fluoride (PVDF). IOP Conference Series: Materials Science and Engineering 409:012021. https://doi.org/10.1088/1757-899x/409/1/012021

Ali SAM, Doherty PJ, Williams DF (1994) The mechanisms of oxidative degradation of biomedical polymers by free radicals. J Appl Polym Sci 51:1389–1398. https://doi.org/10.1002/app.1994.070510805

Almroth BC, Åström L, Roslund S, Petersson H, Johansson M, Persson N (2017) Quantifying shedding of synthetic fibers from textiles; a source of microplastics released into the environment. Environ Sci Pollut Res 25(2):1191–1199. https://doi.org/10.1007/s11356-017-0528-7

Amato-Lourenço LF, Carvalho-Oliveira R, Júnior GR, Galvão LDS, Ando RA, Mauad T (2021) Presence of airborne microplastics in human lung tissue. J Hazard Mater 416:126124. https://doi.org/10.1016/j.jhazmat.2021.126124

Andrady AL (2011) Microplastics in the marine environment. Mar Pollut Bull 62(8):1596–1605. https://doi.org/10.1016/j.marpolbul.2011.05.030

Bain CC, Schridde A (2018) Origin, differentiation, and function of intestinal macrophages. Front Immunol 9:2733. https://doi.org/10.3389/fimmu.2018.02733

Barboza LGA, Lopes C, Oliveira P, Bessa F, Otero V, Henriques B, Raimundo J, Caetano M, Vale C, Guilhermino L (2020) Microplastics in wild fish from north East Atlantic Ocean and its potential for causing neurotoxic effects, lipid oxidative damage, and human health risks associated with ingestion exposure. Sci Total Environ 717:134625. https://doi.org/10.1016/j.scitotenv.2019.134625

Barnes DKA, Galgani F, Thompson RC, Barlaz MA (2009) Accumulation and fragmentation of plastic debris in global environments. Philos Trans R Soc B 364(1526):1985–1998. https://doi.org/10.1098/rstb.2008.0205

Beachell HC, Nemphos SP (1959) Oxidative degradation of polymers in presence of ozone. In: Advances in chemistry series, pp 168–175. https://doi.org/10.1021/ba-1959-0021.ch026

Beranová M, Wasserbauer R, Vanćurová D, Štifter M, Očenášková J, Mára M (1990) Effect of cytochrome P-450 inhibition and stimulation on intensity of polyethylene degradation in microsomal fraction of mouse and rat livers. Biomaterials 11:521–524. https://doi.org/10.1016/0142-9612(90)90070-7

Bermúdez R, Swarzenski PW (2021) A microplastic size classification scheme aligned with universal plankton survey methods. MethodsX 8:101516. https://doi.org/10.1016/j.mex.2021.101516

Bischoff F (1972) Organic polymer biocompatibility and toxicology. Clin Chem 18(9):869–894. https://doi.org/10.1093/clinchem/18.9.869

Boucher J, Friot D (2017) Primary microplastics in the oceans: a global evaluation of sources. https://doi.org/10.2305/iucn.ch.2017.01.en

Bouwmeester H, Hollman P, Peters R (2015) Potential health impact of environmentally released micro- and Nanoplastics in the human food production chain: experiences from Nanotoxicology. Environ Sci Technol 49(15):8932–8947. https://doi.org/10.1021/acs.est.5b01090

Braun T, Ehrlich L, Henrich W, Koeppel S, Lomako I, Schwabl P, Liebmann B (2021) Detection of microplastic in human placenta and meconium in a clinical setting. Pharmaceutics 13(7):921. https://doi.org/10.3390/pharmaceutics13070921

Burgener K, Bhamla MS (2021) A polymer-based technique to remove pollutants from soft contact lenses. Contact Lens and Anterior Eye 44(3):101335. https://doi.org/10.1016/j.clae.2020.05.004

Burkhart JE, Piacitelli C, Schwegler-Berry D, Jones WG (1999) ENVIRONMENTAL STUDY OF NYLON FLOCKING PROCESS. Journal of Toxicology and Environmental Health, Part A: Current Issues 57(1):1–23. https://doi.org/10.1080/009841099157836

Campanale C, Massarelli C, Savino I, Locaputo V, Uricchio VF (2020) A detailed review study on potential effects of microplastics and additives of concern on human health. International Journal of Environmental Research and Public Health (Online) 17(4):1212. https://doi.org/10.3390/ijerph17041212

Carwile JL, Luu H, Bassett LS, Driscoll DA, Yuan C, Chang J, Ye X, Calafat AM, Michels KB (2009) Polycarbonate bottle use and urinary bisphenol a concentrations. Environ Health Perspect 117(9):1368–1372. https://doi.org/10.1289/ehp.0900604

Cella C, La Spina R, Méhn D, Fumagalli F, Ceccone G, Valsesia A, Gilliland D (2022) Detecting micro- and Nanoplastics released from food packaging: challenges and analytical strategies. Polymers 14(6):1238. https://doi.org/10.3390/polym14061238

Chiu H, Xia T, Lee Y, Chen CW, Tsai JC, Wang YJ (2015) Cationic polystyrene nanospheres induce autophagic cell death through the induction of endoplasmic reticulum stress. Nanoscale 7(2):736–746. https://doi.org/10.1039/c4nr05509h

Cole M, Lindeque PK, Halsband C, Galloway TS (2011) Microplastics as contaminants in the marine environment: a review. Mar Pollut Bull 62(12):2588–2597. https://doi.org/10.1016/j.marpolbul.2011.09.025

Comăniţă E, Ghinea C, Roşca M, Simion IM, Petraru M, Gavrilescu M (2015) Environmental impacts of polyvinyl chloride (PVC) production process. IEEE

Conti GO, Ferrante M, Bannı M, Favara C, Nicolosi I, Cristaldi A, Zuccarello P (2020) Micro- and nano-plastics in edible fruit and vegetables. The first diet risks assessment for the general population. Environ Res 187:109677. https://doi.org/10.1016/j.envres.2020.109677

Cooper DM, Loxham M (2019) Particulate matter and the airway epithelium: the special case of the underground? Eur Respir Rev 28(153):190066. https://doi.org/10.1183/16000617.0066-2019

Corradini F, Meza P, Eguiluz R, Casado F, Lwanga EH, Geissen V (2019) Evidence of microplastic accumulation in agricultural soils from sewage sludge disposal. Sci Total Environ 671:411–420. https://doi.org/10.1016/j.scitotenv.2019.03.368

Cox K, Covernton GA, Davies HL, Dower JF, Juanes F, Dudas SE (2019) Human consumption of microplastics. Environ Sci Technol 53(12):7068–7074. https://doi.org/10.1021/acs.est.9b01517

Danso D, Chow J, Streit WR (2019) Plastics: environmental and biotechnological perspectives on microbial degradation. Appl Environ Microbiol 85(19):e01095–e01019. https://doi.org/10.1128/aem.01095-19

Das A, Mahanwar PA (2020) A brief discussion on advances in polyurethane applications. Adv Ind Eng Polym Res 3(3):93–101. https://doi.org/10.1016/j.aiepr.2020.07.002

Dehghani S, Moore F, Akhbarizadeh R (2017) Microplastic pollution in deposited urban dust, Tehran metropolis, Iran. Environ Sci Pollut Res 24(25):20360–20371. https://doi.org/10.1007/s11356-017-9674-1

Deng Y, Zhang Y, Lemos B, Ren H (2017) Tissue accumulation of microplastics in mice and biomarker responses suggest widespread health risks of exposure. Sci Rep 7:46687. https://doi.org/10.1038/srep46687

Desai PR, Patlolla RR, Singh M (2010) Interaction of nanoparticles and cell-penetrating peptides with skin for transdermal drug delivery. Mol Membr Biol 27(7):247–259. https://doi.org/10.3109/09687688.2010.522203

Dessì C, Okoffo ED, O'Brien J, Gallen M, Samanipour S, Kaserzon S, Rauert C, Wang X, Thomas KV (2021) Plastics contamination of store-bought rice. J Hazard Mater 416:125778. https://doi.org/10.1016/j.jhazmat.2021.125778

Détrée C, Gallardo-Escárate C (2018) Single and repetitive microplastics exposures induce immune system modulation and homeostasis alteration in the edible mussel Mytilus galloprovincialis. Fish Shellfish Immunol 83:52–60. https://doi.org/10.1016/j.fsi.2018.09.018

Diaz-Basantes MF, Conesa JA, Fullana A (2020) Microplastics in honey, beer, milk and refreshments in Ecuador as emerging contaminants. Sustain For 12(14):5514. https://doi.org/10.3390/su12145514

Domenech J, Marcos R (2021) Pathways of human exposure to microplastics, and estimation of the total burden. Curr Opin Food Sci 39:144–151. https://doi.org/10.1016/j.cofs.2021.01.004

Dong C, Chen C, Chen Y, Chen H, Lee J, Lin C (2020) Polystyrene microplastic particles: in vitro pulmonary toxicity assessment. J Hazard Mater 385:121575. https://doi.org/10.1016/j.jhazmat.2019.121575

Dris R, Gaspéri J, Saad M, Mirande C, Tassin B (2016) Synthetic fibers in atmospheric fallout: a source of microplastics in the environment? Mar Pollut Bull 104(1–2):290–293. https://doi.org/10.1016/j.marpolbul.2016.01.006

Eo S, Hong SH, Song YK, Han GM, Shim WJ (2019) Spatiotemporal distribution and annual load of microplastics in the Nakdong River, South Korea. Water Research 160:228–237. https://doi.org/10.1016/j.watres.2019.05.053

Eriksen M, Thiel M, Prindiville M, Kiessling T (2018) Microplastic: what are the solutions? In: Wagner M, Lambert S (eds) Freshwater microplastics . The handbook of environmental chemistry, vol 58. Springer, Cham. https://doi.org/10.1007/978-3-319-61615-5_13

Estrela FN, Guimarães ATB, Da Costa Araújo AP, Silva FG, Da Luz TM, Silva AM, Pereira PS, Malafaia G (2021) Toxicity of polystyrene nanoplastics and zinc oxide to mice. Chemosphere 271:129476. https://doi.org/10.1016/j.chemosphere.2020.129476

Fadare OO, Okoffo ED, Olasehinde EF (2021) Microparticles and microplastics contamination in African table salts. Mar Pollut Bull 164:112006. https://doi.org/10.1016/j.marpolbul.2021.112006

Fadare OO, Wan B, Guo L, Zhao L (2020) Microplastics from consumer plastic food containers: are we consuming it? Chemosphere 253:126787. https://doi.org/10.1016/j.chemosphere.2020.126787

Fendall LS, Sewell MA (2009) Contributing to marine pollution by washing your face: microplastics in facial cleansers. Mar Pollut Bull 58(8):1225–1228. https://doi.org/10.1016/j.marpolbul.2009.04.025

Filho WL, Havea PH, Balogun A, Boenecke J, Maharaj A, Haapio MO, Hemstock SL (2019) Plastic debris on Pacific Islands: ecological and health implications. Sci Total Environ 670:181–187. https://doi.org/10.1016/j.scitotenv.2019.03.181

Fischer M, Goßmann I, Scholz-Böttcher BM (2019) Fleur de Sel—an interregional monitor for microplastics mass load and composition in European coastal waters? J Anal Appl Pyrolysis 144:104711. https://doi.org/10.1016/j.jaap.2019.104711

Forte M, Iachetta G, Tussellino M, Carotenuto R, Prisco M, De Falco M et al (2016) Polystyrene nanoparticles internalization in human gastric adenocarcinoma cells. Toxicol In Vitro 31:126–136

Frias J, Pagter E, Nash R, O'Connor IA, Carretero O, Filgueiras AV, Viñas L, Gago J, Antunes J, Bessa F, Sobral P, Goruppi A, Tirelli V, Pedrotti ML, Suaria G, Aliani S, Lopes C, Raimundo J, Caetano M, Gerdts G (2018) Standardised protocol for monitoring microplastics in sediments may 2018. HAL (Le Centre Pour La Communication Scientifique Directe). https://doi.org/10.13140/rg.2.2.36256.89601/1

Galvão A, Aleixo M, De Pablo H, Lopes C, Raimundo J (2020) Microplastics in wastewater: microfiber emissions from common household laundry. Environ Sci Pollut Res 27(21):26643–26649. https://doi.org/10.1007/s11356-020-08765-6

Gigault J, Halle AT, Baudrimont M, Pascal PY, Gauffre F, Phi TL, El Hadri H, Grassl B, Reynaud S (2018) Current opinion: what is a nanoplastic? Environ Pollut 235:1030–1034. https://doi.org/10.1016/j.envpol.2018.01.024

Grainger J, Konkel JE, Zangerle-Murray T, Shaw TN (2017) Macrophages in gastrointestinal homeostasis and inflammation. Pflugers Arch - Eur J Physiol 469(3–4):527–539. https://doi.org/10.1007/s00424-017-1958-2

Grieve M, Biermann T (1997) The population of coloured textile fibres on outdoor surfaces. Sci Justice 37(4):231–239. https://doi.org/10.1016/s1355-0306(97)72196-8

Gündoğdu S (2018) Contamination of table salts from Turkey with microplastics. Food Additives & Contaminants: Part A 35(5):1006–1014. https://doi.org/10.1080/19440049.2018.1447694

Gündoğdu S, Çevik C, Ataş NT (2020) Stuffed with microplastics: microplastic occurrence in traditional stuffed mussels sold in the Turkish market. Food Biosci 37:100715. https://doi.org/10.1016/j.fbio.2020.100715

Harms IK, Diekötter T, Troegel S, Lenz M (2021) Amount, distribution and composition of large microplastics in typical agricultural soils in northern Germany. Sci Total Environ 758:143615. https://doi.org/10.1016/j.scitotenv.2020.143615

Hengstler JG, Foth H, Gebel T, Kramer P, Lilienblum W, Schweinfurth H, Völkel W, Wollin K, Gundert-Remy U (2011) Critical evaluation of key evidence on the human health hazards of exposure to bisphenol a. Crit Rev Toxicol 41(4):263–291. https://doi.org/10.3109/10408444.2011.558487

Hernández EG, Nowack B, Mitrano DM (2017) Polyester textiles as a source of microplastics from households: a mechanistic study to understand microfiber release during washing. Environ Sci Technol 51(12):7036–7046. https://doi.org/10.1021/acs.est.7b01750

Hernandez LM, Xu EG, Larsson HCE, Tahara R, Maisuria VB, Tufenkji N (2019) Plastic teabags release billions of microparticles and nanoparticles into tea. Environ Sci Technol 53(21):12300–12310. https://doi.org/10.1021/acs.est.9b02540

Ho BT, Roberts TK, Lucas S (2017) An overview on biodegradation of polystyrene and modified polystyrene: the microbial approach. Crit Rev Biotechnol 38(2):308–320. https://doi.org/10.1080/07388551.2017.1355293

Horvatits T, Tamminga M, Liu B, Sebode M, Carambia A, Fischer L, Püschel K, Huber S, Fischer EK (2022) Microplastics detected in cirrhotic liver tissue. EBioMedicine 82:104147. https://doi.org/10.1016/j.ebiom.2022.104147

Hou Z, Meng R, Bao J, Lai T, Qu R, Hao S, Jia D, Wang B (2022) Distinct accumulation of nanoplastics in human intestinal organoids. Sci Total Environ 838:155811. https://doi.org/10.1016/j.scitotenv.2022.155811

Hüffer T, Weniger A, Hofmann T (2018) Sorption of organic compounds by aged polystyrene microplastic particles. Environ Pollut 236:218–225. https://doi.org/10.1016/j.envpol.2018.01.022

Hurley MJ, Gottuk DT, Hall JR Jr, Harada K, Kuligowski ED, Puchovsky M, Torero JL, Watts JM Jr, Wieczorek C (2016) SFPE handbook of fire protection engineering. Springer-Verlag, New York, NY, USA

Ibrahim YS, Anuar ST, Azmi AA, Khalik WMAWM, Lehata S, Hamzah S, Ismail D, Feei Z, Dzulkarnaen A, Zakaria Z, Mustaffa N, Sharif SET, Lee YY (2020) Detection of microplastics in human colectomy specimens. JGH Open 5(1):116–121. https://doi.org/10.1002/jgh3.12457

Jenner LC, Rotchell JM, Bennett RT, Cowen ME, Tentzeris V, Sadofsky LR (2022) Detection of microplastics in human lung tissue using μFTIR spectroscopy. Sci Total Environ 831:154907. https://doi.org/10.1016/j.scitotenv.2022.154907

Jin Y, Liang L, Tu W, Luo T, Fu Z (2019) Impacts of polystyrene microplastic on the gut barrier, microbiota and metabolism of mice. Sci Total Environ 649:308–317. https://doi.org/10.1016/j.scitotenv.2018.08.353

Jin Y, Xia J, Pan Z, Yang J, Wang Q, Fu Z (2018) Polystyrene microplastics induce microbiota dysbiosis and inflammation in the gut of adult zebrafish. Environ Pollut 235:322–329. https://doi.org/10.1016/j.envpol.2017.12.088

Kabir MS, Wang H, Luster-Teasley S, Zhang L, Zhao R (2023) Microplastics in landfill leachate: sources, detection, occurrence, and removal. Environmental Science and Ecotechnology 16:100256. https://doi.org/10.1016/j.ese.2023.100256

Kalpana S, Priyadarshini S, Leena MM, Moses J, Anandharamakrishnan C (2019) Intelligent packaging: trends and applications in food systems. Trends Food Sci Technol 93:145–157. https://doi.org/10.1016/j.tifs.2019.09.008

Kang G, Cao Y (2014) Application and modification of poly(vinylidene fluoride) (PVDF) membranes—a review. J Membr Sci 463:145–165. https://doi.org/10.1016/j.memsci.2014.03.055

Kankanige D, Babel S (2020) Smaller-sized micro-plastics (MPs) contamination in single-use PET-bottled water in Thailand. Sci Total Environ 717:137232. https://doi.org/10.1016/j.scitotenv.2020.137232

Kedzierski M, Lechat B, Sire O, Maguer GL, Tilly VL, Bruzaud S (2020) Microplastic contamination of packaged meat: occurrence and associated risks. Food Packag Shelf Life 24:100489. https://doi.org/10.1016/j.fpsl.2020.100489

Kemona A, Piotrowska M (2020) Polyurethane recycling and disposal: methods and prospects. Polymers 12(8):1752. https://doi.org/10.3390/polym12081752

Koerner GR, Koerner RM (2018) Polymeric geomembrane components in landfill liners. In: *Elsevier eBooks*, pp 313–341. https://doi.org/10.1016/b978-0-12-407721-8.00017-6

Kosuth M, Mason SA, Wattenberg EV (2018) Anthropogenic contamination of tap water, beer, and sea salt. PLoS One 13(4):e0194970. https://doi.org/10.1371/journal.pone.0194970

Krug HF, Wick P (2011) Nanotoxicology: an interdisciplinary challenge. Angew Chem Int Ed 50(6):1260–1278. https://doi.org/10.1002/anie.201001037

Kulkarni RK (1965) Brief review of biochemical degradation of polymers. Polym Eng Sci 5:227–230. https://doi.org/10.1002/pen.760050404

Kumar B, Löschel LA, Imhof HK, Löder MGJ, Laforsch C (2021) Analysis of microplastics of a broad size range in commercially important mussels by combining FTIR and Raman spectroscopy approaches. Environ Pollut 269:116147. https://doi.org/10.1016/j.envpol.2020.116147

Kumar R, Manna C, Padha S, Verma A, Sharma P, Dhar A, Ghosh A, Bhattacharya P (2022) Micro(nano)plastics pollution and human health: How plastics can induce carcinogenesis to humans? Chemosphere 298:134267. https://doi.org/10.1016/j.chemosphere.2022.134267

Kutralam-Muniasamy G, Pérez-Guevara F, Elizalde I, Shruti V (2020) Branded milks—are they immune from microplastics contamination? Sci Total Environ 714:136823. https://doi.org/10.1016/j.scitotenv.2020.136823

Lebreton L, Van Der Zwet J, Damsteeg JW, Slat B, Andrady AL, Reisser J (2017) River plastic emissions to the world's oceans. Nat Commun 8(1):15611. https://doi.org/10.1038/ncomms15611

Lee J, Jeong S, Chae K (2021) Discharge of microplastics fibres from wet wipes in aquatic and solid environments under different release conditions. Sci Total Environ 784:147144. https://doi.org/10.1016/j.scitotenv.2021.147144

Lee Y, Cho J, Sohn J, Kim H (2023) Health effects of microplastic exposures: current issues and perspectives in South Korea. Yonsei Med J 64(5):301. https://doi.org/10.3349/ymj.2023.0048

Lehmann A, Leifheit EF, Gerdawischke M, Rillig MC (2021) Microplastics have shape- and polymer-dependent effects on soil aggregation and organic matter loss—an experimental and meta-analytical approach. Microplast Nanoplas 1(1):7. https://doi.org/10.1186/s43591-021-00007-x

Leslie H, Van Velzen M, Brandsma S, Vethaak D, García-Vallejo JJ, Lamoree M (2022) Discovery and quantification of plastic particle pollution in human blood. Environ Int 163:107199. https://doi.org/10.1016/j.envint.2022.107199

Li Z, Li Q, Li R, Zhou J, Wang G (2020) The distribution and impact of polystyrene nanoplastics on cucumber plants. Environ Sci Pollut Res 28(13):16042–16053. https://doi.org/10.1007/s11356-020-11702-2

Lian J, Liu W, Meng L, Wu J, Zeb A, Cheng L, Lian Y, Sun H (2021) Effects of microplastics derived from polymer-coated fertilizer on maize growth, rhizosphere, and soil properties. J Clean Prod 318:128571. https://doi.org/10.1016/j.jclepro.2021.128571
Liang L, Wan Z, Luo T, Fu Z, Jin Y (2018) Polystyrene microplastics induce gut microbiota dysbiosis and hepatic lipid metabolism disorder in mice. Sci Total Environ 631–632:449–458. https://doi.org/10.1016/j.scitotenv.2018.03.051
Liepins R, Pearce EM (1976) Chemistry and toxicity of flame retardants for plastics. Environ Health Perspect 17:55. https://doi.org/10.2307/3428609
Lin S, Zhang H, Wang C, Su XL, Song Y, Wu P et al (2022) Metabolomics reveal nanoplastic-induced mitochondrial damage in human liver and lung cells. Environ Sci Technol 56:12483–12493
Lithner D, Larsson Å, Dave G (2011) Environmental and health hazard ranking and assessment of plastic polymers based on chemical composition. Sci Total Environ 409(18):3309–3324. https://doi.org/10.1016/j.scitotenv.2011.04.038
Liu C, Li J, Zhang Y, Wang L, Deng J, Gao Y, Lu Y, Zhang J, Sun H (2019a) Widespread distribution of PET and PC microplastics in dust in urban China and their estimated human exposure. Environ Int 128:116–124. https://doi.org/10.1016/j.envint.2019.04.024
Liu Z, Yu P, Cai M, Wu D, Zhang M, Chen M, Zhao Y (2019b) Effects of microplastics on the innate immunity and intestinal microflora of juvenile Eriocheir sinensis. Sci Total Environ 685:836–846. https://doi.org/10.1016/j.scitotenv.2019.06.265
Liu X, Zhao Y, Dou J, Hou Q, Cheng J, Jiang X (2022) Bioeffects of Inhaled Nanoplastics on Neurons and Alteration of Animal Behaviors through Deposition in the Brain. Nano letters 22(3):1091–1099. https://doi.org/10.1021/acs.nanolett.1c04184
Lu K, Qiao R, An H, Zhang Y (2018) Influence of microplastics on the accumulation and chronic toxic effects of cadmium in zebrafish (Danio rerio). Chemosphere 202:514–520
Maddah HA (2016) Polypropylene as a promising plastic: a review. Am J Polym Sci 6(1):1–11. https://doi.org/10.5923/j.ajps.20160601.01
Mason SA, Welch VG, Neratko J (2018) Synthetic polymer contamination in bottled water. Front Chem 6:407. https://doi.org/10.3389/fchem.2018.00407
Mattsson K, Johnson EV, Malmendal A, Linse S, Hansson L, Cedervall T (2017) Brain damage and behavioural disorders in fish induced by plastic nanoparticles delivered through the food chain. Sci Rep 7(1):11452. https://doi.org/10.1038/s41598-017-10813-0
Meng X, Zhang J, Wang W, Gonzalez-Gil G, Vrouwenvel Vrouwenvelder JS, Li Z (2022) Effects of nano- and microplastics on kidney: physicochemical properties, bioaccumulation, oxidative stress and immunoreaction. Chemosphere 288:132631. https://doi.org/10.1016/j.chemosphere.2021.132631
Mohajerani A, Burnett L, Smith JG, Markovski S, Rodwell G, Rahman T, Kurmus H, Mirzababaei M, Arulrajah A, Horpibulsuk S, Maghool F (2020) Recycling waste rubber tyres in construction materials and associated environmental considerations: a review. Resour Conserv Recycl 155:104679. https://doi.org/10.1016/j.resconrec.2020.104679
Monti DM, Guarnieri D, Napolitano G, Piccoli R, Netti PA, Fusco S, Arciello A (2015) Biocompatibility, uptake and endocytosis pathways of polystyrene nanoparticles in primary human renal epithelial cells. J Biotechnol 193:3–10. https://doi.org/10.1016/j.jbiotec.2014.11.004
Moore CJ (2008) Synthetic polymers in the marine environment: a rapidly increasing, long-term threat. Environ Res 108(2):131–139. https://doi.org/10.1016/j.envres.2008.07.025
Nalbone L, Cincotta F, Giarratana F, Ziino G, Panebianco A (2021) Microplastics in fresh and processed mussels sampled from fish shops and large retail chains in Italy. Food Control 125:108003. https://doi.org/10.1016/j.foodcont.2021.108003
Nisticò R (2020) Polyethylene terephthalate (PET) in the packaging industry. Polym Test 90:106707. https://doi.org/10.1016/j.polymertesting.2020.106707
O'Brien S, Rauert C, Ribeiro F, Okoffo ED, Burrows SD, O'Brien J, Wang X, Wright S, Thomas KV (2023) There's something in the air: a review of sources, prevalence and behaviour of microplastics in the atmosphere. Sci Total Environ 874:162193. https://doi.org/10.1016/j.scitotenv.2023.162193

Oliveri Conti G, Ferrante M, Banni M, Favara C, Nicolosi I, Cristaldi A, Fiore M, Zuccarello P (2020) Microand nano-plastics in edible fruit and vegetables. The first diet risks assessment for the general population. Environmental research 187:109677. https://doi.org/10.1016/j.envres.2020.109677

Pac F, Andrey D, Ericksen B, De Moraes Peixoto R, Bm C, Ambühl ME, Jb D, Dubascoux S, Zbinden P, Panchaud A, Poitevin E (2021) Detection and characterization of small-sized microplastics (≥ 4 μm) in Milk products. Sci Rep 11(1):24046. https://doi.org/10.1038/s41598-021-03458-7

Paget V, Dekali S, Kortulewski T, Grall R, Gamez C, Blazy K, Aguerre-Chariol O, Chevillard S, Braun A, Rat P et al (2015) Specific uptake and Genotoxicity induced by polystyrene Nanobeads with distinct surface chemistry on human lung epithelial cells and macrophages. PLoS One 10 e0123297. https://doi.org/10.1371/journal.pone.0123297

Park T, Lee S, Lee M, Lee JK, Lee S, Zoh K (2020) Occurrence of microplastics in the Han River and riverine fish in South Korea. Sci Total Environ 708:134535. https://doi.org/10.1016/j.scitotenv.2019.134535

Paxton N, Allenby MC, Lewis P, Woodruff MA (2019) Biomedical applications of polyethylene. Eur Polym J 118:412–428. https://doi.org/10.1016/j.eurpolymj.2019.05.037

Prata JC (2018) Airborne microplastics: consequences to human health? Environ Pollut 234:115–126. https://doi.org/10.1016/j.envpol.2017.11.043

Rackaityte E, Lynch SV (2020) The human microbiome in the 21st century. Nat Commun 11(1):5256. https://doi.org/10.1038/s41467-020-18983-8

Ragusa A, Svelato A, Santacroce C, Catalano P, Notarstefano V, Carnevali O, Papa F, Rongioletti M, Baiocco F, Draghi S, D'Amore E, Rinaldo D, Matta M, Giorgini E (2021) Plasticenta: first evidence of microplastics in human placenta. Environ Int 146:106274. https://doi.org/10.1016/j.envint.2020.106274

Ramsperger A, Jasinski J, Völkl M, Witzmann T, Meinhart M, Jérôme V, Kretschmer WP, Freitag R, Senker J, Fery A, Kress H, Scheibel T, Laforsch C (2022) Supposedly identical microplastic particles substantially differ in their material properties influencing particle-cell interactions and cellular responses. J Hazard Mater 425:127961. https://doi.org/10.1016/j.jhazmat.2021.127961

Rasmussen LA, Lykkemark J, Andersen TR, Vollertsen J (2023) Permeable pavements: a possible sink for Tyre wear particles and other microplastics? Sci Total Environ 869:161770. https://doi.org/10.1016/j.scitotenv.2023.161770

Rezaei M, Riksen M, Sirjani E, Sameni A, Geissen V (2019) Wind erosion as a driver for transport of light density microplastics. Sci Total Environ 669:273–281. https://doi.org/10.1016/j.scitotenv.2019.02.382

Ribeiro F, Okoffo ED, O'Brien J, Fraissinet-Tachet S, O'Brien S, Gallen M, Samanipour S, Kaserzon S, Mueller JF, Galloway TS, Thomas KV (2020) Quantitative analysis of selected plastics in high-commercial-value Australian seafood by pyrolysis gas chromatography mass spectrometry. Environ Sci Technol 54(15):9408–9417. https://doi.org/10.1021/acs.est.0c02337

Rillig MC, Ingraffia R, De Souza Machado AA (2017) Microplastic incorporation into soil in agroecosystems. Front Plant Sci 8:305719. https://doi.org/10.3389/fpls.2017.01805

Rochman CM, Brookson CB, Bikker J, Djuric N, Earn A, Bucci K, Athey SN, Huntington AK, McIlwraith HK, Munno K, De Frond H, Kolomijeca A, Erdle LM, Grbić J, Bayoumi MS, Borrelle SB, Wu T, Santoro S, Werbowski LM, Hung C (2019) Rethinking microplastics as a diverse contaminant suite. Environ Toxicol Chem 38(4):703–711. https://doi.org/10.1002/etc.4371

Sait STL, Sørensen L, Kubowicz S, Vike-Jonas K, González SV, Asimakopoulos AG, Booth AM (2021) Microplastic fibres from synthetic textiles: environmental degradation and additive chemical content. Environ Pollut 268:115745. https://doi.org/10.1016/j.envpol.2020.115745

Santillo D, Miller K, Johnston P (2017) Microplastics as contaminants in commercially important seafood species. Integrated environmental assessment and management 13(3):516–521. https://doi.org/10.1002/ieam.1909

Scheurer ME, Bigalke M (2018) Microplastics in Swiss floodplain soils. Environ Sci Technol 52(6):3591–3598. https://doi.org/10.1021/acs.est.7b06003

Schneider M, Stracke F, Hansen S, Schaefer UF (2009) Nanoparticles and their interactions with the dermal barrier. Dermatoendocrinol 1(4):197–206. https://doi.org/10.4161/derm.1.4.9501

Schraufnagel DE (2020) The health effects of ultrafine particles. Exp Mol Med 52(3):311–317. https://doi.org/10.1038/s12276-020-0403-3

Schwabl P, Köppel S, Königshofer P, Bucsics T, Trauner M, Reiberger T, Liebmann B (2019) Detection of various microplastics in human stool. Ann Intern Med 171:453. https://doi.org/10.7326/M19-0618

Shi X, Chen Z, Wu L, Wei W, Ni B (2023) Microplastics in municipal solid waste landfills: detection, formation and potential environmental risks. Current Opinion in Environmental Science & Health 31:100433. https://doi.org/10.1016/j.coesh.2022.100433

Simón D, Borreguero AM, de Lucas A, Rodríguez JF (2018) Recycling of polyurethanes from laboratory to industry, a journey towards the sustainability. Waste Manag 76:147–171

Sin LT, Tueen BS (2023) Plastics and environmental sustainability issues. In: *Elsevier eBooks*, pp 1–43. https://doi.org/10.1016/b978-0-12-824489-0.00006-4

Shan S, Zhang Y, Zhao H, Zeng T, Zhao X (2022) Polystyrene nanoplastics penetrate across the blood-brain barrier and induce activation of microglia in the brain of mice. Chemosphere 298:134261. https://doi.org/10.1016/j.chemosphere.2022.134261

Soltani NS, Taylor MP, Wilson SP (2021) Quantification and exposure assessment of microplastics in Australian indoor house dust. Environ Pollut 283:117064. https://doi.org/10.1016/j.envpol.2021.117064

Song S, Van Dijk F, Eck G, Wu X, Bos S, Boom D, Kooter IM, Wardenaar R, Spierings DC, Cole MA, Salvati A, Gosens R, Melgert BN (2022) Inhalable textile microplastic fibers impair lung repair. European Respiratory Society 8(69). https://doi.org/10.1183/23120541.lsc-2022.69

Sparks C, Awe A, Maneveld J (2021) Abundance and characteristics of microplastics in retail mussels from Cape Town, South Africa. Mar Pollut Bull 166:112186. https://doi.org/10.1016/j.marpolbul.2021.112186

Steinmetz Z, Wollmann C, Schaefer MR, Buchmann C, David J, Tröger J, Muñoz K, Frör O, Schaumann GE (2016) Plastic mulching in agriculture. Trading short-term agronomic benefits for long-term soil degradation? Sci Total Environ 550:690–705. https://doi.org/10.1016/j.scitotenv.2016.01.153

Stock V, Laurisch C, Franke J, Dönmez MH, Voß L, Böhmert L, Braeuning A, Sieg H (2021) Uptake and cellular effects of PE, PP, PET and PVC microplastic particles. Toxicol In Vitro 70:105021. https://doi.org/10.1016/j.tiv.2020.105021

Tamargo A, Molinero N, Reinosa J, Alcolea-Rodriguez V, Portela R, Bañares MÁ, Fernández J, Moreno-Arribas MV (2022) PET microplastics affect human gut microbiota communities during simulated gastrointestinal digestion, first evidence of plausible polymer biodegradation during human digestion. Sci Rep 12(1):528. https://doi.org/10.1038/s41598-021-04489-w

Thompson RC, Olsen YS, Mitchell RP, Davis A, Rowland SJ, John A, McGonigle DF, Russell AE (2004) Lost at sea: where is all the plastic? Science 304(5672):838. https://doi.org/10.1126/science.1094559

Tympa L, Katsara K, Moschou PN, Kenanakis G, Papadakis V (2021) Do microplastics enter our food chain via root vegetables? A Raman based spectroscopic study on Raphanus sativus. Materials 14(9):2329. https://doi.org/10.3390/ma14092329

Van Cauwenberghe L, Janssen CR (2014) Microplastics in bivalves cultured for human consumption. Environ Pollut 193:65–70. https://doi.org/10.1016/j.envpol.2014.06.010

Vancamelbeke M, Vermeire S (2017) The intestinal barrier: a fundamental role in health and disease. Expert Rev Gastroenterol Hepatol 11(9):821–834. https://doi.org/10.1080/17474124.2017.1343143

Vandenberg LN, Luthi D, Quinerly DA (2017) Plastic bodies in a plastic world: multi-disciplinary approaches to study endocrine disrupting chemicals. J Clean Prod 140:373–385

Veneman WJ, Spaink HP, Brun NR, Bosker T, Vijver MG (2017) Pathway analysis of systemic transcriptome responses to injected polystyrene particles in zebrafish larvae. Aquat Toxicol 190:112–120. https://doi.org/10.1016/j.aquatox.2017.06.014

Vethaak A, Leslie H (2016) Plastic debris is a human health issue. Environ Sci Technol 50(13):6825–6826. https://doi.org/10.1021/acs.est.6b02569

Völkl M, Jérôme V, Weig A, Jasinski J, Meides N, Strohriegl P, Scheibel T, Freitag R (2022) Pristine and artificially-aged polystyrene microplastic particles differ in regard to cellular response. Journal of Hazardous Materials 435:128955. https://doi.org/10.1016/j.jhazmat.2022.128955

Wang J, Li J, Liu S, Li H, Chen X, Peng C, Zhang P, Liu X (2021) Distinct microplastic distributions in soils of different land-use types: a case study of Chinese farmlands. Environ Pollut 269:116199. https://doi.org/10.1016/j.envpol.2020.116199

Wasserbauer R, Beranová M, Vancurová D, Doležel B (1990) Biodegradation of polyethylene foils by bacterial and liver homogenates. Biomaterials 11:36–40. https://doi.org/10.1016/0142-9612(90)90049-V

Weithmann N, Möller J, Löder MGJ, Piehl S, Laforsch C, Freitag R (2018) Organic fertilizer as a vehicle for the entry of microplastic into the environment. Sci Adv 4(4):eaap8060. https://doi.org/10.1126/sciadv.aap8060

Winkler A, Santo N, Ortenzi MA, Bolzoni E, Bacchetta R, Tremolada P (2019) Does mechanical stress cause microplastic release from plastic water bottles? Water Res 166:115082. https://doi.org/10.1016/j.watres.2019.115082

Wright S, Kelly FJ (2017) Plastic and human health: a micro issue? Environ Sci Technol 51(12):6634–6647. https://doi.org/10.1021/acs.est.7b00423

Wu P, Lin S, Cao G, Wu J, Jin H, Wang C, Wong MH, Yang Z, Cai Z (2022) Absorption, distribution, metabolism, excretion and toxicity of microplastics in the human body and health implications. J Hazard Mater 437:129361. https://doi.org/10.1016/j.jhazmat.2022.129361

Wu S, Wu M, Tian D, Qiu L, Li T (2019) Effects of polystyrene microbeads on cytotoxicity and transcriptomic profiles in human Caco-2 cells. Environmental Toxicology (Print) 35(4):495–506. https://doi.org/10.1002/tox.22885

Wünsch JR (2000) Polystyrene: synthesis, production and applications. iSmithers Rapra Publishing

Yang X, Wang H, Zhang L, Kong L, Chen Y, He Q, Li L, Grossart H, Ju F (2021) Marine algae facilitate transfer of microplastics and associated pollutants into food webs. Sci Total Environ 787:147535. https://doi.org/10.1016/j.scitotenv.2021.147535

Yang Y, Xie E, Du Z, Peng Z, Han Z, Li L, Zhao R, Qin Y, Xue M, Li F, Hua K, Yang X (2023) Detection of various microplastics in patients undergoing cardiac surgery. Environ Sci Technol 57(30):10911–10918. https://doi.org/10.1021/acs.est.2c07179

Yurtsever M (2019) Tiny, shiny, and colorful microplastics: are regular glitters a significant source of microplastics? Mar Pollut Bull 146:678–682. https://doi.org/10.1016/j.marpolbul.2019.07.009

Zhang C, Wang J, Zhou A, Ye Q, Feng Y, Wang Z, Wang S, Xu G, Zou J (2021) Species-specific effect of microplastics on fish embryos and observation of toxicity kinetics in larvae. J Hazard Mater 403:123948. https://doi.org/10.1016/j.jhazmat.2020.123948

Zhang S, Yang X, Gertsen H, Peters P, Salánki T, Geissen V (2018) A simple method for the extraction and identification of light density microplastics from soil. Sci Total Environ 616–617:1056–1065. https://doi.org/10.1016/j.scitotenv.2017.10.213

Zhou G, Wu Q, Tang P, Chen C, Cheng X, Wei X, Ma J, Liu B (2023) How many microplastics do we ingest when using disposable drink cups? J Hazard Mater 441:129982. https://doi.org/10.1016/j.jhazmat.2022.129982

Microplastics in Landfill and Leachate

13

Mahbub Alam, Maisha Binte Sultan, Amit Hasan Anik, Shabiha Hossain, Alamgir Kabir, MD Tanvir Hasnine, and Qiuyan Yuan

Abstract

This chapter comprehensively examines the current literature on microplastics (MPs) in landfills and leachate, focusing on sources, occurrence, characteristics, fate, transport mechanisms, environmental implications, and removal techniques. Landfills serve as significant repositories for plastic waste globally, sourced from municipal and industrial activities. Plastic waste undergoes intricate degradation processes within landfills, yielding MPs through physical, biological, and chemical pathways. MPs are distributed throughout landfill bodies and leachate due to dynamic landfill conditions. Leachate, a byproduct of waste decomposition, serves as a critical pathway for MPs and other contaminants into surrounding ecosystems. Characterization studies reveal diverse MP compositions in landfill leachate, with variations in size, shape, color, and polymer types. Groundwater

M. Alam (✉)
Department of Environmental Science, Bangladesh University of Professionals (BUP), Dhaka, Bangladesh

Center for Environmental Nanoscience and Risk, Department of Environmental Health Sciences, Arnold School of Public Health, University of South Carolina, Columbia, SC, USA
e-mail: mahbub@email.sc.edu; mahbub.alam@bup.edu.bd

M. B. Sultan · A. H. Anik · A. Kabir
Department of Environmental Science, Bangladesh University of Professionals (BUP), Dhaka, Bangladesh

S. Hossain
Department of Environmental Science, Bangladesh University of Professionals (BUP), Dhaka, Bangladesh

Department of Geography and Sustainability, University of Tennessee, Knoxville, TN, USA

M. T. Hasnine · Q. Yuan
Department of Civil Engineering, University of Manitoba, Winnipeg, MB, Canada

A. Kumar, V. Singh (eds.), *Microplastics Pollution and its Remediation*,
https://doi.org/10.1007/978-981-97-4068-0_13

quality and ecological health are impacted by the destiny and transit of MPs, which are influenced by several environmental factors. The environmental implications of MPs in landfill leachate include physical and chemical effects, posing risks to terrestrial and aquatic organisms. Removal techniques face challenges due to MPs' small size and diverse characteristics, with methods like filtration, oxidation, and membrane bioreactors showing varied efficacy. This chapter underscores the urgency of addressing MP pollution in landfills and leachate, advocating for continued research and advanced removal technologies to mitigate environmental risks.

Keywords

Landfilled microplastics · Leachate · Occurrence · Fate and transport · Environmental implications · Removal techniques

13.1 Introduction

Pollution from plastic is a major worldwide issue. In the year 2022, global plastics output grew rapidly, surging to 400.3 Mt (million tonnes) (Plastics Europe 2023). By 2025, it is anticipated that the world would produce 600 Mt (Arkin et al. 2019; Geyer et al. 2017). Nearly 40% of applications for plastics include wrapping, construction and building, automobiles, electronics and devices, homes, sports, and farming (Arkin et al. 2019; Kirstein et al. 2021). According to Arkin et al. (2019), polypropylene (PP), polyethylene (PE), and polyvinyl chloride (PVC) are the three most often used plastics. Most types of plastic are poorly degradable due to their durable properties, which are triggered by adding plasticizers (e.g., phthalates) and other additives, but rather breakdown into smaller pieces over time (Kirstein et al. 2021). The plastics that are present in the environment may be divided into five categories: megaplastics (>50 cm), macroplastics (>5 to 50 cm), mesoplastics (≥5 mm to 5 cm), microplastics (MPs; ≥1 μm to <5 mm), and nanoplastics (NPs; ≥1 nm to <1 μm) (Gigault et al. 2018; Anik et al. 2021). MPs are a type of pollution that has attracted a lot of interest recently and are thought to be a rising issue worldwide (Gigault et al. 2018; He et al. 2019; Zhang et al. 2020).

MPs are incredibly durable and extensively dispersed in many media related to the environment, as reported by National Oceanic and Atmospheric Administration (NOAA) (2018), Karim et al. (2019), Anik et al. (2021), and Rahman et al. (2023). Granules, fibers, pieces, and pellets are examples of MPs that are present in the surroundings and mostly consist of many types of artificial polymers (Ryan et al. 2009; Karim et al. 2019). MPs are further divided into primary and secondary categories. Primary MPs are intentionally produced, while secondary MPs are released when big plastics break down (GESAMP 2016; Lehtiniemi et al. 2018). However, MPs observed in the environment are caused by the breakdown of big polymers and the disposal of consumer goods, wastewater discharge, and waste from industry (Afrin et al. 2020; Anik et al. 2021).

Every ecological media, including freshwater, saltwater, soil, sediment, beaches, and the air, has documented the presence and quantity of MPs (Hirai et al. 2011; Claessen et al. 2013; Free et al. 2014; Law and Thompson 2014; Coppock et al. 2017). Following their release into the environment, the wind may carry MPs and eventually find their way to enter the aquatic ecosystem (Lambert et al. 2014; Jambeck et al. 2015). The majority of MPs are washed out of the terrestrial environment via aquatic channels including rivers and estuaries, as well as by industrial and agricultural outputs. Eventually, MPs sink into the ocean (Barletta et al. 2019; Ajith et al. 2020). The world's uppermost waters contain around 24.4 trillion pieces, or 578,000 t, of MPs (Isobe et al. 2021). According to Zhang et al. (2020), recent research has demonstrated the existence of MPs in the air and atmospheric deposition in distant, rural, and urban areas. The average quantity of air MPs from both dry and wet deposition has been determined to be 118 particles/m^2/day in Paris, France, 275 particles/m^2/day in Dongguan, China, and 313 particles/m^2/day in Hamburg, Germany, respectively (Dris et al. 2015, 2016; Cai et al. 2017; Klein and Fischer 2019). Zhou et al. (2017) report that in Yantai, a Chinese coastal city, the atmospheric MP deposition flux reached a high of 602 particles/m^2/day.

The current plastic waste management relies on recycling, incineration, and deposition in landfills. According to Anik et al. (2021) review, 300 Mt of plastic waste is created annually, with just 9% of that waste being recycled, 12% is burned, and the other 79% being either removed from the environment or dumped in landfills. Leachate from plastic debris in landfills causes additional degradation and releases MPs into the soil. The complicated process of plastics degrading in landfills is influenced by a number of variables, including temperature, pH, alkalinity, bacterial activity, polymer type and age, and methods of weathering (Akbay and Ozdemir 2016; Venkatesh et al. 2021). A research by He et al. (2019) looked for MP evidence in landfill leachate. They reported finding 17 distinct kinds of MPs in the leachate samples, ranging in size from 100 to 1000 μm and with average concentrations of 0.42 to 24.58 items/L. MPs in the form of CA (cellulose acetate), LDPE (low-density polyethylene), and HDPE (high-density polyethylene) were discovered by Afrin et al. (2020) during their investigation of landfilled soil samples. They identified three categories of MPs, which were films, fragments, and fibers. Thus, MPs are released when plastic debris is disposed of in a landfill and thereafter end up in the terrestrial environment (Afrin et al. 2020). Furthermore, shattered plastic waste from dumps may enter aquatic bodies through leachate pond leaks and outflows (Rahman et al. 2023). MPs are transported from landfill sites to surrounding ecosystems with the help of rainfall and winds (Golwala et al. 2021). MPs not only enter into the environment but also carry toxic metals with them (Hodson et al. 2017). MP is considered as Trojan horse for trace metals (i.e., Pb, Cu, Cd, Zn, Ti, Fe, As, Be, Bi, Cr, In, Th, Sn, etc.), but there are few studies on their adsorption mechanism (Hildebrandt et al. 2021; Fan et al. 2021). Soil organisms have the ability to absorb MPs that have accumulated in soil with increasing concentrations (Huerta Lwanga et al. 2016, 2017). Plant growth and soil characteristics can be impacted by MPs (de Souza Machado et al. 2018). Additionally, MPs produce a number of harmful substances, including PBDEs (polybrominated diphenyl ethers), PAHs (polycyclic aromatic hydrocarbons), and phthalate (Wardrop et al. 2016). Several studies, including those by Farrell et al. (2013) and Setala et al. (2014), have demonstrated the trophic transfer of MPs. This transfer mechanism may serve as a pathway for MP ingestion by various species, as highlighted by Santana et al. (2018).

The lack of suitable methodology makes it difficult to define MPs. The Fourier transform-infrared (FTIR), Raman spectroscopy, Transmission electron microscopy (TEM), Atomic force microscopy (AFM), and Energy-disperse X-ray spectroscopy (EDS) are a few promising techniques that can be used to characterize MPs (Nascimento et al. 2018; Schwaferts et al. 2020; Afrin et al. 2020). FTIR and Raman spectroscopy are the most used methods for chemically identifying MPs (Kappler et al. 2016). Furthermore, to remove MPs from different environmental compartments, efficient removal procedures are desperately needed in addition to MP identification (Padervand et al. 2020; Anik et al. 2021). Chemical treatments, biological removal and ingestion, and physical sorption and filtration are among the removal techniques for MPs (Padervand et al. 2020). According to Padervand et al. (2020), traditional activated sludge technologies are less efficient in removing MPs from influents in water treatment facilities than membrane bioreactors when adsorption and filtration are combined. However, eliminating MPs from leachate and landfills is a difficult task that calls for a thorough comprehension of the underpinning systems and procedures (Kabir et al. 2023).

There are still significant study gaps that need to be filled, even if the level of knowledge at this time indicates that a substantial amount of study has already been done to evaluate the MPs' origins, distribution, destiny, impacts, identification, and removal strategies. These include: (1) a lack of standard procedures for the identification, characterization, and elimination of MPs in various environmental media, such as landfill and leachate; (2) insufficient data on the distribution and fate of MPs, particularly in landfill and leachate; (3) limited information and evidence on landfilled MP's environmental health effects. Therefore, the goal of this book chapter is to thoroughly examine the most recent research on the origins, prevalence, traits, and environmental effects of microplastics (MPs) in landfills and leachate. The present characterization and removal methods from landfills and leachate, as well as the fate and transport mechanisms of landfilled MPs, are also evaluated in the present study. Moreover, this study also seeks to draw further attention to researchers globally to delve deeper into the exploration of landfilled MPs and their environmental health impacts, thereby creating opportunities for further in-depth research in this field.

13.2 Sources and Generation of Microplastics

13.2.1 Sources and Generation of Microplastics in Landfill

Landfills represent crucial hubs for plastic waste management, accommodating a substantial portion of the world's plastic debris. Estimates suggest that landfills store between 21% and 42% of the global plastics generated, indicating their significant role in plastic waste disposal (He et al. 2019). The influx of plastic waste into landfills stems primarily from diverse sources, including municipal waste collection and unauthorized dumping (Barnes et al. 2009). This plastic waste encompasses a wide array of materials, ranging from everyday packaging to industrial-grade plastics, forming a substantial component of landfill solid waste (Barnes et al. 2009). Plastic waste encounters a complicated sequence of changes once it is placed in landfills, which are influenced by environmental elements such as moisture, exposure to sunlight, and the activity of bacteria (Barnes et al. 2009). These procedures cause plastics to deteriorate, which eventually produces MPs. Plastic components' chains of polymers are broken down by chemical processes such as photooxidation and hydrolysis, which makes them more prone to breaking into smaller fragments (Zhao and Lou 2016). Furthermore, the physical degradation of plastics is facilitated by the mechanical forces used during waste compression and the heterogeneous character of landfill waste, which amplifies the creation of MPs, in landfill settings (Su et al. 2019).

The production and trade of textiles represent significant contributors to the primary MP pollution, as highlighted by recent research (Liu et al. 2020). The textile industry produces a whopping 42 Mt of artificial fibers annually, of which MPs polyester makes up an astounding 80% (Xu et al. 2020). Remarkably, as part of textile manufacturing waste, a higher percentage of microsynthetic fibers than natural fibers end up in landfills (Xu et al. 2020). Furthermore, tires are a principal source of MPs pollution because, during the manufacturing process, they generate large amounts of powdered form and debris that range in size from nanometers in size to micrometers; the majority of these particles are then dumped in landfills (Galafassi et al. 2019). According to Su et al. (2019), the major source of MPs in landfills, notably second-generation MPs, is the disintegration of plastic debris, as evidenced by the irregular forms and fragmented structures seen in waste.

In addition to the primary MPs generated within landfills, these sites also receive direct inputs of primary and secondary MPs from external sources. For instance, disposed sludge containing MPs and other contaminants directly contributes to the MPs load in landfills (Yang et al. 2015). Moreover, waste in landfills has a far greater quantity of MPs than leachate, thus highlighting landfills' importance as major sinks for MPs (Puthcharoen and Leungprasert 2019). Landfills operate as sinks for MPs, but they also supply the surroundings with MPs. MPs from landfills have been found in surrounding waterways, sediments, and aquatic life, demonstrating how waste habitats are linked to larger ecosystems (Carr et al. 2016). This emphasizes the necessity of all-encompassing management plans to stop MPs contamination before it starts and lessen its negative effects on the ecosystem.

13.2.2 Sources and Generation of Microplastics in Landfill Leachate

Landfill leachate primarily contains MPs originating from the dumped waste and treatment plant of wastewater residuals, including slurry and fat, oil, and grease

(FOG) (Golwala et al. 2021). Kabir et al. (2023) state that secondary MPs are produced through various mechanical processes when plastic waste is deposited in landfills. Figure 13.1 illustrates the generation of secondary MPs from macroplastics in landfills. Additionally, wastewater is a major source of MPs because, during purification, MPs become lodged in solid debris and FOG (Sun et al. 2019). Thus, the disposal of FOG and sludge formed from sewage in landfills may contribute to the elevated concentration of MPs in the leachate (Iyare et al. 2020).

In landfills, trash passes through many treatment phases, such as methanogenesis, acid production and hydrolysis, aerobic biological degradation, a shift to anaerobic circumstances, and ultimate maturation and stability (US EPA 2006). Every one of these phases quickens the decomposition of plastics, producing secondary MPs (Hou et al. 2021). The breakdown and fragmentation of plastic garbage in landfills happen more quickly throughout each of these phases. Temperatures rise from 35 °C to above 70 °C during the aerobic breakdown phase due to an increase in microbial activity (Silva et al. 2021). Improved temperature, moisture content, and oxygen consumption in the landfill promote the production of hydrogen peroxide groups in the polymeric chain of carbon by aerobic microbial activity (Chamas et al. 2020). According to Yousif et al. (2023), the oxidative degradation of plastics sets off a series of processes that start with the creation of hydroperoxide groups. As a result of oxidative deterioration, cracks and fractures appear on plastic surfaces (Hou et al. 2021). These cracked plastics disintegrate under physical pressures, for instance, mechanical shear and heat stress, generating fragments and MPs (Shen et al. 2022). Arhant et al. (2019) reported that during the anaerobic acidification phase, hydrolysis accelerates the breakdown and disintegration of plastics, leading to an increased formation of secondary MPs.

Moreover, microbeads in pharmaceuticals and personal care products (PPCPs) such shampoo products, bath gels, cosmetics, sunblock, face masks, and makeup are among the numerous human activities linked to mass-produced MPs/NPs that

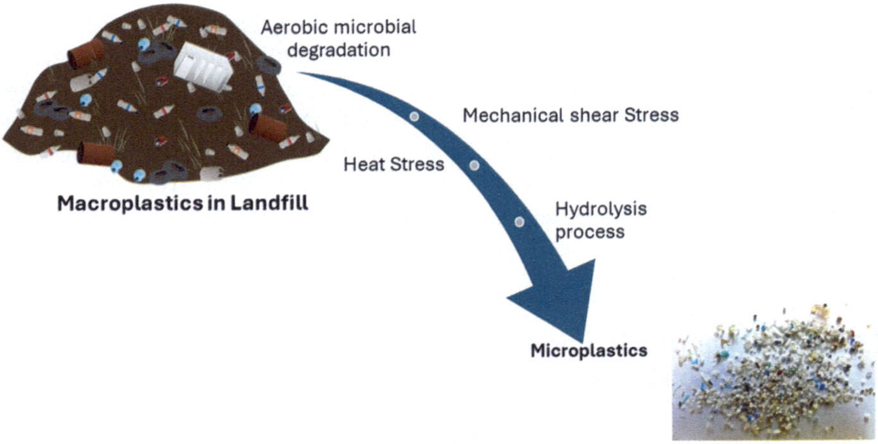

Fig. 13.1 Generation of secondary MPs from microplastic in landfills

contribute to the buildup of MPs in landfills. In the study by Kabir et al. (2023), primary MPs are also present in leachate from landfills via waste generated by businesses and facilities that manage these materials. The most prevalent MPs discovered in untreated leachate, according to Sun et al. (2021), were polyethylene and polypropylene, both of which are extensively used in consumer goods, including packaging and containers, as well as construction.

13.3 Occurrence and Characterization of Microplastics in Landfill and Leachate

13.3.1 Occurrence of Microplastics in Landfill and Leachate

In industrialized and growing nations worldwide, the most popular methods for getting rid of solid waste are landfills and open-pit dumping. Each year, open dumpsites receive vast quantities of municipal waste (Chamanee et al. 2023). Even while landfills are sometimes thought of as part of the infrastructure for managing garbage, improper management and uncontrolled treatment may greatly increase the danger of environmental contamination. Around 90% of landfills lack infrastructure or function as open dumping sites, reflecting a common approach to management of waste, especially in countries with low to middle incomes, particularly in Asia (Kumarathilaka et al. 2017). Approximately 20% of MSW, or 42% of the global plastic waste output, is deposited in landfills. Plastic products make up a significant amount of garbage at these sites (Wan et al. 2022).

Landfilling represents a prolonged process spanning decades, characterized by intricate and dynamic variations in system circumstances such as pH, oxygen levels, and temperature (Su et al. 2019). Within landfills, plastics undergo degradation through exposure to various treatment stages during landfill stabilization, resulting in the leachate and landfill sites' many biological, chemical, and physical reactions that produce MPs (Kabir et al. 2023; Singh et al. 2023). The plastics accumulated in open dump sites or landfills experience harsh conditions, including anaerobic environments, temperature variations, pH shifts, physical stress, sunlight exposure, thermal degradation, microbial degradation, oxidation reactions, and mechanical scratching. Together, these actions help plastic waste in landfills weather from macro to micro/nano scales, leading to MPs formation (Dehal et al. 2023; Silva et al. 2021; He et al. 2019). Consequently, landfills are perceived as significant repositories of MPs, encompassing minute plastic granules from cosmetics and small fragments resulting from macroplastic breakdown (Su et al. 2019). These minute particles or fibers have the potential to disperse from landfills into surrounding environments via airborne pathways, posing increased risks in instances of inadequate management (Anik et al. 2021; Rahman et al. 2024a).

One major environmental threat is leachate, a liquid byproduct produced when rainfall interacts with degraded waste materials in landfills (Wijekoon et al. 2022). Following rainfall events, landfills can produce substantial volumes of leachate, which may follow various pathways, such as infiltration into the ground, surface runoff, evapotranspiration, or intrusion into groundwater (Abdel-Shafy et al. 2024). The amount and component of leachate mostly depend on the rates at which water seeps into landfills, as well as the disintegration and deterioration processes that take place there (Costa et al. 2019). Recognized as a potent source of contamination, landfill leachate detrimentally impacts both surface and groundwater quality (Wang et al. 2023). Plaid leachate is a significant conduit for environmental pollutants, with an array of pollutants encompassing inorganic and organic substances derived from soil, bits of microplastic, groundwater, freshwater bodies, and landfills (Silva et al. 2021). Furthermore, the presence of MPs in leachate may contribute to the spread of other contaminants, worsening the effects on the habitats that receive the discharge of leachate (Rahman et al. 2023). Nevertheless, MPs abundance within landfills and leachate exhibits considerable variability attributable to diverse sampling methodologies, analytical techniques, and leachate treatment procedures (Kabir et al. 2023). Additionally, the complex composition of waste within landfills significantly influences the concentration and types of MPs present in leachate. Notably, considerable variation in MP concentrations has also been observed in wastewater (Su et al. 2019). Reported MP concentrations in raw and treated wastewater range from 1 to 3169 items/L and from 0.0007 to 125 items/L, respectively (Gatidou et al. 2019). Interestingly, MPs abundance in raw wastewater appears lower than in treated landfill leachate. This discrepancy may be because smaller MPs and nanoplastics are more common in landfill strata than in effluent due to the breakdown of plastic and MPs. However, such smaller particles may remain undetected by current methodologies, leading to potential underestimation of MP concentrations based solely on number counts (Sun et al. 2021).

On the contrary, landfills' age or operational status can impact the abundances of MP detected in landfill leachate. For example, Su et al. (2019) observed that younger landfills exhibited higher average abundances of MPs (8–10 items/L) compared to older landfills, where the number of MPs in leachate was lower (4 items/L). The increasing use of plastic items over the years may cause this tendency. Indeed, global plastic production has surged by 20-fold over the last five decades, resulting in an estimated production of 9200 Mt of plastic, a significant portion of which, exceeding 6900 Mt, has ended up in landfills or has contributed to environmental pollution (Walker and Fequet 2023). Plastic trash makes up a substantial amount of solid waste in landfills—between 2.95% and 21.76%—which means plastic waste accumulates more in functioning landfills than in aged or inactive ones (Zhou et al. 2014). As plastic waste accumulates within landfills, it undergoes fragmentation over time, giving rise to secondary MPs that subsequently permeate into leachate. However, microbial degradation of polymers within landfills could also contribute to the reduced MP concentration observed in leachate from aged landfills (Park and Kim 2019). Furthermore, the limited detection of smaller particles in recent studies may partially explain this phenomenon. Due to analytical constraints, most studies analyzing MPs in leachate have focused on particle sizes ranging from 50 to 5000 μm, with smaller MPs or NPs often overlooked and undetected within older landfill sites (Kabir et al. 2023).

Research indicates spatial disparities in the abundance of MP found in the landfill leachate, potentially linked to regional disparities in plastic production (Silva et al. 2021). For instance, a study focusing on European nations identified Germany as Europe's highest plastic waste-generating country, producing an annual total of around 14.5 Mt of plastic waste (Plastic Pollution by Country 2024 n.d.). In contrast, China generates around 59.1 Mt of plastic waste annually (Kabir et al. 2023). Measurements of MP concentrations in China ranged widely from 0.4 to 235.4 items/L (He et al. 2019; Su et al. 2019; Sun et al. 2021). In India, investigations at the Chennai landfill revealed MP abundances varying from 2 to 80 items/L (Kabir et al. 2023; Bharath et al. 2021). Notably, at a major landfill site in Bangladesh, MPs identified in leachate, freshwater, and underground water with average quantities of 270 ± 13, 2090 ± 100, and 647 ± 22 pieces/L, respectively (Rahman et al. 2024b). Comparatively, MP concentration ranges in leachate from Nordic countries, including Iceland, Norway, and Finland, were notably lower, ranging from 0 to 4.5 items/L (Van Praagh et al. 2019). Moreover, Southeast European nations like Serbia, Bosnia, and Herzegovina have an average MP concentration that is three times greater compared to that of Nordic nations (Kabir et al. 2023). This discrepancy is largely attributed to higher levels of plastic waste generation in developing nations like China, India, Bangladesh, and Southeast European countries, where comprehensive waste management practices, including proper waste sorting during collection and landfill and leachate management, may not be consistently implemented as in developed countries. The abundance of MPs found in different landfills throughout the globe is shown in Table 13.1.

13.3.2 Identification and Characterization of Microplastics in Landfill and Leachate

Several scientific investigations have documented the occurrence of MPs in landfill and leachate exhibiting a range of shades, including red, orange, yellow, brown, gray, white, blue, and green (Table 13.1). The coloring of MPs is influenced by the colors of their parent materials and the duration of their persistence, as noted by Bharath et al. (2021). Transparent MPs often come from plastic disposables such as bags, plates, and cups or from companies that use them as raw materials (Dey et al. 2021). On the other hand, colored MPs are susceptible to weathering, even though their breakdown is often caused by larger, colored plastic objects. (Liu et al. 2020). Light-colored MPs occur in landfill leachate—such as translucent, white, or yellowish particles—indicating prolonged deterioration resulting from weathering processes (Corcoran 2020). The size of MPs is typically determined by their greatest length (He et al. 2019). Across various regions worldwide, the most common size range of MPs detected in landfill leachate falls between 100 and 1000 μm (He et al. 2019). However, the size distribution of MPs in leachate may be quite diverse, spanning from 20 μm to 5 mm. The method employed for detection has a substantial impact on the range of sizes that are detected (Obbard 2018).

Table 13.1 Occurrences, characterization, and identification techniques of MPs in landfill and leachate

Country	Sampling location	Abundances of MPs	Characterization of MPs				Identification technique	References
			Size	Shape	Color	Polymer type		
Bangladesh	Amin Bazar Landfill site, Savar	270 ± 13, 2090 ± 100, and 647 ± 22 MP/L	0.1–0.5 mm	Fragment	Blue, red, and transparent	LDPE, HDPE, PETE, Polypropylene (PP), Nitrile, Nylon	FTIR	Rahman et al. (2024a, b)
India	Landfill, Chennai	2–80 items/L	<500 μm	Nylon (70%), pellets (18%), foam (6%), fragments (3%), fibers/PVC (2%), and polythene (1%)	White (38%), black (27%), green (8%), red (18%), blue (6%), and yellow (2%)	Polyethylene (PE) Polystyrene (PS)	Stereo microscope ATR-FTIR SEM-EDX	Bharath et al. (2021)
India	Jawahar Nagar landfill, Hyderabad	9–21 items/L		Fiber (83%), followed by fragment (11%), film (3%), and foam (2%)	Yellow (35%), transparent (16%), purple (15%), blue (11%), pink (8%), white (6%), black (3%), green (2%), and orange (1%)	LDPE, PP, PET, CA, and Nitrile	Stereo microscope ATR-FTIR SEM-EDX	Sekar and Sundaram (2023)

China	Landfills at Shanghai, Wuxi, Suzhou, Changzhou	0.42–24.58 items/L	100–1000 µm	Lines, flakes, fragments, pellets, and foams		PE and PP	ATR-FTIR Micro-FTIR	He et al. (2019)
China	Landfill leachate at Suzhou	235.4 ± 17.1 items/L	10–500 µm	Bead, fiber, film, and fragment	Transparent or yellowish, red, pink, purple, black, blue, and brown	PE, PP, PA, PVA, PVB, PMP, PAA	Micro-Raman spectrometer Olympus BXFM optical microscope	Sun et al. (2021)
China	Untreated and treated leachates from landfill, Hong Kong	49.0 ± 24.3 to 507.6 ± 37.3 items/L	100–500 µm	Fragments or filaments	Transparent or yellow	PET, PE	Stereomicroscope µFTIR spectroscopy	Qin et al. (2024)
China	Waste disposal center in Shanghai	(129 ± 54) to (1288 ± 184) MP items/L	20–100 µm	Film, fiber, fragment, flake, floc	Translucent, light, dark	PE and PP were the most frequently present polymer types	FTIR microscope	Zhang et al. (2024)
Indonesia	Galuga landfills, Bogor City	80,640 ± 604.80 items/day	500–1000 µm	Fiber, fragment, granule, foam		PE, PP, PS, PET, polyester, and cellophane	FTIR µFTIR spectroscopy	Cordova and Riani (2021)

(continued)

Table 13.1 (continued)

Country	Sampling location	Abundances of MPs	Characterization of MPs				Identification technique	References
			Size	Shape	Color	Polymer type		
Indonesia	Randegan landfill, Mojokerto City	9.00 ± 0.85 particles/L	350–<1000 μm (64.44%), 100–350 μm (31.11%), and 1000–5000 μm (4.45%)	Fiber (64.44%), fragment (28.89%), and film (6.67%)	Black color (53.33%)		PTFE membrane Dissecting microscope FTIR spectrometer	Trihadiningrum et al. (2023)
France	Coastal landfill at Dollemard Coast	244 items/L	<20–500 μm	Films, fragments, microbeads, or fibers	Dark, red, yellow, blue, white, or green	PS = 40%, PA = 18% and PET = 14% were the most detected polymers	Stereomicroscope μ-Raman spectroscopy	Kazour et al. (2019)
Finland	Turku, Salo, Lahti	0.16–1.10 items/L	50–5000 μm			PE, PET, PS, PP, PVC, PA	FT-IR spectroscopy	van Praagh et al. (2018)
Finland	Helsinki	0.080 and 0.261 fibers/L, 0.002 and 0.017 items/L	>1 mm				Stereomicroscope FTIR	Kilponen (2016)
Thailand	Landfill at Takhian Tia	20.90 ± 4.96 Items/1 kg dryweight		Granules 47%, films 28%, fibers 17%, irregular 8%		PE, PP, PET	FTIR	Puthcharoen and Leungprasert (2019)

Iran	Landfill site at Bushehr port	79.2 items/L	0.45–5000 μm	Fragment, fiber, film, angular	Red, brown, white, transparent, black, blue, green	Nylon (80%), PP (14%), PS (4%), PC (2%)	Micro-Raman spectroscopy Binocular microscope	Mohammadi et al. (2022)
Iran	Hamadan landfill	16,940 to 76,513 items/kg dry soil	0.45–25 μm (59.32%), 25 μm to 5 mm (40.68%)	Fiber most abundant (71%)	Black, colorless/ white, red, blue, green	LDPE (35%) and HDPE (33%) polymers	Stereo microscope (ATR-FTIR) spectroscopy	Rahmani et al. (2023)
Serbia	Jagodina Bijeljina Bosnia Herzegovina Serbia Novi Sad	0.64–2.16 mg/L	150 μm	Fiber, fragment, foam			Stereomicroscope	Narevski et al. (2021)

Various morphologies of MPs, including fibers, fragments, films, granulates, circles, crumbs, foams, lines, pellets, and beads, have been identified in landfill leachate. Among these forms, foams and fragments are notably prevalent in landfill leachate globally (Table 13.1), as they can easily traverse through waste layers and enter leachate during precipitation events (Zhang et al. 2021). The distinct shapes of MPs serve as indicators of their origins (Metz et al. 2020). For instance, films, commonly derived from plastic sacks and other materials for packaging, are prevalent in landfill leachate. Granules and spheres often originate from items such as microbeads, bottles, and food boxes (Puthcharoen and Leungprasert 2019). Resin pellets are primarily discovered in the vicinity of industrial zones, whereas fragments and foams are prevalent in areas near fishing ports (Antunes et al. 2018). Within household wastewater and treatment facilities, textile fibers from laundered garments and microbeads from personal care items comprise a substantial proportion of the detected MPs (Tang et al. 2020; Ziajahromi et al. 2017). Hence, it is feasible to deduce their probable origins by examining the types of microplastic shapes present in landfill leachate.

Several landfill leachates have been found to contain over 25 different types of polymers. Reviews of the literature indicate that HDPE, LDPE, PP, PVC, polystyrene (PS), and polyethylene terephthalate (PET) are among the plastic polymers most frequently detected in landfill leachate globally. Recognizing the polymeric composition can assist in pinpointing the sources and origins of MPs. Polyethylene (PE) emerges as the most extensively used polymer globally, with applications spanning various products, including personal care items and packaging (Montazer et al. 2020). Rigid plastic products and furniture, including carpets, fishing gear, and textile floor coverings, are made from PP (Genis 2017). Meanwhile, PS is predominantly used in the assembly and packing of single-use products like cutlery, containers for meals, and cups for coffee (Letcher 2020). PVC finds widespread usage across multiple industries owing to its exceptional flexibility in construction, waterproofing, apparel, toys, and sports equipment (Turner and Filella 2021).

Determining the physical (color, shape, and size) and chemical (polymeric chemistry) characteristics of MPs from landfill leachate is essential for thoroughly characterizing these particles. However, employing a single identification technique for fully characterizing MP complicated matrices, like leachate from landfills, poses challenges. Consequently, a combination of analytical methods is often employed. Here, Table 13.1 summarizes MPs' characterization and existing identification techniques based on different particle sizes in the landfills. Visual identification with the naked eye provides a straightforward means of physical characterization, particularly for MPs between the dimensions of 2 and 5 mm (Heo et al. 2013). However, there are restrictions on this approach's size and susceptibility to inaccuracies, making identifying MPs within the intricate environmental context of landfill leachate challenging. Stereomicroscopes are commonly employed in leachate analysis to enumerate and classify physical attributes such as microplastic size, shape, color, and surface characteristics. However, errors can also be committed using stereoscopic microscopes, particularly when working with materials with many biological stuffs in them. Hidalgo-Ruz et al. (2012) reported error rates of up to 70% when solely relying on a stereomicroscope for identification, with error rates escalating as particle size decreases. Fibrous MPs or transparent particles less than 100 μm pose challenges for identification with stereoscopic microscopes (Song et al. 2015), making the reliance on stereomicroscopes alone unreliable for studying leachate, given the prevalence of such particles in landfills. Scanning electron microscopy (SEM) may be used to improve the analysis of complex environmental materials by providing high-resolution pictures of microplastic particles. SEM makes it easier to distinguish MPs from organic particles and characterize their surface appearance. The number of samples that may be analyzed is nevertheless limited by the cost, duration, and labor required for sample preparation and inspection of SEM-based techniques (Shim et al. 2017; Dehghani et al. 2017).

The main method for chemically characterizing MPs in landfill leachate is Fourier transform infrared (FTIR) spectroscopy. FTIR spectroscopy makes it possible to identify certain polymer varieties and detect plastic particles by providing access to a large polymeric spectra database. FTIR spectroscopy is used in several modes based on the size range that is being studied. He et al. (2019), for instance, used Micro-FTIR (μ-FTIR) for particles smaller than 1 mm and Attenuated Total Reflectance (ATR) mode (ATR-FTIR) for MPs bigger than 1 mm (Rolsky et al. 2020). However, Raman spectroscopy provides an alternate method for analyzing very small MPs (less than 1 μm), determining the sample's polymer composition by looking at the polarity of the chemical bonds inside the particle. Through the use of a laser beam to illuminate the particle and analyze the frequencies of backscattered light, this method generates a distinct spectrum for each polymer based on the makeup of its molecules and constituent atoms (Löder and Gerdts 2015; Schymanski et al. 2018). Nonetheless, the precision of results can be affected by possible interference from extraneous wavelengths and fluorescent released by colorants and dyes (Lenz et al. 2015). Elements smaller than 20 μm can be detected using Micro-Raman spectroscopy (Schymanski et al. 2018). For instance, Kazour et al. (2019) employed Raman spectroscopy to identify MPs ranging from 20 to 500 μm in surface water samples near a coastal landfill.

13.4 Fate and Transport of Microplastics in Landfill Leachate

MPs from landfill leachate can enter the surrounding ecosystem through various natural processes, such as wind dispersion, flooding, leaching, and runoff, as depicted in Fig. 13.2 (Wojnowska-Baryła et al. 2022). Physical barriers are constructed to stop such leaks in hygienic and properly managed landfills, greatly lowering the danger of leachate pollution. On the other hand, there are no such barriers in open landfills. Therefore, leachate can readily discharge into the environment. Leachate eventually seeps into the saturated groundwater aquifer through the soil, forming a plume that descends into the unsaturated zone. Leachate's high salinity and density cause it to sink into groundwater first, restricting transverse or lateral mixing (Vithanage et al. 2014). Groundwater contamination by MPs primarily originates from landfills, with buried plastics identified as the source of 90% of detected

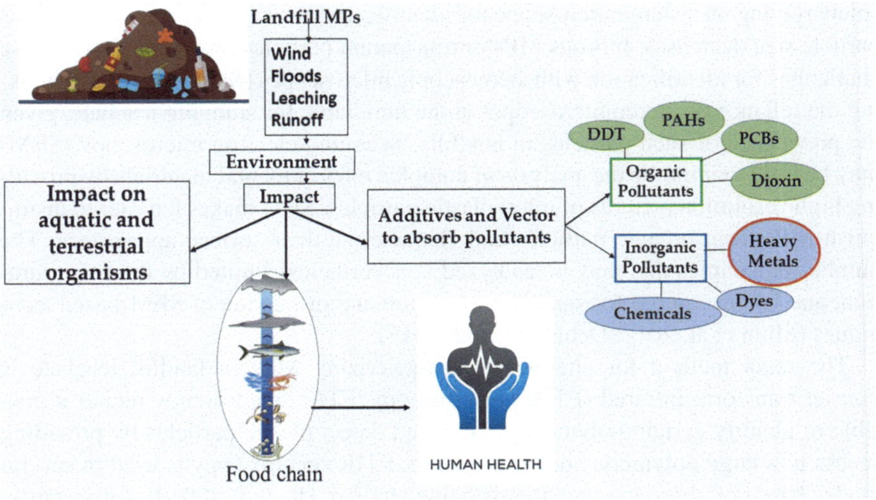

Fig. 13.2 Fate and effects of landfill MPs

MPs (Singh and Bhagwat 2022; Bharath et al. 2021). MPs' ecological effect is mostly determined by the amount and toxicity of MPs found in leachate as well as the penetrability and composition of the geological layers that control both vertical as well as horizontal propagation (Wali et al. 2021). Through alternating dry and wet periods, leachate swiftly infiltrates nearby groundwater, leading to the contamination of surface water reservoirs connected to groundwater reservoirs with MPs (Upadhyay and Bajpai 2021). Furthermore, in order to facilitate the spread of pollutants across aquatic systems, leachate accumulating at landfill peripheries is conveyed by the natural land slope and drawn directly into nearby water bodies through surface and subsurface drainage channels. Because of increased surface runoff during rainy seasons, pollution levels rise (Abdul Maulud et al. 2021). Leachate from landfills stands as a notable contributor of MPs in freshwater ecosystems, with considerable daily emissions expected into these environments (Golwala et al. 2021; Cordova and Riani 2021). The characteristics of MPs, outflow patterns, and the dumpsite's closeness to aquatic environments greatly influence how MPs migrate from leachate into water systems (Priya et al. 2022).

According to Bradney et al. (2019), MPs have the capacity to gather contaminants from nearby sediment or aquatic habitats. Through processes including hydrophobic partitioning, bonding of hydrogen, π–π connections, static attraction, and pore-filling, their hydrophobic nature and large surface area lead to considerable sorption affinities with a variety of organic contaminants (Hartmann et al. 2017). In accordance with recent studies, MPs are important for the transport of heavy metals such as Cr, Cu, Zn, Cd, As, and Pb, with their mobility affected by the physicochemical characteristics of these metals (Gao et al. 2021; Fig. 13.2). The scope of MPs to bind with heavy metals is influenced by a number of variables, including temperature, salinity, pH, and the amount of dissolved organic matter in aqueous solutions (Sewwandi et al. 2022). Compared to bigger plastic particles, MPs have been found to have a higher propensity to bond with organic contaminants (Tang et al. 2020). Several investigations have demonstrated that a wide range of organic pollutants can bind to MPs, including dioxin-like substances, hydrophilic organic compounds, PCBs, PAHs, medicines, polybrominated diphenyl ethers (PBDEs), and antibiotics (Zhang et al. 2019; Godoy et al. 2019; Wang et al. 2020; Fig. 13.2). Furthermore, metal-driven co-selection of multi-antibiotic-resistant microorganisms has been linked to MPs (Imran et al. 2019). The identification of unique patterns of antibiotic release associated with landfill age by analyzing the presence of heavy metals, antibiotics, and antimicrobial resistance genes (ARGs) in leachate and landfill trash from sites of different ages (Wu et al. 2017). In laboratory experiments, Su et al. (2019) demonstrated MPs' function in landfill leachate as vectors for antibiotic-resistant genes, showing selective enrichment ratios ranging from 5.7 to 1000.0 folds, with increased enrichment potential observed over time. Moreover, MPs assist in creating a unique habitat called a plastisphere, in which microorganisms live on the surface of plastics and may increase the concentration of pathogens and ARGs in adjacent landfill leachate (Shen et al. 2022).

Furthermore, according to Song et al. (2023), MPs have been observed to strengthen the ecological risks linked with antibiotics by magnifying the collective impacts of copper, tetracyclines, and sulfonamides. Future research ought to examine how MPs and antibiotics interact, as well as how these interactions affect the environment by examining things like the formation and colonization of ARGs on MP surfaces (Song et al. 2023). Additives used to increase plastic flexibility during production, such as polybrominated diphenyl ethers, bisphenol A (BPA), phthalates, and nonylphenol, can leak out of MPs when polymers break down over time (Campanale et al. 2020). These chemicals, many of which operate as endocrine disruptors, have the potential to interact harmfully with existing contaminants due to their hydrophobic nature and vast surface area (Ullah et al. 2023). Bisphenol A (BPA) is frequently found in leachate from municipal solid waste (MSW) dumps; values can range from 1.3 to 17,200 μg/L (Chamanee et al. 2023). In a similar vein, BPA concentrations in landfill leachate ranged from 0.70 to 2.72 mg/L, according to Narevski et al. (2021). However, the function of MPs removed from leachate as harmful pollutant transporters has not yet been studied.

13.5 Environmental Implications of Microplastics in Landfill Leachate

When MPs get into other parts of the ecosystem, they can seriously harm living things. Studies on the toxicity of MPs are still in their infancy, despite continuous study. Nonetheless, there are two types of harmful impacts that MPs exposure can have on organisms: chemical and physical (Wojnowska-Baryła et al. 2022). Particle size, shape, and concentration all have an impact on the physical consequences; on the other hand, the dangerous compounds produced by MPs are responsible for the chemical effects. MPs can contain additives, polymers, raw materials sourced from plastics, and environmental chemicals (Wojnowska-Baryła et al. 2022). Hahladakis et al. (2018) note that these substances include flame retardants, additives, plasticizing agents, antioxidants, UV stabilizers, oils, dyes, and flame-resistant compounds. Many of these substances have hazardous qualities and can contaminate the environment.

MPs possess a high surface-area-to-volume ratio, rendering them effective sorbents for organic pollutants and toxic metals (Fig. 13.2). These adsorbed pollutants can enter organisms and, under favorable conditions, be released, impacting growth, development, reproduction, and even causing death (Huang et al. 2021). MPs can induce oxidative stress reactions, affect pigment production, enzyme activities, endocrine and metabolic functions, and exhibit cytotoxic, neurotoxic, and genotoxic effects (Huang et al. 2021). Furthermore, these impacts can accumulate and amplify through the food chain from individuals to populations, communities, and ecosystems, exacerbating environmental degradation (Ma et al. 2020).

Aquatic animals can uptake MPs, with differences observed between species and life stages. Additionally, various shapes and surface characteristics of MPs can cause different levels of mechanical damage to the intestine (Vroom et al. 2017). In terrestrial animals, MPs have been found to accumulate in earthworm intestines, leading to reduced growth rates and increased mortality rates (Wang et al. 2019). MPs can endanger human health by building up along the food chain. Additionally, MPs intake via the atmosphere, particularly through dust inhalation, may expose humans to particles, with infants and workers being particularly vulnerable (Huang et al. 2020).

Additionally, in the plastisphere bacteria grow on the plastic surface (Shen et al. 2022). Pathogens and ARGs in landfill leachate may be enhanced as a result. Additionally, by amplifying the combined effects of certain chemicals, MPs may increase the environmental dangers linked to antibiotics (Song et al. 2023). Future research needs to concentrate on understanding the interactions between antibiotics and MPs and the growth and colonization of ARGs on MP surface (Song et al. 2023).

13.6 Removal Techniques of Microplastics in Landfills and Leachate

In landfill environments, degraded plastics transition into MPs via landfill leachate, which then disperses into soil, aquatic ecosystems, and organisms through trophic transfer (Santana et al. 2018). Removing MPs from landfill leachate poses technical challenges owing to their diverse characteristics and tiny size (Xu et al. 2020). MP might be removed using a variety of methods, each having pros and cons of its own, such as sequenced batch reactors, membrane bioreactors, oxidation, coagulation/flocculation, activated carbon, elimination, vaporization, filtering, aeration, and reverse osmosis process (Table 13.2). In some cases, due to the heterogeneity in size, advanced filtration techniques may also struggle to capture MPs consistently. Leachate treatment methods may reduce MP concentrations by 3–100% depending on the technique used (Kabir et al. 2023). Research shows that compared to untreated

Table 13.2 MPs removal mechanisms, working conditions, limitations, and efficiencies of various current approaches

Type	Approach	Mechanism	Working condition	Limitation	Particle size	Removal rate	References
Physical	Membrane filtration	Uses a semipermeable membrane to filter water and isolate MPs by blocking certain particles according to size	• Steady flow conditions with proper pretreatment to prevent clogging of the membrane pores	• Fouling can reduce filtration efficiency over time • Requires regular maintenance or replacement of membranes	Nanometers to micrometers	~90%	Ganiyu et al. (2015)
	Density-based separation using salt/ethanol	Relies on differences in the density of particles to separate them from a liquid stream	• Controlled flow rates and settling times • Sufficient residence time for particles	• Variability in the density of MPs leading to incomplete separation or loss of target particles	Depending on how the system is set up and runs, the particles can range in size from bigger silt and debris to microscopic MPs	It has shown efficacy in enhancing MP extraction compared to alternative digestion methods	Li et al. (2019), Herrera et al. (2018)
	Dissolved air floatation	Involves the introduction of fine air bubbles into a liquid stream, which attach to particles, reducing their effective density and causing them to float to the surface for removal	• Adequate contact time between the air bubbles and particles • Systems with stable influent characteristics • Proper chemical dosing for coagulation and flocculation	• Treating highly turbid or heavily contaminated water streams	Tens of micrometers to millimeters	>90%	Hou et al. (2021), Rao and Leja (2013)

(continued)

Table 13.2 (continued)

Type	Approach	Mechanism	Working condition	Limitation	Particle size	Removal rate	References
	Sand bed filtration	Relies on the physical straining and adsorption of particles as water passes through a bed of sand	• Well-graded sand media • Controlled flow rates • Regular backwashing to prevent clogging	• May not effectively remove very fine particles or dissolved contaminants	Tens of micrometers to millimeters in size	70–99%, based on the operational environment and particle size diversity	Enfrin et al. (2019), van Praagh et al. (2018)
Chemical	Electro-coagulation	Metal hydroxide coagulants, or sludge blankets for MP adhesion, are created in the water stream when metal ions (Fe^{2+} and Al^{3+}) released from electrodes react with hydroxide	• Appropriate pH • Conductivity • Temperature conditions • Careful control of current density and electrode configuration	• High energy consumption • Electrode fouling • The need for frequent maintenance	Sub-micron to tens of micrometers in size	90–90.24% efficacy of elimination	Perren et al. (2018)
	Advanced oxidation process	Involve the formation of extremely reactive hydroxyl radicals (OH) or another oxidizing agent through chemical reactions that degrade MPs present in the wastewater	• Efficient mixing and contact time between the oxidants and contaminants are crucial for optimal performance	• High energy consumption • Specialized equipment • Form potentially harmful by-products	Nanometers to millimeters	Chlorpyrifos, dieldrin, and diuron were among the pesticides discovered in raw landfill leachates, and their clearance rates ranged from 98% to 99%	Miao et al. (2020), Jiang et al. (2021)

Biological	Bio-flocculation by activated sludge	Involves the aggregation and capture of microplastic particles suspended in landfill leachate by microbial flocs present in activated sludge	• Optimal temperature and pH • Nutrient levels, to support microbial activity and bio-flocculant production	• Availability of suitable microbial species and nutrients • Variations in leachate composition and characteristics	Micrometers to millimeters in size	~70–80%	Cheng et al. (2021)
	Sequencing batch reactor (SBR)	Remove microplastics (MPs) from landfill leachate through a cyclic process involving alternating stages of filling, reaction, settling, and decanting	• Proper mixing • Sufficient settling time	• Fluctuations in leachate composition • Variations in influent characteristics • The presence of inhibitory substances	Micrometers to millimeters	>80–90%	Zhang et al. (2021)
	Membrane bioreactor (MBR)	Utilize a combination of biological treatment and membrane filtration to remove microplastics (MPs) from landfill leachate	• Stable influent characteristics • Proper nutrient levels • Adequate aeration	• Membrane fouling	Micrometers to millimeters	99.9% removal efficiency	Zhang et al. (2021), Anik et al. (2021)

waste, MP abundances following sand bedding filters, SRB, and engineered soil filtering are often lower. This emphasizes the need for sophisticated technology to efficiently extract MPs from landfill leachate (van Praagh et al. 2018; Wang et al. 2016; Dolar et al. 2016; Kattel et al. 2016).

Physical techniques for MP removal encompass various approaches characterized by source materials, effluent composition, facility capacity, and separation principles by considering factors such as size, density, hydrophobicity (Poerio et al. 2019; Hernandez et al. 2017). Divergent filtration methods including membrane, sand, gravity, ultrafiltration, and reverse osmosis are employed to eliminate MPs, however, sand bed, gravity, and artificial soil filtration may not capture MPs effectively (Sun et al. 2019). Integration with other approaches like membrane bioreactors has enhanced MP separation by achieving a remarkable 99.9% removal rate from wastewater. Moreover, density-based separation techniques leverage solutions like salt or ethanol showing efficacy in enhancing MP extraction compared to alternative digestion methods (Li et al. 2019; Herrera et al. 2018). Contrariwise, hydrophobicity-driven separation utilizes interfacial tensions in flotation techniques. While flotation with oil is common, challenges arise from potential transformations during oil removal. On the contrary, dissolved air flotation generates smaller bubbles through cavitation, producing micro- or nanobubbles, which enhance the chances of removing the light MPs, making it an effective method (Hou et al. 2021; Rao and Leja 2013; Zimmerman et al. 2008). Soil and sand bed filtration demonstrate effective MP removal, which is influenced by pore size, with rapid sand filtration utilizing layers of anthracite, silica, and gravel that can remove up to 99% suspended solids (Enfrin et al. 2019; van Praagh et al. 2018).

Chemical removal procedures such as coagulation with materials such as polyacrylamide (PAM), Fe- and Al-based salts, and $FeCl_3{\cdot}6H_2O$ effectively capture MPs by causing them to form bigger clusters (Anik et al. 2021). However, MP's interactions with these drugs may reduce therapy efficacy. Electrocoagulation uses metal ions (Fe^{2+} and Al^{3+}) to generate metal hydroxide coagulants, which adhere to sludge blankets and effectively remove about 90% of MPs in wastewater treatment facilities. Another new technology is the advanced oxidation process (AOP), which employs hydroxyl and sulfate-free radicals to mineralize macropollutants in landfill leachate (Li et al. 2022). Some processes of AOP include ozone-based, hydrogen peroxide-based, Fenton and Fenton-like photo-Fenton, and photocatalysis processes. Electro Fenton-like technology and photocatalytic degradation methods utilizing materials like TiO_2 and BiOCl-X have been shown to effectively degrade MPs by altering their surface morphology and molecular weight through oxidation (Miao et al. 2020; Jiang et al. 2021). While constructive wetland (CW) alone may not be highly effect in capturing the MPs, the use of reacting mediums, such as powdered iron or zero-valent iron (ZVI), can improve the removal of hydrophobic organic micropollutant's (OMP's) inbuilt CW (Carranza-Diaz et al. 2014; Tang et al. 2015; Huang et al. 2017; Chiemchaisri et al. 2015). This integrated method, which uses these reactive media in CWs, involves physical processes like adsorption and chemical reactions.

MPs in landfill leachate have been successfully reduced by biological treatment techniques, such as bio-flocculation by activated sludge and degradation by organisms that consume plastic (Cheng et al. 2021). Studies indicate high removal efficiencies of 100% by integrating both sequencing batch reactors (SBR), which helps to settle high-density MPs with biomass and membrane bioreactors (MBR) by effectively trapping them in the membrane pores (Zhang et al. 2021). Pharmacies have been successfully removed from wastewater by combining advanced oxidation processes (AOPs) with membrane separation (Ganiyu et al. 2015). Another developing method for the removal of MP is the microbial breakdown of polymers in landfill leachate, which involves microalgae like *Scenedesmus dimorphus* and *Anabaena spiroides* as well as bacteria like *Bacillus* species. As a source of nutrition, these bacteria are able to efficiently break down a variety of polymers, including PE, PS, PET, and PP (Anik et al. 2021; Bahl et al. 2021; Chia et al. 2020; Kumar et al. 2017).

The difference between separate removal methods, their mechanism, limitations, removal rate, and best working conditions have been tabulated in (Table 13.2). According to this, MBRs and AOPs offer promising solutions to address the complex challenge of MP pollution in landfill leachate, contributing to environmental protection efforts. Moreover, some integrated approaches like reactive media in CW, combining both SBR and MBR, and integration of AOP with membrane filtration have also exhibited a promising start, and more research is needed in this sector depending on its complexity.

13.7 Concluding Remarks

In conclusion, our thorough investigation of MPs in landfills and leachate clarifies the many threats that these contaminants bring to human health and environmental quality. Investigating their origins and production first, it is clear that wastewater treatment procedures and landfill operations have a major role in the buildup of MPs in leachate and landfills. The identification and description of MPs show how commonplace these particles are in garbage settings, as well as how different their physical characteristics are. Comprehending the destination and modes of transportation of microplastics (MPs) in landfill leachate elucidates the several avenues by which these contaminants may disperse and affect adjacent ecosystems. The environmental ramifications also highlight how urgently mitigating techniques are needed to address the threats that MPs represent to terrestrial and marine life as well as human populations. Finally, insights into removal techniques provide valuable guidance for managing microplastic pollution in landfills and leachate, offering promising avenues for future research and practical implementation. Overall, this chapter underscores the complex nature of MPs pollution in landfill environments and emphasizes the significance of making coordinated efforts to lessen its harmful effect on the environment and public health.

Competing Interest The authors declare they have no conflicting interests.

References

Abdel-Shafy HI, Ibrahim AM, Al-Sulaiman AM, Okasha RA (2024) Landfill leachate: sources, nature, organic composition, and treatment: an environmental overview. Ain Shams Eng J 15(1):102293. https://doi.org/10.1016/j.asej.2023.102293

Abdul Maulud KN, Fitri A, Wan Mohtar WHM, Wan Mohd Jaafar WS, Zuhairi NZ, Kamarudin MKA (2021) A study of spatial and water quality index during dry and rainy seasons at Kelantan River Basin, Peninsular Malaysia. Arab J Geosci 14(2):1–19

Afrin S, Uddin MK, Rahman MM (2020) Micro-plastics contamination in the soil from Urban Landfill site, Dhaka, Bangladesh. Heliyon 6(11):e05572. https://doi.org/10.1016/j.heliyon.2020.e05572

Ajith N, Arumugam S, Parthasarathy S, Manupoori S, Janakiraman S (2020) Global distribution of MPs and its impacts on marine environment—a review. Environ Sci Pollut Res. https://doi.org/10.1007/s11356-020-09015-5

Akbay IK, Ozdemir T (2016) Monomer migration and degradation of polycarbonate via UV-C irradiation within aquatic and atmospheric environments. J Macromol Sci A 53(6):340–345

Anik AH, Hossain S, Alam M, Sultan MB, Hasnine MT, Rahman MM (2021) MPs pollution: a comprehensive review on the sources, fates, effects, and potential remediation. Environ Nanotechnol Monitor Manag 16:100530. https://doi.org/10.1016/j.enmm.2021.100530

Antunes J, Frias J, Sobral P (2018) MPs on the Portuguese Coast. Mar Pollut Bull 131:294–302. https://doi.org/10.1016/j.marpolbul.2018.04.025

Arhant M, Le Gall M, Le Gac P-Y, Davies P (2019) Impact of hydrolytic degradation on mechanical properties of PET—towards an understanding of MPs formation. Polym Degrad Stab 161:175–182. https://doi.org/10.1016/j.polymdegradstab.2019.01.021

Arkin C, Azoulay D, Caterbow A et al (2019) Plastic Atlas: facts and figures about the world of synthetic polymers, 2nd edn. Heinrich Böll Foundation, Berlin. ISBN 978-3-86928-211-4

Bahl S, Dolma J, Singh JJ, Sehgal S (2021) Biodegradation of plastics: a state of the art review. Mater Today Proc 39:31–34

Barletta M, Lima ARA, Costa MF (2019) Distribution, sources and consequences of nutrients, persistent organic pollutants, metals and micro-plastics in South American estuaries. Sci Total Environ 651:1199–1218. https://doi.org/10.1016/j.scitotenv.2018.09.276

Barnes DK, Galgani F, Thompson RC, Barlaz M (2009) Accumulation and fragmentation of plastic debris in global environments. Philos Trans R Soc B Biol Sci 364(1526):1985–1998

Bharath KM, Natesan U, Vaikunth R, Praveen Kumar R, Ruthra R, Srinivasalu S (2021) Spatial distribution of microplastic concentration around landfill sites and its potential risk on groundwater. Chemosphere 277:130263. https://doi.org/10.1016/j.chemosphere.2021.130263

Bradney L, Wijesekara H, Palansooriya KN, Obadamudalige N, Bolan NS, Ok YS, Rinklebe J, Kim K-H, Kirkham M (2019) Particulate plastics as a vector for toxic trace-element uptake by aquatic and terrestrial organisms and human health risk. Environ Int 131:104937

Cai L, Wang J, Peng J, Tan Z, Zhan Z, Tan X, Chen Q (2017) Characteristic of MPs in the atmospheric fallout from Dongguan city, China: preliminary research and first evidence. Environ Sci Pollut Res 24:24928–24935. https://doi.org/10.1007/s11356-017-0116-x

Campanale C, Massarelli C, Savino I, Locaputo V, Uricchio VF (2020) A detailed review study on potential effects of MPs and additives of concern on human health. Int J Environ Res Public Health 17(4):1212

Carr SA, Liu J, Tesoro AG (2016) Transport and fate of microplastic particles in wastewater treatment plants. Water Res 91:174–182

Carranza-Diaz O, Schultze-Nobre L, Moeder M, Nivala J, Kuschk P, Koeser H (2014) Removal of selected organic micropollutants in planted and unplanted pilotscale horizontal flow constructed wetlands under conditions of high organic load. Ecol Eng 71:234–245

Chamanee G, Sewwandi M, Wijesekara H, Vithanage M (2023) Global perspective on MPs in landfill leachate: occurrence, abundance, characteristics, and environmental impact. Waste Manag 171:10–25. https://doi.org/10.1016/j.wasman.2023.08.011

Chamas A, Moon H, Zheng J, Qiu Y, Tabassum T, Jang JH, Abu-Omar M, Scott SL, Suh S (2020) Degradation rates of plastics in the environment. ACS Sustain Chem Eng 8(9):3494–3511. https://doi.org/10.1021/acssuschemeng.9b06635
Cheng YL, Kim JG, Kim HB, Choi JH, Tsang YF, Baek K (2021) Occurrence and removal of MPs in wastewater treatment plants and drinking water purification facilities: a review. Chem Eng J 410:128381
Chia WY, Tang DYY, Khoo KS, Lup ANK, Chew KW (2020) Nature's fight against plastic pollution: algae for plastic biodegradation and bioplastics production. Environ Sci Ecotechnol 4:100065
Chiemchaisri C, Chiemchaisri W, Witthayaphirom C (2015) Remediation of MSW landfill leachate by permeable reactive barrier with vegetation. Water Sci Technol 71(9):1389–1397
Claessen M, Cauwenberghe LV, Vandegehuchte MB, Janssen CR (2013) New techniques for the detection of MPs in sediments and field collected organisms. Mar Pollut Bull 70:227–233
Coppock RL, Cole M, Lindeque PK, Queiros AM, Galloway TS (2017) A smallscale, portable method for extracting MPs from marine sediments. Environ Pollut 230:829–837
Corcoran PL (2020) Degradation of MPs in the environment. In: Rocha-Santos T, Costa M, Mouneyrac C (eds) Handbook of MPs in the environment. Springer International Publishing, Cham. https://doi.org/10.1007/978-3-030-10618-8_10-1
Cordova MR, Riani E (2021) Micro- and mesoplastics release from the Indonesian municipal solid waste landfill leachate to the aquatic environment: case study in Galuga landfill area, Indonesia. Mar Pollut Bull 163:111986. https://doi.org/10.1016/j.marpolbul.2021.111986
Costa AM, De Souza Marotta Alfaia RG, Campos JC (2019) Landfill leachate treatment in Brazil—an overview. J Environ Manag 232:110–116. https://doi.org/10.1016/j.jenvman.2018.11.006
de Souza Machado AA, Kloas W, Zarfl C (2018) MPs as an emerging threat to terrestrial ecosystems. Glob Chang Biol 2018(24):1405–1416
Dehal A, Prajapati A, Patil M, Kumar AR (2023) A review on MPs in landfill leachate: formation, occurrence, detection, and removal techniques. Chem Ecol 40(1):65–101. https://doi.org/10.1080/02757540.2023.2290183
Dehghani S, Moore F, Akhbarizadeh R (2017) Microplastic pollution in deposited urban dust, Tehran metropolis, Iran. Environ Sci Pollut Res 24:20360–20371. https://doi.org/10.1007/s11356-017-9674-1
Dey A, Dhumal CV, Sengupta P, Kumar A, Pramanik NK, Alam T (2021) Challenges and possible solutions to mitigate the problems of single-use plastics used for packaging food items: a review. J Food Sci Technol 58(9):3251–3269
Dolar D, Košutić K, Strmecky T (2016) Hybrid processes for treatment of landfill leachate: coagulation/UF/NF-RO and adsorption/UF/NF-RO. Sep Purif Technol 168:39–46. https://doi.org/10.1016/j.seppur.2016.05.016
Dris R, Gasperi CJ, Rocher AV, Saad BM, Renault N, Tassin B (2015) Microplastic contamination in an urban area: a case study in Greater Paris. Environ Chem 12:592–599
Dris R, Gasperi J, Saad M, Mirande C, Tassin B (2016) Synthetic fibers in atmospheric fallout: a source of MPs in the environment? Mar Pollut Bull 104:290–293. https://doi.org/10.1016/j.marpolbul.2016.01.006
Enfrin M, Dumée LF, Lee J (2019) Nano/MPs in water and wastewater treatment processes—origin, impact and potential solutions. Water Res 161:621–638
Fan T, Zhao J, Chen Y, Wang M et al (2021) Coexistence and adsorption properties of heavy metals by polypropylene MPs. Adsorp Sci Technol 2021:1–12. https://doi.org/10.1155/2021/4938749
Farrell M, Kuhn TK, Macdonald LM, Maddern TM et al (2013) Microbial utilisation of biochar derived carbon. Sci Total Environ 465:288–297
Free CM, Jensen OP, Mason SA, Eriksen M, Williamson NJ, Boldgiv B (2014) High-levels of microplastic pollution in a large, remote, mountain lake. Mar Pollut Bull 85(1):156–163
Galafassi S, Nizzetto L, Volta P (2019) Plastic sources: a survey across scientific and grey literature for their inventory and relative contribution to MPs pollution in natural environments, with an emphasis on surface water. Sci Total Environ 693:133499

Ganiyu SO, Van Hullebusch ED, Cretin M, Esposito G, Oturan MA (2015) Coupling of membrane filtration and advanced oxidation processes for removal of pharmaceutical residues: a critical review. Sep Purif Technol 156:891–914. https://doi.org/10.1016/j.seppur.2015.09.059

Gao X, Hassan I, Peng Y, Huo S, Ling L (2021) Behaviors and influencing factors of the heavy metals adsorption onto MPs: a review. J Clean Prod 319:128777. https://doi.org/10.1016/j.jclepro.2021.128777

Gatidou G, Arvaniti OS, Stasinakis AS (2019) Review on the occurrence and fate of MPs in Sewage Treatment Plants. J Hazard Mater 367:504–512. https://doi.org/10.1016/j.jhazmat.2018.12.081

Genis A (2017) Analysis of the global and Russian markets of polypropylene and of its main consumption areas. Russ J Gen Chem 87(9):2137–2150

GESAMP (2016) In: Kershaw PJ, Rochman CM (eds) Sources, fate and effects of MPs in the marine environment: part two of a global assessment, vol 93. IMO/FAO/UNESCO-IOC/UNIDO/WMO/IAEA/UN/UNEP/UNDP Joint Group of Experts on the Scientific Aspects of Marine Environmental Protection, London, p 220

Geyer R, Jambeck JR, Law KL (2017) Production, use, and fate of all plastics ever made. Sci Adv 3(7):e1700782

Gigault J, Halle AT, Baudrimont M et al (2018) Current opinion: what is a nanoplastic? Environ Pollut 235:1030–1034

Godoy V, Blázquez G, Calero M, Quesada L, Martín-Lara MA (2019) The potential of MPs as carriers of metals. Environ Pollut 255:113363. https://doi.org/10.1016/j.envpol.2019.113363

Golwala H, Zhang X, Iskander SM, Smith AL (2021) Solid waste: an overlooked source of MPs to the environment. Sci Total Environ 769:144581. https://doi.org/10.1016/j.scitotenv.2020.144581

Hahladakis JN, Velis CA, Weber R, Iacovidou E, Purnell P (2018) An overview of chemical additives present in plastics: migration, release, fate and environmental impact during their use, disposal and recycling. J Hazard Mater 344:179–199

Hartmann NB, Rist S, Bodin J, Jensen LH, Schmidt SN, Mayer P, Meibom A, Baun A (2017) MPs as vectors for environmental contaminants: exploring sorption, desorption, and transfer to biota. Integr Environ Assess Manag 13(3):488–493. https://doi.org/10.1002/ieam.1904

He P, Chen L, Shao L, Zhang H, Lü F (2019) Municipal solid waste (MSW) landfill: a source of MPs?—evidence of MPs in landfill leachate. Water Res 159:38–45

Heo NW, Hong SH, Han GM et al (2013) Distribution of small plastic debris in cross-section and high strandline on Heungnam beach, South Korea. Ocean Sci J 48:225–233. https://doi.org/10.1007/s12601-013-0019-9

Hernandez LM, Yousefi N, Tufenkji N (2017) Are there nanoplastics in your personal care products? Environ Sci Technol Lett 4:280–285. https://doi.org/10.1021/acs.estlett.7b00187

Herrera A, Garrido-Amador P, Martinez I, Samper MD, Lopez-Martinez J, Gomez M (2018) Novel methodology to isolate MPs from vegetal-rich samples. Mar Pollut Bull 129:61–69

Hidalgo-Ruz V, Gutow L, Thompson RC, Thiel M (2012) MPs in the marine environment: a review of the methods used for identification and quantification. Environ Sci Technol 46(6):3060–3075. https://doi.org/10.1021/es2031505

Hildebrandt L, Nack FL, Zimmermann T, Pröfrock D (2021) MPs as a Trojan horse for trace metals. J Hazard Mater Lett 2:100035. ISSN 2666-9110. https://doi.org/10.1016/j.hazl.2021.100035

Hirai H, Takada H, Ogata Y, Yamashita R, Mizukawa K et al (2011) Organic micropollutants in marine plastics debris from the open ocean and remote and urban beaches. Mar Pollut Bull 62:1683–1692

Hodson ME, Duffus-Hodson CA, Clark A, Prendergast-Miller MT, Thorpe KL (2017) Plastic bag derived-MPs as a vector for metal exposure in terrestrial invertebrates. Environ Sci Technol 51(8):4714–4721

Hou L, Kumar D, Yoo CG, Gitsov I, Majumder EL-W (2021) Conversion and removal strategies for MPs in wastewater treatment plants and landfills. Chem Eng J 406:126715

Huang L, Liu G, Dong G, Wu X, Wang C, Liu Y (2017) Reaction mechanism of zerovalent iron coupling with microbe to degrade tetracycline in permeable reactive barrier (PRB). Chem Eng J 316:525–533

Huang Y, Qing X, Wang W, Han G, Wang J (2020) Mini-review on current studies of airborne MPs: analytical methods, occurrence, sources, fate and potential risk to human beings. Trends Analyt Chem 125:115821
Huang D, Tao J, Cheng M, Deng R, Chen S, Yin L, Li R (2021) MPs and nanoplastics in the environment: macroscopic transport and effects on creatures. J Hazard Mater 407:124399. https://doi.org/10.1016/j.jhazmat.2020.124399
Huerta Lwanga E, Gertsen H, Gooren H, Peters P, Salanki T, van der Ploeg M, Besseling E, Koelmans AA, Geissen V (2016) MPs in the terrestrial ecosystem: implications for lumbricus terrestris (Oligochaeta, lumbricidae). Environ Sci Technol 50(5):2685–2691
Huerta Lwanga E, Vega JM, Quej VK, Chi JDLA, Cid LSD, Chi C, Segura GE, Gertsen H, Salanki T (2017) Field evidence for transfer of plastic debris along a terrestrial food chain. Sci Rep 7:1–7
Imran M, Das KR, Naik MM (2019) Co-selection of multi-antibiotic resistance in bacterial pathogens in metal and microplastic contaminated environments: an emerging health threat. Chemosphere 215:846–857. https://doi.org/10.1016/j.chemosphere.2018.10.114
Isobe A, Azuma T, Cordova MR et al (2021) A multilevel dataset of microplastic abundance in the world's upper ocean and the Laurentian Great Lakes. Microplastics Nanoplastics 1(1):16. https://doi.org/10.1186/s43591-021-00013-z
Iyare PU, Ouki SK, Bond T (2020) MPs removal in wastewater treatment plants: a critical review. Environ Sci Water Res Technol 6(10):2664–2675
Jambeck J, Geyer R, Wilcox C, Siegler TR, Perryman M, Andrady AL, Law KL (2015) Plastic waste inputs from land into the ocean. Science 347(6223):768–771
Jiang R, Lu G, Yan Z, Liu J, Wu D, Wang Y (2021) Microplastic degradation by hydroxy-rich bismuth oxychloride. J Hazard Mater 405:124247
Kabir MS, Wang H, Luster-Teasley S, Zhang L, Zhao R (2023) MPs in landfill leachate: sources, detection, occurrence, and removal. Environ Sci Ecotechnol 16:100256. https://doi.org/10.1016/j.ese.2023.100256
Kappler A, Fischer D, Oberbeckmann S, Schernewski G, Labrenz M, Eichhorn K, Voit B (2016) Analysis of environmental MPs by vibrational microspectroscopy: FTIR, Raman or both? Anal Bioanal Chem 408:8377–8391. https://doi.org/10.1007/s00216-016-9956-3
Karim EM, Sanjee AS, Mahmud S, Shaha M, Moniruzzaman M, Das CK (2019) MPs pollution in Bangladesh: current scenario and future research perspective. Chem Ecol. https://doi.org/10.1080/02757540.2019.1688309
Kattel E, Dulova N, Trapido M (2016) Hazardous waste landfill leachate treatment by combined chemical and biological techniques. Desalin Water Treat 57:13236
Kazour M, Terki S, Rabhi K, Jemaa S, Khalaf G, Amara R (2019) Sources of MPs pollution in the marine environment: importance of wastewater treatment plant and coastal landfill. Mar Pollut Bull 146:608–618. https://doi.org/10.1016/j.marpolbul.2019.06.066
Kilponen J (2016) MPs and harmful substances in urban runoffs and landfill leachates: possible emission sources to marine environment. Bachelor's thesis in environmental engineering, Lahti University of Applied Sciences
Kirstein IV, Hensel F, Gomiero A, Iordachescu L et al (2021) Drinking plastics?—quantification and qualification of MPs in drinking water distribution systems by µFTIR and Py-GCMS. Water Res 188:116519. ISSN 0043-1354. https://doi.org/10.1016/j.watres.2020.116519
Klein M, Fischer EK (2019) Microplastic abundance in atmospheric deposition within the Metropolitan area of Hamburg, Germany. Sci Total Environ 685:96–103. https://doi.org/10.1016/j.scitotenv.2019.05.405
Kumar RV, Kanna G, Elumalai S (2017) Biodegradation of polyethylene by green photosynthetic microalgae. J Bioremed Biodegr 8(381):2
Kumarathilaka P, Wijesekara H, Bolan N, Kunhikrishnan A, Vithanage M (2017) Phytoremediation of landfill leachates. In: Ansari A, Gill S, Gill R, Lanza GR, Newman L (eds) Phytoremediation. Springer, Cham. https://doi.org/10.1007/978-3-319-52381-1_17
Lambert S, Sinclair CJ, Boxall AB (2014) Occurrence, degradation and effect of polymer-based materials in the environment. Rev Environ Contamin Toxicol 227:1–53

Law KL, Thompson RC (2014) MPs in the seas. Science 345(6193):144–145

Lehtiniemi M, Hartikainen S, Näkki P, Engström-Öst J, Koistinen A, Setälä O (2018) Size matters more than shape: ingestion of primary and secondary MPs by small predators. Food Webs 17:e00097. https://doi.org/10.1016/j.fooweb.2018.e00097

Lenz R, Enders K, Stedmon CA, Mackenzie D, Nielsen TG (2015) A critical assessment of visual identification of marine microplastic using Raman spectroscopy for analysis improvement. Mar Pollut Bull 100(1):82–91. https://doi.org/10.1016/j.marpolbul.2015.09.026

Letcher TM (2020) Introduction to plastic waste and recycling. In: Plastic waste and recycling. Elsevier, Amsterdam

Li QL, Wu JT, Zhao XP, Gu XY (2019) Separation and identification of MPs from soil and sewage sludge. Environ Pollut 254:113076

Li S, Yang Y, Zheng H, Zheng Y, Jing T, Ma J, Nan J, Leong YK, Chang JS (2022) Advanced oxidation process based on hydroxyl and sulfate radicals to degrade refractory organic pollutants in landfill leachate. Chemosphere 297, 134214. https://doi.org/10.1016/j.chemosphere.2022.134214

Liu P, Zhan X, Wu X, Li J, Wang H, Gao S (2020) Effect of weathering on environmental behavior of MPs: properties, sorption and potential risks. Chemosphere 242:125193

Löder MGJ, Gerdts G (2015) Methodology used for the detection and identification of MPs—a critical appraisal. In: Bergmann M, Gutow L, Klages M (eds) Marine anthropogenic litter. Springer, Cham, pp 201–227. https://doi.org/10.1007/978-3-319-16510-3_8

Ma H, Pu S, Liu S, Bai Y, Mandal S, Xing B (2020) MPs in aquatic environments: toxicity to trigger ecological consequences. Environ Pollut 261:114089

Metz T, Koch M, Lenz P (2020) Quantification of MPs: which parameters are essential for a reliable inter-study comparison? Mar Pollut Bull 157:111330

Miao F, Liu Y, Gao M, Yu X, Xiao P, Wang M et al (2020) Degradation of polyvinyl chloride MPs via an electro-Fenton-like system with a TiO_2/graphite cathode. J Hazard Mater 399:123023

Mohammadi A, Malakootian M, Dobaradaran S, Hashemi M, Jaafarzadeh N (2022) Occurrence, seasonal distribution, and ecological risk assessment of MPs and phthalate esters in leachates of a landfill site located near the marine environment: Bushehr Port, Iran as a case. Sci Total Environ 842:156838

Montazer Z, Habibi Najafi MB, Levin DB (2020) Challenges with verifying microbial degradation of polyethylene. Polymers 12(1):123

Narevski AC, Novaković MI, Petrović MZ, Mihajlović IJ, Maoduš NB, Vujić GV (2021) Occurrence of bisphenol A and MPs in landfill leachate: lessons from South East Europe. Environ Sci Pollut Res 28(31):42196–42203. https://doi.org/10.1007/s11356-021-13705-z

Nascimento LF, Lima JF, Filho PCS, Serra OA (2018) Effect of lanthanum loading on nanosized CeO_2-ZnO solid catalysts supported on cordierite for diesel soot oxidation. J Environ Sci 73:58–68

NOAA (National Oceanic and Atmospheric Administration) (2018) What are MPs? https://oceanservice.noaa.gov/facts/MPs.html

Obbard RW (2018) MPs in Polar Regions: the role of long range transport. Curr Opin Environ Sci Health 1:24–29

Padervand M, Lichtfouse E, Robert D, Wang C (2020) Removal of MPs from the environment. A review. Environ Chem Lett 18:807–828. https://doi.org/10.1007/s10311-020-00983-1

Park SY, Kim CG (2019) Biodegradation of micro-polyethylene particles by bacterial colonization of a mixed microbial consortium isolated from a landfill site. Chemosphere 222:527–533. https://doi.org/10.1016/j.chemosphere.2019.01.159

Perren W, Wojtasik A, Cai Q (2018) Removal of Microbeads from Wastewater Using Electrocoagulation. ACS Omega 3 (3), 3357-3364. https://doi.org/10.1021/acsomega.7b02037

Plastic Pollution by Country 2024 (n.d.). https://worldpopulationreview.com/country-rankings/plastic-pollution-by-country

Plastics Europe (2023) Plastics—the fast Facts 2023. https://plasticseurope.org/knowledge-hub/plastics-the-fast-facts-2023/

Poerio T, Piacentini E, Mazzei R (2019) Membrane processes for microplastic removal. Molecules 24:4148. https://doi.org/10.3390/molecules24224148

Priya K, Renjith K, Joseph CJ, Indu M, Srinivas R, Haddout S (2022) Fate, transport and degradation pathway of MPs in aquatic environment—a critical review. Reg Stud Mar Sci 56:102647
Puthcharoen A, Leungprasert S (2019) Determination of MPs in soil and leachate from the landfills. Thai Environ Eng J 33(3):39–46
Qin Z, Siddiqui MA, Xin X, Mou J, Varjani S, Chen G, Lin CSK (2024) Identification of MPs in raw and treated municipal solid waste landfill leachates in Hong Kong, China. Chemosphere 351:141208. https://doi.org/10.1016/j.chemosphere.2024.141208
Rahman MM, Sultan MB, Alam M (2023) MPs and adsorbed micropollutants as emerging contaminants in landfill: a mini review. Curr Opin Environ Sci Health 31:100420. https://doi.org/10.1016/j.coesh.2022.100420
Rahman M, Toha M, Sikder S (2024a) Micro/nanoplastics pollution from landfill sites: a comprehensive review on the formation, distribution, effects and potential mitigation. In: Technical landfills and waste management. Springer, pp 247–265. https://doi.org/10.1007/978-3-031-52633-6_10
Rahman MA, Haque MM, Tareq SM (2024b) Abundance and characteristics of MPs in the landfill leachate of Amin Bazar, Dhaka: a potential risk to aquatic environments. Phys Chem Earth A/B/C 134:103573. https://doi.org/10.1016/j.pce.2024.103573
Rahmani A, Brojerdi M, Seid-Mohammadi A, Shabanloo A, Zabihollahi S, Zafari D (2023) Abundance and characteristics of MPs in different zones of waste landfill site: a case study of Hamadan, Iran. Case Stud Chem Environ Eng 8:100494. https://doi.org/10.1016/j.cscee.2023.100494
Rao SR, Leja J (2013) Surface chemistry of froth flotation: reagents and mechanisms. Springer
Rolsky C, Kelkar V, Driver EM, Halden RU (2020) Municipal sewage sludge as a source of MPs in the environment. Curr Opin Environ Sci Health 14:16–22. https://doi.org/10.1016/j.coesh.2019.12.001
Ryan PG, Moore CJ, van Franeker JA, Moloney CL (2009) Monitoring the abundance of plastic debris in the marine environment. Philos Trans R Soc Lond Ser B Biol Sci 364:1999–2012
Santana MFM, Moreira FT, Pereira CDS (2018) Continuous exposure to MPs does not cause physiological effects in the cultivated mussel *Perna perna*. Arch Environ Contam Toxicol 74:594–604
Schwaferts C, Sogne V, Welz R, Meier F, Klein T, Niessner R, Elsner M, Ivleva NP (2020) Nanoplastic analysis by online coupling of Raman microscopy and field-flow fractionation enabled by optical tweezers. Anal Chem 92(8):5813–5820
Schymanski D, Goldbeck C, Humpf H, Fürst P (2018) Analysis of MPs in water by micro-Raman spectroscopy: release of plastic particles from different packaging into mineral water. Water Res 129:154–162. https://doi.org/10.1016/j.watres.2017.11.011
Sekar V, Sundaram B (2023) Preliminary evidence of MPs in landfill leachate, Hyderabad, India. Process Saf Environ Prot 175:369–376. https://doi.org/10.1016/j.psep.2023.05.070
Setala O, Lehtinen VF, Lehtiniemi M (2014) Ingestion and transfer of MPs in the planktonic food web. Environ Pollut 185:77–83
Sewwandi M, Amarathunga A, Wijesekara H, Mahatantila K, Vithanage M (2022) Contamination and distribution of buried MPs in Sarakkuwa beach ensuing the MV X-press pearl maritime disaster in Sri Lankan Sea. Mar Pollut Bull 184:114074. https://doi.org/10.1016/j.marpolbul.2022.114074
Shen M, Xiong W, Song B, Zhou C, Almatrafi E, Zeng G, Zhang Y (2022) MPs in landfill and leachate: occurrence, environmental behavior and removal strategies. Chemosphere 305:135325. https://doi.org/10.1016/j.chemosphere.2022.135325
Shim WJ, Hong SH, Eo S (2017) Identification methods in microplastic analysis: a review. Anal Methods 9(9):1384–1391. https://doi.org/10.1039/c6ay02558g
Silva AL, Prata JC, Duarte AC, Soares AM, Barceló D, Rocha-Santos T (2021) MPs in landfill leachates: the need for reconnaissance studies and remediation technologies. Case Stud Chem Environ Eng 3:100072. https://doi.org/10.1016/j.cscee.2020.100072
Singh S, Bhagwat A (2022) MPs: a potential threat to groundwater resources. Groundwater Sustain Dev 19:100852

Singh S, Malyan SK, Maithani C, Kashyap S, Tyagi VK, Singh R, Malhotra S, Sharma M, Kumar A, Panday BK, Pandey R (2023) MPs in landfill leachate: occurrence, health concerns, and removal strategies. J Environ Manag 342:118220. https://doi.org/10.1016/j.jenvman.2023.118220
Song YK, Hong SH, Jang M, Han GM, Rani M, Lee J, Shim WJ (2015) A comparison of microscopic and spectroscopic identification methods for analysis of MPs in environmental samples. Mar Pollut Bull 93(1–2):202–209. https://doi.org/10.1016/j.marpolbul.2015.01.015
Song L, Yang S, Gong Z, Wang J, Shi X, Wang Y, Zhang R, Wu Y, Wager YZ (2023) Antibiotics and antibiotic-resistant genes in municipal solid waste landfills: current situation and perspective. Curr Opin Environ Sci Health 31:100421. https://doi.org/10.1016/j.coesh.2022.100421
Su Y, Zhang Z, Wu D, Zhan L, Shi H (2019) Occurrence of MPs in landfill systems and their fate with landfill age. Water Res 164:114968. https://doi.org/10.1016/j.watres.2019.114968
Sun J, Dai X, Wang Q, van Loosdrecht MC, Ni B-J (2019) MPs in wastewater treatment plants: detection, occurrence and removal. Water Res 152:21–37
Sun J, Zhu Z, Li W, Yan X, Wang L, Zhang L, Jin J, Dai X, Ni B (2021) Revisiting MPs in landfill leachate: unnoticed tiny MPs and their fate in treatment works. Water Res 190:116784. https://doi.org/10.1016/j.watres.2020.116784
Tang X, Wang S, Yang Y, Tao R, Dai Y, Dan A, Li L (2015) Removal of six phthalic acid esters (PAEs) from domestic sewage by constructed wetlands. Chem Eng J 275:198–205
Tang N, Liu X, Xing W (2020) MPs in wastewater treatment plants of Wuhan, Central China: abundance, removal, and potential source in household wastewater. Sci Total Environ 745:141026. https://doi.org/10.1016/j.scitotenv.2020.141026
Trihadiningrum Y, Wilujeng SA, Tafaqury R, Radita DR, Radityaningrum AD (2023) Evidence of MPs in leachate of Randegan landfill, Mojokerto City, Indonesia, and its potential to pollute surface water. Sci Total Environ 874:162207. https://doi.org/10.1016/j.scitotenv.2023.162207
Turner A, Filella M (2021) Polyvinyl chloride in consumer and environmental plastics, with a particular focus on metal-based additives. Environ Sci Processes Impacts 23(9):1376–1384
Ullah S, Ahmad S, Guo X, Ullah S, Ullah S, Nabi G, Wanghe K (2023) A review of the endocrine disrupting effects of micro and nano plastic and their associated chemicals in mammals. Front. Endocrinol. 13:1084236. https://doi.org/10.3389/fendo.2022.1084236Upadhyay K, Bajpai S (2021) MPs in landfills: a comprehensive review on occurrence, characteristics and pathways to the aquatic environment. Nat Environ Pollut Technol 20:9
US EPA (2006) Landfill bioreactor performance: second interim report. Outer Loop Recycling & Disposal Facility, Louisville
van Praagh M, Hartman C, Brandmyr E (2018) MPs in landfill leachates in the Nordic countries. Nordic Council of Ministers, Denmark
Van Praagh M, Hartman C, Brandmyr E (2019) MPs in landfill leachates in the Nordic countries. TemaNord. https://doi.org/10.6027/tn2018-557
Venkatesh S, Mahboob S, Govindarajan M, Al-Ghanim KA, Ahmed Z, Al-Mulhm N, Gayathri R, Vijayalakshmi S (2021) Microbial degradation of plastics: sustainable approach to tackling environmental threats facing big cities of the future. J King Saud Univ Sci 33(3):101362. ISSN 1018-3647. https://doi.org/10.1016/j.jksus.2021.101362
Vithanage M, Wijesekara S, Siriwardana A, Mayakaduwa SS, Ok YS (2014) Management of municipal solid waste landfill leachate: a global environmental issue. Environ Deterior Human Health. https://doi.org/10.1007/978-94-007-7890-0_11
Vroom RJE, Koelmans AA, Besseling E, Halsband C (2017) Aging of MPs promotes their ingestion by marine zooplankton. Environ Pollut 231:987–996
Wali SU, Umar KJ, Abor IG (2021) Modelling flow and fate of contaminants in groundwater using a version of the five steady-state pollutant transport models: a status-review. J Electron Computer Netw Appl Math 1(01):1–30. ISSN: 2799-1156
Walker TR, Fequet L (2023) Current trends of unsustainable plastic production and micro(nano) plastic pollution. TrAC Trends Anal Chem 160:116984. https://doi.org/10.1016/j.trac.2023.116984

Wan Y, Chen X, Liu Q, Hu H, Wu C, Xue Q (2022) Informal landfill contributes to the pollution of MPs in the surrounding environment. Environ Pollut 293:118586. https://doi.org/10.1016/j.envpol.2021.118586

Wang G, Lu G, Zhao J, Yin P, Zhao L (2016) Evaluation of toxicity and estrogenicity of the landfill-concentrated leachate during advanced oxidation treatment: chemical analyses and bioanalytical tools. Environ Sci Pollut R 23:16015–16024

Wang H-T, Ding J, Xiong C, Zhu D, Li G, Jia X-Y, Zhu Y-G, Xue X-M (2019) Exposure to MPs lowers arsenic accumulation and alters gut bacterial communities of earthworm *Metaphire californica*. Environ Pollut 251:110–116

Wang Z, Sedighi M, Lea-Langton A (2020) Filtration of microplastic spheres by biochar: removal efficiency and immobilisation mechanisms. Water Res 184:116165

Wang L, Wang H, Huang Q, Yang C, Wang L, Liu Z, Zhou Q, Wang T, Ning C (2023) MPs in landfill leachate: a comprehensive review on characteristics, detection, and their fates during advanced oxidation processes. Water 15(2):252. https://doi.org/10.3390/w15020252

Wardrop P, Shimeta J, Nugegoda D, Morrison P, Miranda AF, Tang M, Clarke B (2016) Chemical pollutants sorbed to ingested microbeads from personal care products accumulate in fish. Environ Sci Technol 50(7):4037–4044

Wijekoon P, Koliyabandara PA, Cooray AT, Lam SS, Athapattu B, Vithanage M (2022) Progress and prospects in mitigation of landfill leachate pollution: risk, pollution potential, treatment and challenges. J Hazard Mater 421:126627. https://doi.org/10.1016/j.jhazmat.2021.126627

Wojnowska-Baryła I, Bernat K, Zaborowska M (2022) Plastic waste degradation in landfill conditions: the problem with MPs, and their direct and indirect environmental effects. Int J Environ Res Public Health 19(20):13223. https://doi.org/10.3390/ijerph192013223

Wu D, Huang XH, Sun JZ, Graham DW, Xie B (2017) Antibiotic resistance genes and associated microbial community conditions in aging landfill systems. Environ Sci Technol 51(21):12859–12867. https://doi.org/10.1021/acs.est.7b03797

Xu Z, Sui Q, Li A, Sun M, Zhang L, Lyu S, Zhao W (2020) How to detect small MPs (20–100 Mm) in freshwater, municipal wastewaters and landfill leachates? A trial from sampling to identification. Sci Total Environ 733:139218. https://doi.org/10.1016/j.scitotenv.2020.139218

Yang G, Zhang GM, Wang HC (2015) Current state of sludge production, management, treatment and disposal in China. Water Res 78:60–73

Yousif E, Ahmed D, Zainulabdeen K, Jawad A (2023) Photo-physical and morphological study of polymers: a review. Phys Chem Res 11(2):409–424. https://doi.org/10.22036/PCR.2022.342751.2105

Zhang X, Zheng M, Yin X, Wang L, Lou Y, Qu L, Liu X, Zhu H, Qiu Y (2019) Sorption of 3,6-dibromocarbazole and 1,3,6,8-tetrabromocarbazole by MPs. Mar Pollut Bull 138:458–463. https://doi.org/10.1016/j.marpolbul.2018.11.055

Zhang Y, Kang S, Allen S, Allen D, Gao T, Sillanpää M (2020) Atmospheric MPs: a review on current status and perspectives. Earth Sci Rev 203:103118. ISSN 0012-8252. https://doi.org/10.1016/j.earscirev.2020.103118

Zhang Z, Su Y, Zhu J, Shi J, Huang H, Xie B (2021) Distribution and removal characteristics of MPs in different processes of the leachate treatment system. Waste Manag 120:240–247. https://doi.org/10.1016/j.wasman.2020.11.025

Zhang L, Zhao W, Liang Z, Cai Z, Yan R, Yu X, Barceló D, Sui Q (2024) New insights into microplastic contamination in different types of leachates: abundances, characteristics, and potential sources. Engineering. https://doi.org/10.1016/j.eng.2024.02.008

Zhao Y, Lou Z (2016) Pollution control and resource recovery: municipal solid wastes at landfill. Butterworth Heinemann, pp 11–92

Zhou C, Fang W, Xu W, Cao A, Wang R (2014) Characteristics and the recovery potential of plastic wastes obtained from landfill mining. J Clean Prod 80:80–86. https://doi.org/10.1016/j.jclepro.2014.05.083

Zhou Q, Tian C, Luo Y (2017) Various forms and deposition fluxes of MPs identified in the coastal urban atmosphere. Chin Sci Bull 62:3902–3909. https://doi.org/10.1360/n972017-00956

Ziajahromi S, Neale PA, Rintoul L, Leusch FD (2017) Wastewater treatment plants as a pathway for MPs: development of a new approach to sample wastewater-based MPs. Water Res 112:93–99

Zimmerman WB, Tesar V, Butler S, Bandulasena HC (2008) Microbubble generation. Recent Pat Biomed Eng 2:1–8. https://doi.org/10.2174/187221208783478598

Environmental Risks and Hazards of Microplastics

14

Priyancka Arora, Aakash Singh, Himanshi Sharma, and Ajay Kumar

Abstract

Microplastics are minute plastic particles that are generated during the manufacturing of commercial goods and the deterioration of large-sized plastic components. This degradation is the result of exposure of plastic materials to external components, like sunlight, ocean waves, and many others. The significant attention microplastics have garnered is due to their potential to cause serious environmental harm. These particles are regarded as contaminants that may be harmful to the environment and the health of animals. Studies have shown that these small plastic fragments release harmful additives, can absorb other substances, and can serve as surfaces for biofilms. As a result, they increase the availability, toxicity, and mobility of numerous hazardous substances by acting as reservoirs for them. Furthermore, when consumed by smaller organisms, microplastics may present a risk of trophic transfer to humans and other higher animals. As they further break down, they can breach cell membranes, heightening associated risks. Exposure to microplastics can impact feeding habits, metabolic functions, reproduction, and behavior. Additionally, they may also alter the physical characteristics of water and soil, endangering ecosystems and living things. Enhancing international efforts to minimize pollution, creating biodegradable polymers and additives, and lowering consumption while encouraging the reuse of plastic are among the suggested remedies. This chapter explores the origins, uptake, and impacts of microplastics on the pedosphere and hydrosphere.

P. Arora · H. Sharma
Department of Biotechnology, Amity University, Mohali, Punjab, India

A. Singh · A. Kumar (✉)
Department of Biotechnology, Faculty of Engineering and Technology, Rama University, Kanpur, Uttar Pradesh, India
e-mail: drajay.fet@ramauniversity.ac.in

A. Kumar, V. Singh (eds.), *Microplastics Pollution and its Remediation*,
https://doi.org/10.1007/978-981-97-4068-0_14

Keywords

Microplastics · Polystyrene · Marine plastic debris · Persistent organic pollutants · Bioaccumulation · Health risks

14.1 Introduction

Microplastics (MPs) are tiny pieces of plastic debris found in the surroundings due to the processing and disintegration of consumer products and industrial scraps. Thomson et al., in 2004, coined the term microplastics and defined it as particles smaller than 5 mm, while they studied the microscopic plastic elements accumulated in water bodies. Since the mid-1900s, the use of plastics has steadily expanded, and their annual manufacturing quantities have now exceeded a billion tons. Because of this extensive use, plastics are widely dispersed throughout the water ecosystem (Eriksen et al. 2014). Plastic is a significant contributor to environmental pollution. Now, microplastics have merged as an additional major factor. Most plastics are utilized, and subsequently, they are disposed away on land. By 2030, the amount of microplastics present in certain marine regions is expected to have doubled. Understanding the material composition, physical properties, applications, transportation, and fragmentation into microplastics and nanoplastics is essential to finding a solution to this worldwide issue. The majority of reported investigations of plastic trash have concentrated on the ocean's surface. However, there is a growing trend toward examining submerged sediments, lakes, rivers, soils, and various ecosystems (Cózar et al. 2014). Not all plastics have the same chemical composition, and as a result, they do not behave similarly in the natural environment. First, we need to consider the composition and diversity of microplastics to understand their behavior, fate, and impacts. There is no efficient way to assess the presence of microplastics in the environment and therefore it needs to be addressed. Plastics are produced using polymeric compounds like polystyrene, polyethylene, polypropylene, polyvinyl chloride, polycarbonate, and polyethylene terephthalate. The physical and chemical characteristics of the materials that make up microplastics determine how they interact with the environment. After this another feature that contributes to the reason for contact of these with biota is the minute size and distinct shapes of the microplastics (beads, fibers, pellets, and films). For instance, compared to beads, fibers are more enduring in animal stomachs. The chemical characteristics of microplastics can also affect their resilience and ability to bind. These are composed of chemically stable compounds, and their interactions are defined by compositional irregularities such as stability, polarity, functional group, and crystallinity. These constituents' properties may transform as they deteriorate. As an illustration, when the material breaks down to create microplastics, the crystallinity of the resulting particles increases significantly. These qualities change both during and after use (Pandey et al. 2022).

Microplastics are categorized into two types based on their origin, i.e., primary microplastics and secondary microplastics. The former ones are specifically made of microplastics that are utilized in industry and household settings. They are used as microbeads in personal hygiene products and as industrial abrasives for sensitive surfaces. These are also utilized as pellets and precursor resins in the production of completed plastic products. Other areas that use these include acrylics, detergents, protective coatings, and drilling fluids for the oil and gas sector. Additionally, medical microplastics may function as a vector. Secondary microplastics, which are found in plastic bottles, tires, sheets, fishing nets, and bags, are vast volumes of plastic that have broken down into minute particles (Barnes et al. 2009). When larger plastics break down during use (such as tire wear particles) or after use, secondary microplastics are created. Compared to primary microplastics, secondary microplastics are much more common. Plastics' atmospheric half-lives can range from a 100 to a 1000 years, contingent upon the type of polymer and the surrounding conditions (Ward et al. 2019). The majority lack scientific evidence supporting the theory that higher air temperatures, frictional forces, and UV exposure cause the majority of plastic disintegration that takes place on land (Shim and Thomposon 2015). The extensive utilization of plastics in our daily routines, along with inadequate waste disposal methods, has caused the buildup of tiny plastic particles known as microplastics in different types of environments such as oceans, rivers, lakes, soils, and atmosphere. The minute size of microplastics allows them to be effortlessly carried by wind, water flow, and animals, leading to their widespread presence worldwide. This chapter focuses on the perilous consequences of micro-sized plastics on the biosphere and its living beings.

14.2 Sources of Microplastics

Microplastics can originate from primary and secondary sources, contributing to the presence of nondegradable plastic particles of various sizes in the environment, due to which there is an occurrence of microplastics (MPs). It is very difficult, if not impossible, to pinpoint the exact sources of microplastics due to their small size. Plastic pellets, paint, wastewater from washing, sewage sludge, artificial turf, rubber roads in cities, plastic running tracks in schools, and tire wear from vehicles are the main primary sources of environmental microplastics. Secondary sources, on the other hand, include large-scale plastic waste from farming, fishing, and other activities, as well as urban waste like plastic bottles and bags. Vehicle tire wear is considered a significant source of microplastics in the environment due to the substantial rise in car numbers on the planet (An et al. 2020). Approximately 21–42% of the plastics produced worldwide are stored in landfills, making them one of the primary repositories of plastic trash. Secondary microplastics can be produced from plastic waste in landfills through complex metabolic processes and physical alterations (Youcai and Ziyang 2016). Microplastics are not isolated pollutants; instead, they can act as carriers of other hazardous adsorption contaminants such as heavy metals, polycyclic aromatic hydrocarbons (PAHs), polychlorinated biphenyls (PCBs), and per- and polyfluoroalkyl substances (PFAS) by interacting with these particles electrostatically or through their hydrophobic molecules (Du et al. 2020).

Microplastics may make it easier for micro-pollutants to enter the environment, which could have detrimental impacts on ecosystem health as well as human health. According to reports, the concentration of pollutants in microplastics can increase by 6–70 times on comparison with the surrounding environment (Alimi et al. 2018). Synthetic fabrics, the degradation of synthetic rubber tires, and urban dust are the primary contributors to microplastics in the air (Wang et al. 2021a, b). Other possible sources include construction supplies, factory emissions, plastic furniture fragments, landfills, traffic debris, waste incineration, tumble dryer exhaust, synthetic materials used in horticulture soils, and sewage sludge used as fertilizer (Dris et al. 2017).

14.3 Sources of Microplastics in Various Ecosystems

14.3.1 Aquatic Environment

Studies have indicated that roughly 35% of the microplastics found in the ocean originate from synthetic fabrics. The multitude of fibers is dispersed upon the cleaning of synthetic fabrics. The number of microfibers released during the laundering process is dependent on various factors like fabric composition, the specific type of machine employed for washing purposes, the age of the garments, and the composition of detergent utilized (De Falco et al. 2018). The origin of microplastics found in aquatic bodies could be from automotive tires, artificial fabric fibers after their laundry, agricultural plastic waste, polishing products, personal care products, and fragments of macroplastics. Household wastewater is also a significant contributor of microplastics in aquatic ecosystems, with plastic pellets, sometimes called nurdles, and fibrous microplastics being key ingredients and these can enter the water bodies through sewage outflows (Wong et al. 2020; Zhang et al. 2018). The main pathway is the river through which plastics enter seas and oceans. Approximately 42% of all pollutants released into the marine environment by rivers are microplastics derived from tires and damage caused on roads. Tires create an array of powders and debris that ranges in size from a few nanometers to a few micrometers. In contrast to the microplastics produced from plastic bottles, fibers, and bags, the production of these from tires is more significant (Lassen et al. 2015). Water obtained from wastewater treatment plants and roadsides also accumulates secondary microplastics. Although wastewater treatment can reduce the amount of microplastics in effluent, it does not completely eliminate them. Furthermore, even after undergoing wastewater treatment, microplastics can enter aquatic ecosystems through various pathways (Devi et al. 2023).

14.3.2 Terrestrial Environment

One of the principal contributors to the microplastic pollution in land is the manufacturing and trading of textiles. The textile business classifies textile fibers into three basic categories: natural fibers, regenerated fibers, and synthetic fibers. Synthetic fibers account for over 170% of the microplastic waste from textiles. The textile sector produces 42 million tons of synthetic fibers a year, around 80% of which are polyesters (PES) (Kelly et al. 2019). Due to the significant harm that synthetic textile fibers have caused to the marine environment, awareness of their impact has increased, resulting in a slight reduction in the emission of microplastics. Nevertheless, this is not the only and the primary reason for microplastic pollution in land ecosystems. Additional sources like the degradation of larger plastic objects, tire ercsion, inadequate waste disposal, and air pollution also play a role in contaminating terrestrial environments with microplastics. The direct emission of microfibers into the air is significantly higher than their release into water (Xu et al. 2020). There are several ways that microplastics can get into soil, few of the ways are the utilization of municipal wastes, surface runoff, irrigation water from farms, mulching sheets, and environmental plastic deposition. Microplastics mainly enters the agricultural soils through the use of plastic mulching, which directly affects crops, enhances soil porosity, influences the metabolic processes of soil organisms, and changes the abundance and diversity of soil microorganisms (Hodson et al. 2017).

14.4 Understanding the Hazards of Microplastic

Lightweight polymers are inexpensive, simple to make, and easy to handle. In the 2010s, the capability to produce plastics increased dramatically. The recent COVID-19 crisis has led to a hike in the need for personal protective equipment (PPE), necessitating a substantial increase in PPE production to keep up with the growing number of COVID-19 patients and the requirements of healthcare systems lacking sufficient equipment. Mittens, plastics, and other PPE kits, in addition to the increase in supplied products due to more people spending time indoors, have all contributed to the usage and production of plastics. High molecular weights of olefin-based polymers have posed a persistent issue as they are not biodegradable, evidenced by polyethylene (PE) and polypropylene (PP). When nonbiodegradable polymers come into contact with the environment, they interact with a variety of microbes, organic compounds, and inorganic chemicals. These small plastic particles may be hazardous, according to research findings (Kim et al. 2022). Plastic, representing plastic bags, is widely recognized for its contribution to global pollution. Examining media headlines on the relationship between plastic and the environment, it is evident that the presence of plastic debris in the environment has, perhaps, been recognized as a sustainability issue since the 1970s. At-home activities promote the use of insulated and airtight plastics that are produced using additives like polystyrene. Other chemical additives, such as dyes, plasticizers, UV inhibitors, and fire-resistant materials, are also present in the plastics polymers. When these polymers disperse in the air during degradation, they are associated with severe health concerns. It was studied in over 100 human patients with malignant lung disease, and 87% of them have polymeric fibers (Hale et al. 2020). This obvious problem has spurred the development of various innovative waste disposal methods and regulations, including the German Law on Circular Economy (Kramm and Völker 2018).

The diversity of microplastic particles and intervening factors renders the conventional technique for environment risk assessment of chemical compounds inadequate for assessing the "safe" or "potentially dangerous" threshold of microplastics in natural ecosystems. Plastic particles come from a variety of sources and do not belong to a single class of material. Microplastics have a wide range of physicochemical characteristics, much like their origins. They vary in size and form, include different chemicals, and have different polymeric compositions, all of which can affect their biological impact. Due to related chemicals such as phthalates and BPA, microplastics can pose a risk, causing physical harm owing to their structure, and leading to unintended consequences when consumed (Wright et al. 2013). The identification of the microplastic problem is challenging due to its multifaceted nature. To adequately address this societal concern, more scientific research is needed to provide conclusive evidence. The same is true regarding the assessment of the risks to human health. Because society perceives microplastics as a threat, action must be taken despite the lack of conclusive proof. This ambiguity can complicate risk management decisions. In light of the possibility that the complexity of the phenomenon may remain incompletely comprehended, the forthcoming research studies should address this ambiguity and manage intricate global dangers (Kramm and Völker 2018).

14.5 Impact of Microplastic on Ecosystem

The toxicity of microplastics has been extensively researched, which has led to the recognition of microplastic contamination as a significant environmental issue of global proportions. Although they are diminutive in size, microplastics are a threat to the environment and may even be harmful to human health.

- The primary concern is the environmental effects of microplastics. Marine life is threatened by these widely dispersed microplastics. The aquatic environment contains plastics of various colors, densities, and shapes. Due to the pervasive nature of microplastic contamination in the surroundings, all forms of biota are vulnerable. Given their micro-size, low volume, and low density in comparison to marine water, microplastics float and are mistaken as food by marine species, these species ingest or absorb them once they reach the sea causing physical harm and impairing their feeding, reproduction, and growth (Napper et al. 2015). Furthermore, microplastics can build up in sediments and affect benthic invertebrates as well as other species that live at the bottom over an extended period. The diverse and functionally significant seabed habitats act as repositories for microplastics (Coppock et al. 2021).
- Microplastic accumulation is responsible for the pollution of the food web. When human-made plastic items are discarded into the environment, they can harm

species, ultimately affecting humans through the food chain (Mao et al. 2022). These are ingested by high-nutrient creatures from low-nutrient organisms whose bodies have already acquired microplastics. They can damage organisms through a variety of pathways once they have entered their bodies. Forty percent of the snails had their gastric wall villi severely destroyed, while the histology of the liver and kidney was unaffected (Song et al. 2019). The gradual increase of microplastics in the terrestrial food chain, from the soil to the dung of earthworms or chickens over time, has been reported (Lwanga et al. 2017).

- Microplastics have the potential to infiltrate the human body via the consumption of fish contaminated by them, presenting a risk to human health and potentially contributing to a decline in biodiversity (de Souza Machado et al. 2018). The possibility of microplastic ingestion and subsequent contamination poses a problem for the food web.
- The detrimental effects of the accumulation of microplastics are also associated with chemical toxicity. These can adsorb toxic chemicals from the surroundings. Incidences of adsorption of certain hydrophobic chemicals, such as persistent organic pollutants (POPs) like polychlorinated biphenyls (PCBs) and pesticides, have been documented with microplastics (Blackburn and Green 2022). These substances have the potential to be released when consumed by organisms, which could have harmful effects and lead to bioaccumulation in the food chain.
- The hazards of microplastics are now evident in human health. Microplastics have been found in recent research to have toxicological effects on marine life, including the accumulation of microplastics in organs due to translocation, blood flow restriction, and inflammation (Blackburn and Green 2022). The possibility of immune toxicity in blood has been linked to MPs and POPs. It was shown that over time, the concentration of microplastics in the terrestrial food chain increased, moving from the soil to the dung of earthworms or chickens (Lwanga et al. 2017). This causes the particles to build up in their tissues and gastrointestinal tracts, endangering their health.
- Microplastics have a serious impact on ecological disturbance. Microplastics have the potential to damage biodiversity and disturb natural processes. On their surfaces, they can encourage the growth of dangerous bacteria, change ecosystems, and have an impact on the cycling of nutrients. Ecosystem services may be disrupted and sensitive species may disappear as a result of these changes, which may have a domino effect on entire ecosystems (Campanale et al. 2020).

Furthermore, under some circumstances, the combination of microplastics and persistent organic pollutants in the environment might have unknown consequences on a variety of species, most of which could be detrimental. Controlling microplastic pollution is crucial for protecting the ecosystem and species because of the growing number of microplastics entering the environment (Mao et al. 2022). The hazards of microplastics are discussed in detail in the upcoming sections.

14.6 Health Concerns Associated with Microplastics

Evidence exists to show that some additives in plastics, such as phthalates, bisphenol A, or dyes, are associated with health risks, tumor-causing or mutagenic. MPs may release chemicals, dyes, and pigments that gather on various surfaces and within water bodies, potentially with uncertain health effects (Gasperi et al. 2018). The most common toxic additive is phthalates, which are used as plasticizers to enhance the fluidity of plastics. The plastic polymer covalently binds with them, leading to their rapid disintegration in the environment, thereby causing pollution. Sources of phthalates from plastic are widespread as they are present in household dust, human urine, and occasionally in breast milk as well (Bamai et al. 2014). PBDEs, or polybrominated diphenyl ethers, are additives found in many commercial items that act as flame retardants. Up to 15% of PBDEs may be integrated into plastic, and they are not chemically bonded, which means they might leak during the process of manufacturing, disposal, and recycling. Microplastics can induce oxidative stress, neurological damage, endocrine disruption, immune system impairment, and other issues upon entering an organism (Mao et al. 2022). They can impact neuronal function and behavior, potentially due to oxidative stress and microglia activation in the brain. In vivo testing in mice has shown polystyrene (PS) particulate matter is associated with inhibition of acetylcholinesterase (AChE), which increases oxidative stress, thereby changing neurotransmitter levels (Deng et al. 2017).

Numerous investigations have demonstrated that there are microplastics in the environment that are easily ingested. Around 16% of all plastic produced worldwide is now composed of plastic textile fibers, whose production has been increasing by more than 6% annually. General wear and washing can cause small fibers from clothes to shed; it has been approximated that a single garment will discharge 1900 fibers into wastewater with every wash (Liu et al. 2019). The main consequences of phthalates are asthma and allergies, which are most prevalent in children due to their weaker immunity. It has also been demonstrated that phthalate exposure has biological impacts during pregnancy and may shorten the length of pregnancy (Bamai et al. 2014). Analogous to phthalates, bisphenol-A (BPA) has also been investigated and found to be a reproductive toxin, linked to unfavorable pregnancy outcomes. Microplastics raise the possibility of platelet-driven events through a process that initiates or expedites platelet aggregation. This effect is influenced by their distinctive characteristics in terms of both size and surface chemistry. Large amounts of the chemical bisphenol A (BPA) are manufactured primarily to produce composite polymers. Water bottles, eyeglasses, shatterproof windows, epoxy coatings on certain metal food containers, bottle lids, and water conduits are among the products that incorporate them. Several studies have examined cytotoxicity, oxidative stress, and cellular viability to assess the detrimental effects of microplastics and their additives on cells. After exposure to varying concentrations of pollutants, human cerebral and epithelial cells displayed oxidative stress with the introduction of microplastics; nonetheless, there was no appreciable decrease in cell viability (Schirinzi et al. 2017).

Primary and secondary toxins are the two categories of toxins found in microplastics.

Primary toxics include phthalates, bisphenol-A (BPA), polyvinyl chloride (PVC), and polybrominated diphenyl ethers (PBDEs). Secondary toxins include POPs, PAHs, antimicrobials, halogenated flame retardants (HFRs), and PCBs. Pesticides such as carbendazim, dipterex, and malathion have the ability to adhere to microplastics. Uninfluenced MPs swallowed by an infected amphipod effectively eliminated organic contaminants. It has been found that MPs and POPs cause immunotoxic effects in blood clams; however, the toxicity seemed to be lessened when the plastics had a larger size of 30 μm as opposed to the smaller 500-nm diameter (Tang et al. 2020). More microplastics are being discovered in the soil, suggesting that plastic mulching sheets are being used in agriculture for extended periods. They represent a significant source of microplastics in arable soil (Huang et al. 2020). Research is currently being conducted to estimate the global quantity of airborne microplastics. When atmospheric microplastic deposition was measured in Central London, the results were 20 times higher than at a more remote site (Wright et al. 2020). In an additional investigation, mice were administered varying dosages of polyethylene microplastics. High MP concentrations resulted in enhanced gut bacterial variety and diversity. All dietary groups had considerably greater levels of interleukin 1a compared to the plain group. Mice given high concentrations of MPs displayed elevated inflammatory responses as well as dysregulated TLR4, AP-1, and IRF5 expression (Li et al. 2020).

14.7 Microplastic in Our Oceans: Risks and Concern

With an increase in understanding of the impacts of plastic particles in the water, there has been a significant rise in global awareness of the issues caused by end-of-life plastic. The hydrosphere is considered the principal reservoir for microplastics due to anthropogenic actions like tourism and wastewater treatment. The wastewater treatment plant outfalls account for approximately seven million microplastic particles daily, which are released into water bodies (Haque and Fan 2023). Aquatic contamination from plastic litter has been a significant sustainability issue in the past few years. Since the 1960s, there has been a wealth of evidence showing that smaller particles of plastic waste in the environment cause different problems than larger ones. Big items contribute to the amount of waste, increasing the likelihood that wildlife may come into contact with them as they break down into microplastics (Shim and Thomposon 2015). Additionally, a wider variety of creatures may consume the trash as particle sizes decrease. This, in turn, makes tiny polymers more accessible and ingestible, particularly for small creatures. Furthermore, because it would be very difficult to physically remove tiny plastic particles from the environment, their persistence in the ecosystem may rise. These microplastics not only stack in the aquatic environment but can also be transported to least accessible areas such as glacial zones. Of the total microplastics collected in water bodies, as much as 70% are sedimented in the waterbeds, while 15% remain dispersed in the water. There is a remarkable variance in the dissemination mechanism of microplastics, which is influenced by the degradation pathways of microplastics, as well as water and wind currents (Haque and Fan 2023). The horizontal and vertical distributions are the two most likely routes for microplastics to disperse throughout the hydrosphere. Water conditions, wind speed, precipitation, and water circulation velocity all affect horizontal dispersion. Depending on their size and density, different types of microplastics can either float or sink. Microplastic movement is impacted by hydrological factors such as river bed morphology, velocity of water flow, depth of the water, and direction of the current. The building of dams and reservoirs in rivers can have an impact on the flow of water and the destiny of microplastics (Xia et al. 2020; Bullard et al. 2021). These are distributed vertically by sinking into water beds, such as riverbeds, ocean floors, and seabeds. The properties of microplastics and the hydrosphere are factors that influence biological fouling and the production of microplastic aggregates. The process of biological fouling includes microorganisms and nutrients adhering to the surface of microplastics, forming biofilms that draw marine life, and sinking to the bottom (Alimi et al. 2018; Kaiser et al. 2017). Upon analysis, it has been revealed the existence of microplastics in many areas of the hydrosphere, including freshwater, rivers, oceans, shorelines, deep seas, and coral reefs. Sediments are a hot location for the buildup of microplastics, and the deep sea is thought to be a global reservoir for them (Haque and Fan 2023). During their development, coral reefs may potentially adsorb microplastics from the nearby waters (Reichert et al. 2022).

14.7.1 Risks and Concerns

Microplastics and the chemical toxicity associated with them are major causes of contamination these days. Microplastics are present in various aquatic organisms, including fish, springtails, shrimp, and oysters (Thompson et al. 2009). This presence can result in various consequences, such as impaired growth, reproductive system dysfunction, problems in hatching, and increased mortality rates. Physical damage associated with the consumption of microplastics includes impairment to the gastrointestinal tract, chemical alteration of enzyme activities, influence on gene modification, and the creation of resistant genes (Cole et al. 2015; Sussarellu et al. 2016). Upon entering the food chain, they cause biomagnification as they move from primary to secondary consumers and finally to humans. It has also been discovered that larger aquatic animals, such as walruses, fur seals, humpback whales, and turtles, contain microplastics in their bodies (van Raamsdonk et al. 2020). The southern and western coastlines of Korea were found to have the highest concentrations of paint resin particles and other microplastic waste in saltwater worldwide. Elevated concentrations of microplastic were discovered in the sea-surface microlayer. Extensive surveillance in near-shore waters, together with polymer compositional profiles, identified a variety of input sources that included shipyards, seafood farms, and sewage outflow (Chae et al. 2015; Song et al. 2015). Research conducted by Sequeira et al. (2020) revealed that microplastics have been ingested by up to

60% of fish from 198 species across 24 nations. It has been discovered that microplastics cause hamstring strain in the planktonic crustacean *Daphnia magna*, which hinders its ability to reproduce effectively (Blackburn and Green 2022).

14.8 Microplastics and Terrestrial Life: Unseen Perils

The study of microplastics is primarily focused on hydrosphere and very less reports are established on soil biota as the sink for microplastics. However, there are many detrimental impacts of microplastics on terrestrial organisms as well, and therefore, it is required to be investigated further. The major contribution of microplastics is given by tourism activities, due to which there is plastic litter accumulation. Other domestic activities also contribute to the same. According to the studies, at least 90 million supermarket bags are used every day, and these bags end up as improperly handled plastic garbage (Li and Richter 2015). Following this, farming practices such as the use of plastic mulches and fertilizers contribute to the bulk of microplastics. Agricultural runoffs transport microplastics to the nearest water bodies, but some microplastics are trapped in the soil. This plastic waste degrades in the presence of the environment and releases microplastics into the ground. It is believed that soil contains 300 million tons of microplastics (Lebreton et al. 2017).

The fate and movement of microplastics in soil are poorly understood. Numerous pathways of microplastic transfer in the soil are illustrated by experimental research. One such process is pedoturbation by soil fauna. This process occurs with the assistance of soil fauna, such as earthworms, soil larvae, and vertebrates (Gabet et al. 2003). Additional pathways for the transportation of microplastics in soil include root migration, tillage operations, and tuber crop harvesting. Microplastics in soil migrate downward due to a variety of factors, including soil characteristics, external influences, and wetting-drying cycles. Microplastics' composition, size, and form all influence how they move through soil (Huerta Lwanga et al. 2016). To fully comprehend how soil characteristics and external influences affect the migration of microplastics in soil, further research is needed. The diagrammatic representation of the effect of microplastics when they enter the soil or land and how it affects the food chain is shown in Fig. 14.1.

The presence of microplastics raises concerns regarding soil ecology and ecosystem dynamics. The significant concentration of microplastics in the soil can reduce water retention and infiltration, which can result in nearly no entry of oxygen into the soil. They also interfere with the rate of evaporation as it affects the permeability of the soil, thereby enhancing the soil moisture content, which modifies the physical characteristics of the soil like bulk density, water retention capacity, porosity of the soil, and soil aggregate (Yu et al. 2023; Wang et al. 2022). Conversely, the clay soil reacts differently with the piling up of microplastic. Due to evaporation, it causes shrinkage and cracking in the clay soil (Zhang et al. 2015). Therefore, microplastics can impact the enzymatic functions within the soil system. MPs can increase zinc bioavailability and earthworm interaction with zinc, but they can also decrease the number of bacteria, have a negative effect on nutrient transfer, and disrupt the cycle

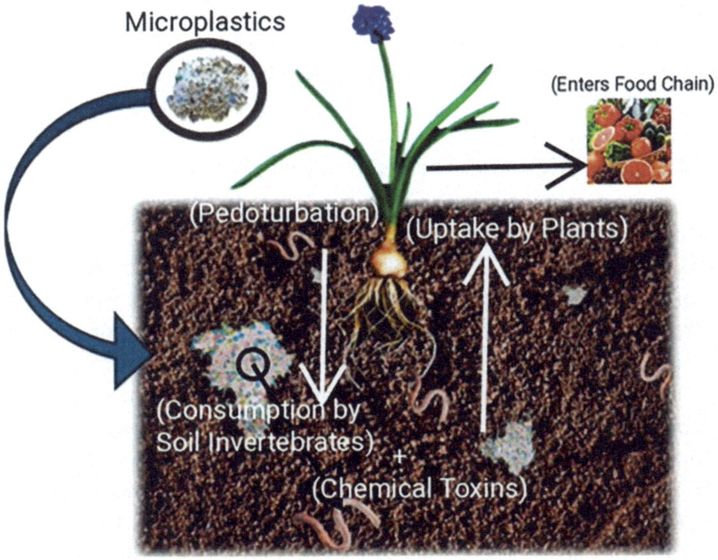

Fig. 14.1 The movement of microplastics along the food chain

of organic carbon (Hodson et al. 2017). The existence of plastic material in the soil has a considerable influence on soil enzymes, which are crucial markers of soil fertility. MPs have a substantial impact on the enzymatic activity in the soil, the surface microbiota, and ecological processes (Huang et al. 2019). Some of the enzymes whose activities are altered due to the accumulation of microplastics are catalases, ureases, fluorescein diacetate hydrolase, and extracellular enzymes (Liu et al. 2017). Microplastics carry POPs in agricultural soil such as polycyclic aromatic hydrocarbons, polychlorinated biphenyls, pesticides, and herbicides (Frias et al. 2010). Microplastics have the potential to serve as transporters of heavy metals present in the soil, either by moving the metals into deeper soil layers or making them accessible for plant uptake (Hodson et al. 2017). Organisms inhabiting the land environment experience negative impacts from the microplastic buildup in the soil. The ingestion of microplastics poses challenges for animals due to their indigestibility, leading to internal accumulation (Peng et al. 2017). Research has shown that mice exposed to soil contaminated with microplastics exhibited reduced metabolic activity and intestinal harm (Jin et al. 2019). While smaller species such as earthworms are capable of breaking down microplastics, this process may compromise their digestive systems and diminish their survival prospects (Sajjad et al. 2022). Because of their tiny size and large surface area, they promote chemotactic factors that boost permeability and prevent macrophage movement once they enter the respiratory system of animals. This could lead to several respiratory illnesses such as chronic inflammation (Atis et al. 2005). Additionally, microplastics are hazardous for plant growth and survival as well. For instance, it is observed that wheat and spring onions grown in soil contaminated with polystyrene, low-density

Table 14.1 Different forms of microplastics and their impact in organisms of terrestrial and aquatic environments

Microplastic form	Environmental impact	Majorly impacted organisms
Microexfoliates	Nonbiodegradable, it does not break down naturally, a single shower can flush around 100,000 microbeads into wastewater (Mason et al. 2016)	Fishes and Filter-feeding animals (Rochman et al. 2015)
Microfibers	Nonbiodegradable and remains in the environment for extended periods, often lasting hundreds of years They can absorb and accumulate toxic chemicals present in the atmosphere (Zubris and Richards 2005)	Mussels, clams, and oysters are particularly vulnerable to microfiber ingestion (Browne et al. 2011) Can be ingested by zooplankton, causing blockages in digestive systems (Cole et al. 2011)
Microplastic flakes	Bioaccumulation in the food chain (Rochman et al. 2013) Can accumulate on shorelines and in sediments, altering the structure which can disrupt important ecological processes, such as the burrowing behavior of benthic organisms (Rochman et al. 2015)	Sea turtles; can lead to intestinal blockages and malnutrition (Schuyler et al. 2016)
Microplastic nurdles	Contribute to water pollution. When spilled into water they can float and disperse widely, contaminating the water column (Koelmans et al. 2016)	Seabirds; they often mistake nurdles for fish eggs and consume them (Provencher et al. 2017)
Foamed microplastics	Chemical pollution; it can leach chemicals into the environment and can also release toxic additives, such as flame retardants (Koelmans et al. 2016)	Birds (Northern Fulmars); particularly seabirds like gulls and albatrosses, are highly impacted (Avery-Gomm et al. 2012)
Microplastic films	Can change the physical and chemical properties of water bodies by lowering oxygen levels, and blocking sunlight (Sussarellu et al. 2016)	Oysters; exposure to polystyrene microplastics impacts their fertility (Sussarellu et al. 2016)

polyethylene, high-density polyethylene, or polypropylene displayed decreased biomass in the roots and shoots (Qi et al. 2018).

Several examples of organisms that are reported to have the impact of various types of microplastics are listed in Table 14.1.

14.9 The Global Proliferation of Microplastics

Microplastics are small particles of plastics that transition from macro-debris to nanoplastics. This suggests that upon fragmentation of microplastics, there is a significant expansion in nanoparticle dumps in the globe. In the last seven decades, there has been a remarkable increase in the global reliance on plastics due to a high yearly growth rate in production. In developed countries, these are mostly disposed of in landfills and these accumulated piles are considered as mismanaged wastes. Sheer and Moss (2011) have reported that medical wastes that were discarded in landfills were washed up on nearby beaches, also referred to as syringe tides, in New York in 1988. The issue of marine plastic contamination is widespread and irreversible. Large amounts of plastics are floating in the ocean, which accounts for nearly 236,000 metric tons (Van Sebille et al. 2015) and significant annual inputs of plastic debris are concerning. It has been documented that 4.8 to 12.7 million tons of plastics are added to the ocean every year, with an estimated tenfold rise in total imports by 2025 (Jambeck et al. 2015). This data excludes the plastic collection beyond the ocean's surface as sampling in deeper waters, sediments, soil, air, and biological systems are not studied much. Recently, large amount of microplastic debris has been located in areas that are least occupied and explored such as Arctic sea ice (Obbard et al. 2014), high mountain peaks, fresh-water lake summits (Neelavannan et al. 2022), oceanic trenches including Mariana Trench (Jamieson et al. 2019), and Antarctic snow (Aves et al. 2022). As plastic pollution increases, research into the ecological disruption brought on by plastic contamination is essential. Strategies for mitigating microplastic pollution need to be implemented on a very serious note considering the dangers associated with it.

14.10 Conclusion

Microplastics are pervasive in ecosystems and across various species, presenting environmental and health concerns. Their unique characteristics, such as buoyancy, resilience, lightweight nature, and distinctive shapes, enable them to travel significant distances on Earth. Because of their notable resistance to degradation, microplastics persist in the environment, leading to their accumulation within ecosystems and entering living beings at the lower trophic levels, such as phytoplankton. As they ascend the trophic levels, they impact various organisms, primarily inducing toxic effects through oxidative stress. The chemical components within microplastics can have detrimental effects on animals, humans, and plants, manifesting in issues like ingestion by marine life, disruption of reproductive systems, economic repercussions on sectors like fishing and tourism, as well as health hazards including respiratory and digestive complications. Since 2014, researchers have classified microplastics as a category of organic contaminants of growing concern. However, the scarcity of experimental data poses challenges in fully evaluating their toxicity in humans. Initiatives are underway to address the reduction of microplastic accumulation in oceans and diminish global plastic consumption. Further research is imperative to comprehend the influence of microplastics on ecosystems and human health, facilitating the formulation and enhancement of mitigation strategies. Comprehensive global efforts are essential to significantly decrease plastic usage. While policies and programs exist to combat microplastic pollution, obstacles such as insufficient awareness, limited resources, and ineffective regulations must be addressed. Immediate action is crucial to curtail plastic consumption, advocate for proper disposal and recycling practices, innovate technologies for microplastic removal, raise public consciousness, and enforce efficient regulations. Collaborative endeavors involving individuals, industries, and governments are vital to effectively combat the menace of microplastics and safeguard the planet for future generations.

References

Alimi OS, Farner Budarz J, Hernandez LM, Tufenkji N (2018) Microplastics and nanoplastics in aquatic environments: aggregation, deposition, and enhanced contaminant transport. Environ Sci Technol 52(4):1704–1724

An L, Liu Q, Deng Y, Wu W, Gao Y, Ling W (2020) Sources of microplastic in the environment. In: He D, Luo Y (eds) Microplastics in terrestrial environments: emerging contaminants and major challenges. Springer, Cham, pp 143–159

Atis S, Tutluoglu B, Levent E, Ozturk C, Tunaci A, Sahin K et al (2005) The respiratory effects of occupational polypropylene flock exposure. Eur Respir J 25(1):110–117

Aves AR, Revell LE, Gaw S, Ruffell H, Schuddeboom A, Wotherspoon NE et al (2022) First evidence of microplastics in Antarctic snow. Cryosphere 16(6):2127–2145

Avery-Gomm S, O'Hara, PD, Kleine L, Bowes V, Wilson LK, Barry KL (2012) Northern fulmars as biological monitors of trends of plastic pollution in the eastern North Pacific. Marine pollution bulletin, 64(9), 1776-1781.

Bamai YA, Araki A, Kawai T, Tsuboi T, Saito I, Yoshioka E et al (2014) Associations of phthalate concentrations in floor dust and multi-surface dust with the interior materials in Japanese dwellings. Sci Total Environ 468:147–157

Barnes DK, Galgani F, Thompson RC, Barlaz M (2009) Accumulation and fragmentation of plastic debris in global environments. Philosophical transactions of the royal society B: biological sciences, 364(1526), 1985–1998.

Blackburn K, Green D (2022) The potential effects of microplastics on human health: what is known and what is unknown. Ambio 51(3):518–530

Browne MA, Crump P, Niven SJ, Teuten E, Tonkin A, Galloway T, Thompson R (2011) Accumulation of microplastic on shorelines worldwide: sources and sinks. Environ Sci Technol 45(21):9175–9179

Bullard JE, Ockelford A, O'Brien P, Neuman CM (2021) Preferential transport of microplastics by wind. Atmos Environ 245:118038

Campanale C, Massarelli C, Savino I, Locaputo V, Uricchio VF (2020) A detailed review study on potential effects of microplastics and additives of concern on human health. Int J Environ Res Public Health 17(4):1212

Chae DH, Kim IS, Kim SK, Song YK, Shim WJ (2015) Abundance and distribution characteristics of microplastics in surface seawaters of the Incheon/Kyeonggi coastal region. Arch Environ Contam Toxicol 69:269–278

Cole M, Lindeque P, Halsband C, Galloway TS (2011) Microplastics as contaminants in the marine environment: a review. Mar Pollut Bull 62(12):2588–2597

Cole M, Lindeque P, Fileman E, Halsband C, Galloway TS (2015) The impact of polystyrene microplastics on feeding, function and fecundity in the marine copepod *Calanus helgolandicus*. Environ Sci Technol 49(2):1130–1137

Coppock RL, Lindeque PK, Cole M, Galloway TS, Näkki P, Birgani H et al (2021) Benthic fauna contribute to microplastic sequestration in coastal sediments. J Hazard Mater 415:125583

Cózar A, Echevarría F, González-Gordillo JI, Irigoien X, Úbeda B, Hernández-León S et al (2014) Plastic debris in the open ocean. Proc Natl Acad Sci 111(28):10239–10244

de Souza Machado AA, Kloas W, Zarfl C, Hempel S, Rillig MC (2018) Microplastics as an emerging threat to terrestrial ecosystems. Glob Chang Biol 24(4):1405–1416

De Falco F, Gentile G, Avolio R, Errico ME, Di Pace E, Ambrogi V Cocca M (2018) Pectin based finishing to mitigate the impact of microplastics released by polyamide fabrics. Carbohydrate polymers, 198, 175–180.

Deng Y, Zhang Y, Lemos B, Ren H (2017) Tissue accumulation of microplastics in mice and biomarker responses suggest widespread health risks of exposure. Sci Rep 7(1):46687
Devi A, Hansa A, Gupta H, Syam K, Upadhyay M, Kaur M et al (2023) Microplastics as an emerging menace to environment: insights into their uptake, prevalence, fate, and sustainable solutions. Environ Res 229:115922
Dris R, Gasperi J, Mirande C, Mandin C, Guerrouache M, Langlois V, Tassin B (2017) A first overview of textile fibers, including microplastics, in indoor and outdoor environments. Environmental pollution, 221, 453–458.
Du C, Liang H, Li Z, Gong J (2020) Pollution characteristics of microplastics in soils in southeastern suburbs of Baoding City, China. Int J Environ Res Public Health 17(3):845
Eriksen M, Lebreton LC, Carson HS, Thiel M, Moore CJ, Borerro JC et al (2014) Plastic pollution in the world's oceans: more than 5 trillion plastic pieces weighing over 250,000 tons afloat at sea. PLoS One 9(12):e111913
Frias JPGL, Sobral P, Ferreira AM (2010) Organic pollutants in microplastics from two beaches of the Portuguese coast. Mar Pollut Bull 60(11):1988–1992
Gabet EJ, Reichman OJ, Seabloom EW (2003) The effects of bioturbation on soil processes and sediment transport. Annu Rev Earth Planet Sci 31(1):249–273
Gasperi J, Wright SL, Dris R, Collard F, Mandin C, Guerrouache M Tassin B (2018) Microplastics in air: are we breathing it in?. Current Opinion in Environmental Science & Health, 1, 1-5.
Hale RC, Seeley ME, La Guardia MJ, Mai L, Zeng EY (2020) A global perspective on microplastics. J Geophys Res Oceans 125(1):e2018JC014719
Haque F, Fan C (2023) Fate and impacts of microplastics in the environment: hydrosphere, pedosphere, and atmosphere. Environments 10(5):70
Hodson ME, Duffus-Hodson CA, Clark A, Prendergast-Miller MT, Thorpe KL (2017) Plastic bag derived-microplastics as a vector for metal exposure in terrestrial invertebrates. Environ Sci Technol 51(8):4714–4721
Huang Y, Zhao Y, Wang J, Zhang M, Jia W, Qin X (2019) LDPE microplastic films alter microbial community composition and enzymatic activities in soil. Environ Pollut 254:112983
Huang Y, Liu Q, Jia W, Yan C, Wang J (2020) Agricultural plastic mulching as a source of microplastics in the terrestrial environment. Environ Pollut 260:114096
Huerta Lwanga E, Gertsen H, Gooren H, Peters P, Salánki T, Van Der Ploeg M et al (2016) Microplastics in the terrestrial ecosystem: implications for *Lumbricus terrestris* (Oligochaeta, Lumbricidae). Environ Sci Technol 50(5):2685–2691
Jambeck JR, Geyer R, Wilcox C, Siegler TR, Perryman M, Andrady A et al (2015) Plastic waste inputs from land into the ocean. Science 347(6223):768–771
Jamieson AJ, Brooks LSR, Reid WD, Piertney SB, Narayanaswamy BE, Linley TD (2019) Microplastics and synthetic particles ingested by deep-sea amphipods in six of the deepest marine ecosystems on Earth. R Soc Open Sci 6(2):180667
Jin Y, Lu L, Tu W, Luo T, Fu Z (2019) Impacts of polystyrene microplastic on the gut barrier, microbiota and metabolism of mice. Sci Total Environ 649:308–317
Kaiser D, Kowalski N, Waniek JJ (2017) Effects of biofouling on the sinking behavior of microplastics. Environ Res Lett 12(12):124003
Kim T, Park K, Hong J (2022) Understanding the hazards induced by microplastics in different environmental conditions. J Hazard Mater 424:127630
Koelmans AA, Bakir A, Burton GA, Janssen CR (2016) Microplastic as a vector for chemicals in the aquatic environment: critical review and model-supported reinterpretation of empirical studies. Environ Sci Technol 50(7):3315–3326
Kramm J, Völker C (2018) Understanding the risks of microplastics: a social-ecological risk perspective. Springer International Publishing, pp 223–237
Lassen C, Hansen SF, Magnusson K, Hartmann NB, Jensen PR, Nielsen TG, Brinch A (2015) Microplastics: occurrence, effects and sources of releases to the environment in Denmark. Danish Environmental Protection Agency
Lebreton LC, Van Der Zwet J, Damsteeg JW, Slat B, Andrady A, Reisser J (2017) River plastic emissions to the world's oceans. Nat Commun 8(1):15611
Li B, Ding Y, Cheng X, Sheng D, Xu Z, Rong Q, Zhang Y (2020) Polyethylene microplastics affect the distribution of gut microbiota and inflammation development in mice. Chemosphere, 244, 125492.

Li ZG, Richter JS (2015) Problem and countermeasure on promoting the plastic bag ban of USA. Appl Mech Mater 768:787–796
Liu H, Yang X, Liu G, Liang C, Xue S, Chen H et al (2017) Response of soil dissolved organic matter to microplastic addition in Chinese loess soil. Chemosphere 185:907–917
Liu K, Wang X, Wei N, Song Z, Li D (2019) Accurate quantification and transport estimation of suspended atmospheric microplastics in megacities: implications for human health. Environ Int 132:105127
Lwanga EH, Gertsen H, Gooren H, Peters P, Salánki T, van der Ploeg M et al (2017) Incorporation of microplastics from litter into burrows of *Lumbricus terrestris*. Environ Pollut 220:523–531
Kelly MR, Lant NJ, Kurr M, Burgess JG (2019) Importance of water-volume on the release of microplastic fibers from laundry. Environmental science & technology, 53(20), 11735–11744.
Mao X, Xu Y, Cheng Z, Yang Y, Guan Z, Jiang L, Tang K (2022) The impact of microplastic pollution on ecological environment: a review. Front Biosci Landmark 27(2):46
Mason SA, Garneau D, Sutton R, Chu Y, Ehmann K, Barnes J et al (2016) Microplastic pollution is widely detected in US municipal wastewater treatment plant effluent. Environ Pollut 218:1045–1054
Napper IE, Bakir A, Rowland SJ, Thompson RC (2015) Characterisation, quantity and sorptive properties of microplastics extracted from cosmetics. Mar Pollut Bull 99(1–2):178–185
Neelavannan K, Sen IS, Lone AM, Gopinath K (2022) Microplastics in the high-altitude Himalayas: assessment of microplastic contamination in freshwater lake sediments, Northwest Himalaya (India). Chemosphere 290:133354
Obbard RW, Sadri S, Wong YQ, Khitun AA, Baker I, Thompson RC (2014) Global warming releases microplastic legacy frozen in Arctic Sea ice. Earth's Future 2(6):315–320
Pandey B, Pathak J, Singh P, Kumar R, Kumar A, Kaushik S, Thakur TK (2022) Microplastics in the ecosystem: an overview on detection, removal, toxicity assessment, and control release. Water 15(1):51
Peng J, Wang J, Cai L (2017) Current understanding of microplastics in the environment: occurrence, fate, risks, and what we should do. Integr Environ Assess Manag 13(3):476–482
Provencher JF, Bond AL, Avery-Gomm S, Borrelle SB, Rebolledo ELB, Hammer S et al (2017) Quantifying ingested debris in marine megafauna: a review and recommendations for standardization. Anal Methods 9(9):1454–1469
Qi Y, Yang X, Pelaez AM, Lwanga EH, Beriot N, Gertsen H et al (2018) Macro- and microplastics in soil-plant system: effects of plastic mulch film residues on wheat (*Triticum aestivum*) growth. Sci Total Environ 645:1048–1056
Reichert J, Arnold AL, Hammer N, Miller IB, Rades M, Schubert P et al (2022) Reef-building corals act as long-term sink for microplastic. Glob Chang Biol 28(1):33–45
Rochman CM, Hoh E, Kurobe T, Teh SJ (2013) Ingested plastic transfers hazardous chemicals to fish and induces hepatic stress. Sci Rep 3(1):1–7
Rochman CM, Tahir A, Williams SL, Baxa DV, Lam R, Miller JT et al (2015) Anthropogenic debris in seafood: plastic debris and fibers from textiles in fish and bivalves sold for human consumption. Sci Rep 5(1):1–10
Sajjad M, Huang Q, Khan S, Khan MA, Liu Y, Wang J et al (2022) Microplastics in the soil environment: a critical review. Environ Technol Innov 27:102408
Schirinzi GF, Pérez-Pomeda I, Sanchís J, Rossini C, Farré M, Barceló D (2017) Cytotoxic effects of commonly used nanomaterials and microplastics on cerebral and epithelial human cells. Environ Res 159:579–587
Schuyler QA, Wilcox C, Townsend KA, Wedemeyer-Strombel KR, Balazs G, van Sebille E, Hardesty BD (2016) Risk analysis reveals global hotspots for marine debris ingestion by sea turtles. Glob Chang Biol 22(2):567–576
Sequeira IF, Prata JC, da Costa JP, Duarte AC, Rocha-Santos T (2020) Worldwide contamination of fish with microplastics: A brief global overview. Marine Pollution Bulletin, 160, 111681.
Sheer R, Moss D (2011) Medical waste on beaches. Environ Mag
Shim WJ, Thomposon RC (2015) Microplastics in the ocean. Arch Environ Contam Toxicol 69:265–268

Song YK, Hong SH, Jang M, Han GM, Shim WJ (2015) Occurrence and distribution of microplastics in the sea surface microlayer in Jinhae Bay, South Korea. Arch Environ Contam Toxicol 69:279–287

Song Y, Cao C, Qiu R, Hu J, Liu M, Lu S et al (2019) Uptake and adverse effects of polyethylene terephthalate microplastics fibers on terrestrial snails (*Achatina fulica*) after soil exposure. Environ Pollut 250:447–455

Sussarellu R, Suquet M, Thomas Y, Lambert C, Fabioux C, Pernet MEJ et al (2016) Oyster reproduction is affected by exposure to polystyrene microplastics. Proc Natl Acad Sci 113(9):2430–2435

Tang Y, Rong J, Guan X, Zha S, Shi W, Han Y et al (2020) Immunotoxicity of microplastics and two persistent organic pollutants alone or in combination to a bivalve species. Environ Pollut 258:113845

Thompson RC, Olsen Y, Mitchell RP, Davis A, Rowland SJ, John AW et al (2004) Lost at sea: where is all the plastic? Science 304(5672):838

Thompson RC, Moore CJ, Vom Saal FS, Swan SH (2009) Plastics, the environment and human health: current consensus and future trends. Philos Trans R Soc B Biol Sci 364(1526):2153–2166

van Raamsdonk LW, van der Zande M, Koelmans AA, Hoogenboom RL, Peters RJ, Groot MJ et al (2020) Current insights into monitoring, bioaccumulation, and potential health effects of microplastics present in the food chain. Food Secur 9(1):72

Van Sebille E, Wilcox C, Lebreton L, Maximenko N, Hardesty BD, Van Franeker JA et al (2015) A global inventory of small floating plastic debris. Environ Res Lett 10(12):124006

Wang Z, Sun C, Li F, Chen L (2021a) Fatigue resistance, re-usable and biodegradable sponge materials from plant protein with rapid water adsorption capacity for microplastics removal. Chem Eng J 415:129006

Wang C, Zhao J, Xing B (2021b) Environmental source, fate, and toxicity of microplastics. J Hazard Mater 407:124357

Wang F, Wang Q, Adams CA, Sun Y, Zhang S (2022) Effects of microplastics on soil properties: current knowledge and future perspectives. J Hazard Mater 424:127531

Ward CP, Armstrong CJ, Walsh AN, Jackson JH, Reddy CM (2019) Sunlight converts polystyrene to carbon dioxide and dissolved organic carbon. Environ Sci Technol Lett 6(11):669–674

Wong JKH, Lee KK, Tang KHD, Yap PS (2020) Microplastics in the freshwater and terrestrial environments: Prevalence, fates, impacts and sustainable solutions. Science of the total environment, 719, 137512.

Wright SL, Thompson RC, Galloway TS (2013) The physical impacts of microplastics on marine organisms: a review. Environ Pollut 178:483–492

Wright SL, Ulke J, Font A, Chan KLA, Kelly FJ (2020) Atmospheric microplastic deposition in an urban environment and an evaluation of transport. Environ Int 136:105411

Xia W, Rao Q, Deng X, Chen J, Xie P (2020) Rainfall is a significant environmental factor of microplastic pollution in inland waters. Sci Total Environ 732:139065

Xu C, Zhang B, Gu C, Shen C, Yin S, Aamir M, Li F (2020) Are we underestimating the sources of microplastic pollution in terrestrial environment? J Hazard Mater 400:123228

Youcai Z, Ziyang L (2016) Pollution control and resource recovery: municipal solid wastes at landfill. Butterworth-Heinemann

Yu Y, Battu AK, Varga T, Denny AC, Zahid TM, Chowdhury I, Flury M (2023) Minimal impacts of microplastics on soil physical properties under environmentally relevant concentrations. Environ Sci Technol 57(13):5296–5304

Zhang ZB, Peng X, Zhou H, Lin H, Sun H (2015) Characterizing preferential flow in cracked paddy soils using computed tomography and breakthrough curve. Soil Tillage Res 146:53–65

Zhang K, Shi H, Peng J, Wang Y, Xiong X, Wu C, Lam PK (2018) Microplastic pollution in China's inland water systems: a review of findings, methods, characteristics, effects, and management. Sci Total Environ 630:1641–1653

Zubris KAV, Richards BK (2005) Synthetic fibers as an indicator of land application of sludge. Environ Pollut 138(2):201–211

Sustainable Management of Microplastics

15

Deeksha Ranjan and Laxmi Singh

Abstract

Concern over the effects of Microplastics (MPLs) on humans and the entire natural world has grown in the last few decades. Particles like these originate from a multitude of sources, including personal hygiene products, soapy products, and the detritus of larger plastic goods. MPLs have serious negative effects on other ecological systems and aquatic organisms, including mortality and intrusion with reproduction processes as well as consumption by underwater creatures. The occurrence of MPLs in the surroundings has an impact on a number of commercial enterprises, including touristry and trawling, which has substantial financial ramifications. Human health hazards associated with microplastic exposure include breathing problems, difficulties with digestion, difficulty falling asleep, weight gain, and a greater chance of being diabetic. Prevention and remediation are equally necessary in the fight against MPLs contamination. Even if rehabilitation is necessary, given the state of technology currently, it could require quite a while to complete. However, prevention is something that can and ought to be undertaken right now. To fully comprehend the possible environmental effect, more studies are required on the various risk of adverse reactions, and the process of leaching.

D. Ranjan (✉)
Department of Applied Sciences and Humanities, Faculty of Engineering and Technology, Rama University, Kanpur, Uttar Pradesh, India

L. Singh
Department of Applied Sciences and Humanities, Faculty of Engineering and Technology, Rama University, Kanpur, Uttar Pradesh, India

Department of Biotechnology and Bioinformatics, School of Life Sciences, University of Hyderabad. Hyderabad, Telangana, India

A. Kumar, V. Singh (eds.), *Microplastics Pollution and its Remediation*,
https://doi.org/10.1007/978-981-97-4068-0_15

Keywords

Microplastics · Environmental · Contamination · Remediation · Adverse · Leaching

15.1 Introduction

Plastic is being found in nature more and more, not just because of its enormous quantity of manufacture but also due to improper handling of the material particularly when it turns into trash (Lebreton and Andrady 2019). Due to the detrimental effects of extensively dispersed MPLs in the ecosystem, there is a worrisome scenario regarding accumulation of plastic (da Costa et al. 2017) (Fig. 15.1). MPLs contamination is a contemporary environmental problem caused by the vast amount of plastic trash in our surroundings and the deterioration of bigger polymers into tiny particles (Geyer et al. 2017).

There are plastic wastes of different sizes and varieties. Microplastics are particles of waste having size 1–5 mm in diameter (Kumar et al. 2020). According to their place of origin, microplastics can also be divided into two groups including primary microplastics, which are introduced into the environment right away, and secondary microplastics, which are produced when bigger plastics like meso-, macro-, and mega plastics degrade in the environment (Jambeck et al. 2015) (Fig. 15.2). MPLs are ubiquitous in the ecosystem because they are so tiny that they are impossible to remove using current technologies (Napper et al. 2015).

The introduction of effective recycling processes and MPLs degrading technologies is essential in achieving an environment free of MPLs. Beginning with coast all the way to the broad sea and ocean across the sea level and from semitropic zone to frigid zone, MPLs are present everywhere (Ivar do Sul et al. 2014), increasingly accessible to sea creatures, lower marine animals, including zooplankters, since having smaller size (Rios-Mendoza et al. 2021). Reports of consumed MPLs have been made for a wide range of taxonomies and trophic levels (Watts et al. 2014). By sustaining studies across a range of inanimate and animate patterns in the ecosystems, scientists can ascertain their emission circumstance, areas of concern and organism exposures (Imhof et al. 2017a, b). Over time, biodegradable substitutes for microplastics have been supplanted in a number of hygiene products, including dental products and facial and beauty products (Fig. 15.3). As per the norms, a face-wash pack might have around four million microbeads in it. MPLs from these products get up in nature because sewage plants are unable to eliminate them when they are discharged into the sewers (Cole et al. 2011).

Microplastics may also potentially enter the ecosystem through fabrics from the textile industry when discharged (de Souza Machado et al. 2018). Merely 9–12% of all plastic waste in the environment can be reused through recycling (Talvitie et al. 2017). The remaining burned, abandoned in open spaces, or placed in landfills. Most plastic waste winds up in streams, creeks, and other water bodies, which all contribute to the ocean ultimately (Nizzetto et al. 2016). Due to their tiny stature, low-density levels, breeze, and stream are the causes of their migration. MPLs and

Fig. 15.1 Pictures showing plastic pollution that will gradually lead to microplastic pollution in the ecosystem. The pictures were originally taken by Manasi Krishna at Ambersut, District Lakhimpur, Uttar Pradesh, India

plankton are of the same size, and hence aquatic life confuses the two. The majority of zooplankton consume MPLs, which they then transfer to a greater food chain hierarchy (Mason et al. 2016). It has been found that zooplankton belonging to over 30 species and 28 taxonomic groupings eat MPLs. Numerous research studies have been carried out on sustainable methods of managing plastic trash, including a decrease in sources, and positive usage through changes in laws and regulations (Kim et al. 2017). Additionally, cutting-edge methods like technological advances, turning the plastic debris to a usable form of energy, converting it into worthwhile components, and using it to make building supplies help to assist the environmentally friendly handling of plastic particles (Mintenig et al. 2019). Several questions have been expressed about MPLs in terrestrial and aquatic ecosystems due to their detrimental impacts on ecosystem functioning as well as the health of humans

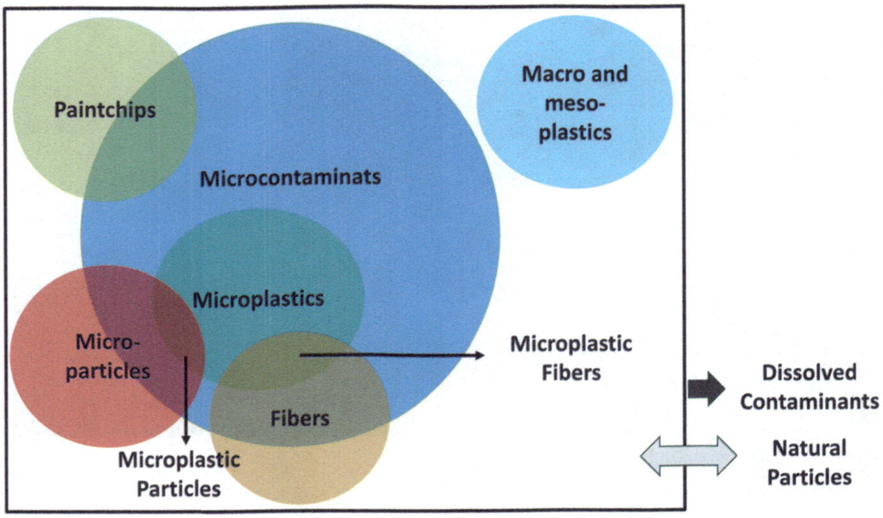

Fig. 15.2 Portrayal of MPLs in a more comprehensive study of microcontaminants

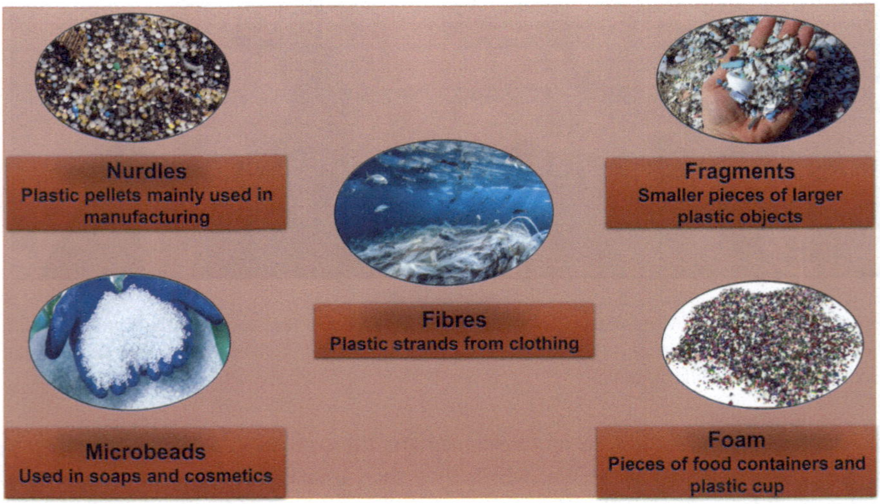

Fig. 15.3 Commonly accessible form of microplastics according to size

(Vandermeersch et al. 2015). In recent times, the management of plastic trash has emerged as a global issue. Thus, attaining the Sustainable Development Goals requires the use of sustainable waste management strategies for plastic (Thornton Hampton et al. 2022). The ecotoxicological effects of sources of plastic trash and concerns about the long-term sustainable handling of MPLs in the environment are initial issues (Lei et al. 2018). Minimizing the environmental impact caused by MPLs can be accomplished by reducing the sources, implementing a regenerative economy into plastics, and managing trash properly (Horton et al. 2017). Suitable planning, waste valuation, and commodities of low value for waste handling methods as well as other feasible substitutes are crucial to minimize MPLs in the surroundings (Free et al. 2014). MPLs are more difficult to track down, manage, and measure in land and aquatic ecosystems (Bradney et al. 2019). These particles like chemical compounds and other inorganic and organic pollutants have the ability to build up across the food supply chain (Blair et al. 2019). Serious ecological consequences are caused by these pollutants, including percolated chemical agents concerning polymers and possibly hazardous components and perpetual compounds bounded to MPLs (Mercy et al. 2023; Groh et al. 2019).

15.2 Source of Microplastics in the Environment

During the production process, plastics are frequently treated with synthetic organofluorine chemical compounds like perfluorohexanesulfonic acid (PFHxS), and perfluorononanoic acid (PFNA), polynuclear aromatics (PNAs), inclusive of acenaphthene, anthracene, naphthalene, acenaphthylene, and others (Brennecke et al. 2016) (Fig. 15.4).

Additionally, metals like arsenic (As), chromium (Cr), and lead (Pb) are also mixed with MPLs and becomes an evidence of serious problem (Sengupta and Dey 2020). These components have the potential to leak into the surroundings (Pauna et al. 2019).

MPLs have the potential to enter via unregulated disposal locations, detritus, and various anthropogenic endeavors, leading to the pollution of ecosystems on land, especially soil (Wright et al. 2013) (Fig. 15.5). Furthermore, aquatic species and soil organisms can more easily consume MPLs due to their petite size (Smiroldo et al. 2019). MPLs deposition throughout the food web causes acute and long-term harmful effects in suspension feeding animals, crustaceans, sea creatures, and terrarium worm (Wang et al. 2019). As a result, creating a clever plan for the environment friendly handling of plastics is crucial. A number of investigations have been carried out to examine MPLs in soil in particular (Jiang 2021). Since the early 1990s, reports of MPLs in water bodies and their detrimental effects have been made, leading to studies centered around MPLs in aquatic ecosystem (da Costa et al. 2018)

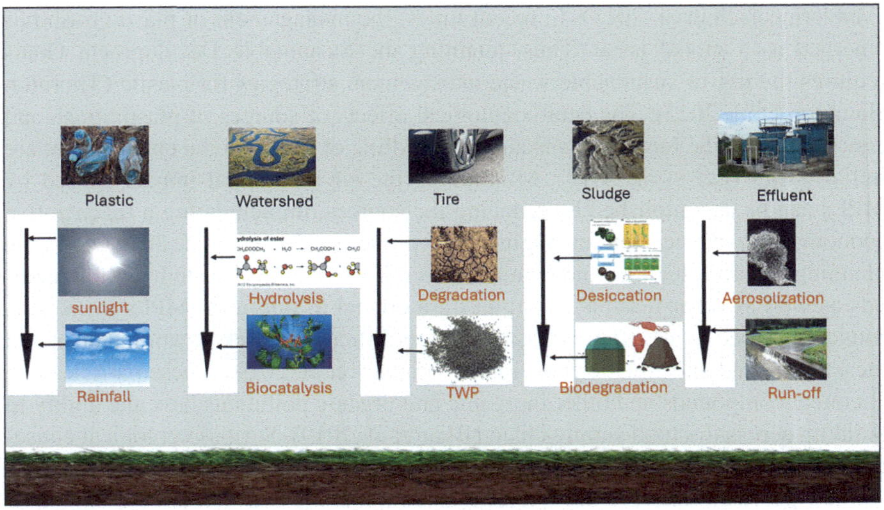

Fig. 15.4 Introduction of MPLs in the ecosystem

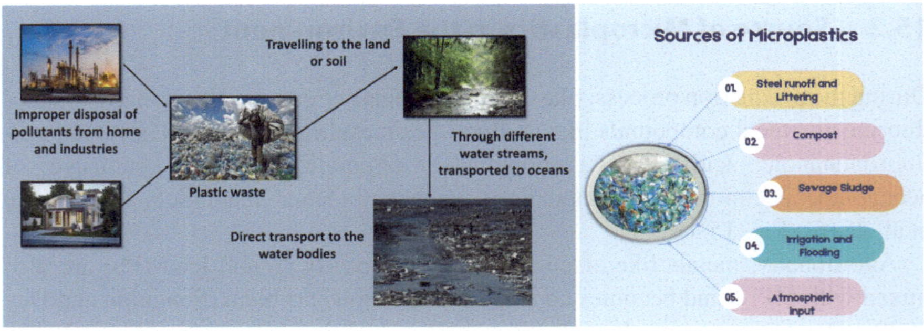

A) B)

Fig. 15.5 Sources of MPLs in the ecosystem

15.3 Chemical Composition of Microplastics

MPLs are comprised of a broad range of compounds and is additionally identified by its chemical composition. Numerous plastic goods use blended substances that contain a variety of polymers in addition to a variety of additives (Rochman et al. 2019) (Fig. 15.6). Research focused on the treatment of wastewater have mentioned about the additives like plasticizing agents, coloring substances, preservatives, and flame retardants, nor adsorbed products like polycyclic aromatic hydrocarbons, polychlorinated biphenyls, dichlorodiphenyltrichloroethane, hazardous metals, and medical supplies and self-care products (Rillig et al. 2021; Zhou et al. 2020).

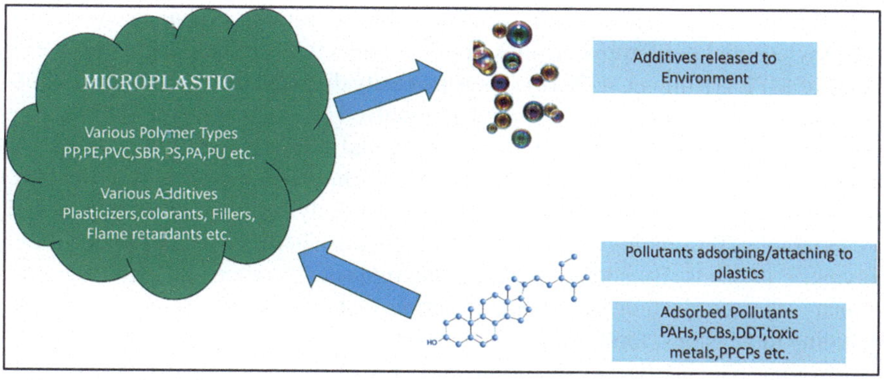

Fig. 15.6 MPLs consist of a broad category of compounds

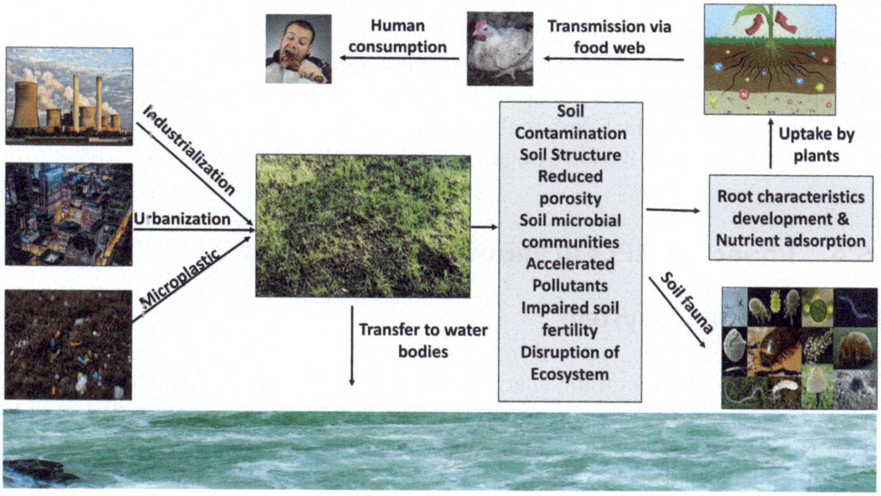

Fig. 15.7 Accumulation of MPLs in the environment

Primary MPLs, like polymer granules, microfibers, and other forms of debris develop as a result of immediate pollution via release of synthetically made materials items for personal use, and other sectors (Schellenberger et al. 2019). According to Mahon et al. (2017), debris from effluent treatment, residences, and waste streams are frequently contaminated with these human-caused environmental MPLs. According to Cole et al. (2011) and Massos and Turner (2017), secondary MPLs are produced when bigger particle sizes of plastic undergo normal natural deterioration due to ultraviolet (UV) rays and biomechanical stresses. In dumpsters and open-pit dumps, inadequate separation of urban construction and demolition waste during its removal raises the risk of pollution (He et al. 2019). Microplastic particles may build up in soils and water bodies as a consequence of the decomposition of MPLs (Fig. 15.7).

MPLs are pollutants that behave comparably to various pollutants in nature. Polyethylene and polypropylene make up more than 90% of the carbon-based elements of MPLs (Rillig et al. 2018). They are easily differentiated from other pollutants, though, as they are typically found undissolved as a separate entity. MPLs may travel extremely far from their point of origin and migrate due to their insoluble nature, which is consistent with the fundamental principles of particulate dynamics. This suggests that studying MPLs can also require studying hydrology geology, and oceanography. Terminology is necessary to distinguish between the different MPLs' polymeric imprints. By adding the term "microplastic" to every aspect of the preliminary characterization and standard-setting that has already been established, including dimensions, form, source, and solvability, it is inferred that they are made of materials like plastic and its component polymeric compounds (Schür et al. 2019). Studies focused on MPLs are intriguing because, in addition to being entirely or partially manufactured, they are resistant to the majority of detrimental environmental processes (Gilani et al. 2023). Such fragments, which might or might not be mistaken for plastics, are the subject of separate scientific investigations and are occasionally discussed in recognized specific domains. Particulates like these encompass paint chips, fibers made of wool and cotton, and spherical carbon-based particulates, a type of black carbon that is unique from other carbons, having somewhat harmful effects in the environment (Frias and Nash 2019).

15.4 Impact of MPLs in Various Ecosystems

15.4.1 On Soil Ecosystem

Consequences of MPLs on the characteristics of soil Physical characteristics. The physical characteristics of soil can be significantly impacted by MPLs (Bilal et al. 2020) (Fig. 15.8). The process of compacting of the soil may result from MPLs accumulation, which decreases the size of pores. Compacted soil can hinder the penetration of water and aeration of the soil, both of which are detrimental to plant growth (Sutanto et al. 2024). MPLs have a large relative exposed area, which contributes to the broad ability to adsorb. As a consequence, they frequently absorb dangerous pollutants with harmful chemical compounds in addition to supplements including, bis(2-ethylhexyl) phthalate, a common substance in the processing of plastics (Godswill and Godspel 2019). The existence, discharge, and mobility trends of compounds as well as other conceivably dangerous components including toxic metals and pervasive hazardous substances present in all varieties of plastics have been well studied in the past (Zeng 2023). MPLs can also take up heavy metals and other contaminants, which can worsen the condition of the soil (Bellasi et al. 2020). These contaminants and chemicals have the ability to build up in the soil, endangering plants, terrestrial creatures, and other living things that rely on good soil (Zhang et al. 2018). As newly recognized perpetual pollutants, MPLs have the potential to gradually alter the composition of soil while having a broad impact on an ecosystem (Zhang et al. 2021a, b).

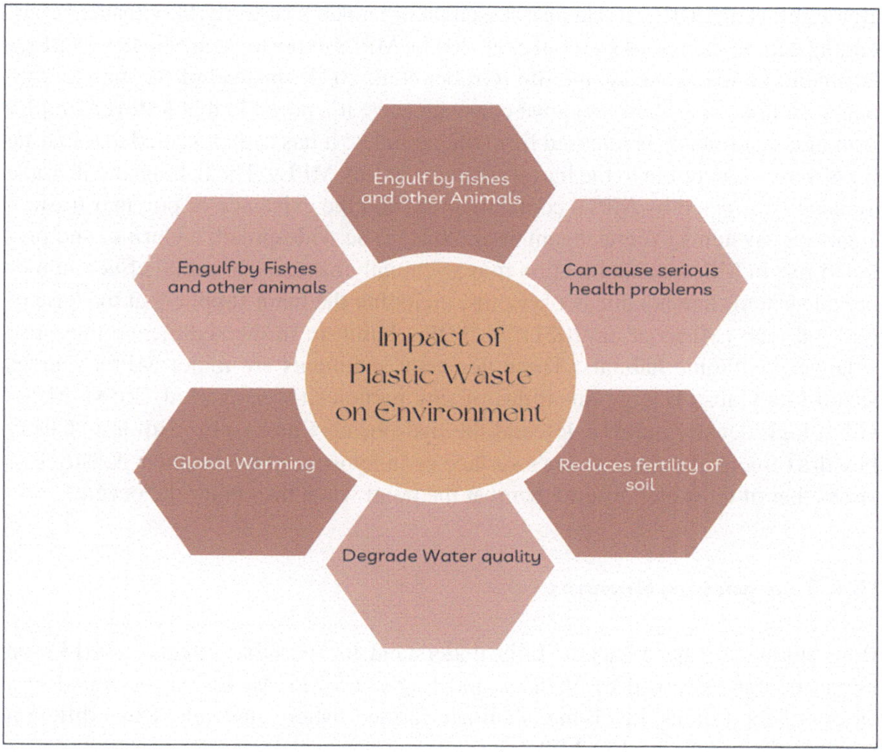

Fig. 15.8 Hazardous effects of MPLs in the environment

15.4.2 Marine Ecosystem

The migration of consumed toxic metals, infectious agents, long-lasting contaminants, etc., could all be greatly impacted by microplastics (Galafassi et al. 2021). Furthermore, microplastics promote the development of microbial biofilms made up of bacteria, fungus, and algae and may also disseminate infections caused by microbes and develop resistance to antibiotics (Zettler et al. 2013; Mao et al. 2020). Because most MPLs fall into the same spectrum of sizes as of food that zooplankton typically feed on, they may inadvertently find their way into the food chain (Aragaw and Mekonnen 2021). Aquatic animals can acquire MPLs through water (Issac and Kandasubramanian 2021). At top levels of the food chain organisms acquire greater MPLs from the prey they consumed, which causes them to amass furthermore MPLs (Trestrail et al. 2020), that results in oxidative overload, disrupted gene regulation, inflammation, and effects on the immune-mediated functions. Furthermore, swallowed plastic may result in intestine obstruction, thus lowering nutritional absorption and modifies hormonal balance (Fendall and Sewell 2009). Lacking uptake of nutrients can lead to a reduction of metabolic stores and insufficient nutrition, which can affect the development and reproducing capability of an organism

(Besseling et al. 2019). It can also lessen an organism's capacity to withstand challenging ecological issues (Vroom et al. 2017). MPLs exposure can also have various detrimental impacts on aquatic life (Gruber et al. 2023). Instead of the plastic fragments itself causing these negative consequences, it's possible that leftover simpler form of contaminants is released from the particles. It has been revealed that marine life is more susceptible to the harmful effects of tiny MPLs. The fishing and its trade are heavily affected by MPLs contamination, and the existence of polymer trash in aquatic ecosystems (Veerasingam et al. 2021). The widespread, enduring, and predicted rise in MPLs contamination may eventually have a substantial effect on biological systems and aquatic ecosystems, including the main suppliers at the base of the food web (Mistri et al. 2021). As MPLs build up in the sediments, they also endanger benthonic habitats. These demersal sediments are major MPLs sources because they absorb large quantities of tiny particles (Mearns et al. 2018). MPLs with a high density quickly descend the hydrohead. Conversely, with low density they drift freely along the water's surface or in hydrohead unless their density surpasses that of the water around them, at the point when they enter the ocean.

15.4.3 Impact on Humans

More studies are necessary to fully understand the possible effects of MPLs on human beings, as noted by Vethaak and Legler (2021). MPLs can be found in a variety of food items, involving, shellfish, farmed fishery, and salt. Three different means of being exposed to MPLs have been identified including consumption of food that contains MPLs, breathing in of air having MPLs (by means of air), and coming into proximity to those tiny bits through the skin (Revel et al. 2018). It is unclear how breathing in the fibrous MPLs affects human health. Most fibers are believed to be expungable from the airway tract, but some can induce inflammation leading to bronchial ulcers (Prata 2018), particularly in people with impaired immune systems. Prior research has indicated that inhaling fibers while working in a factory may increase the risk of certain types of cancer (Laranjo et al. 2015). Moreover, studies on workers who handle nylon flocks have revealed a greater incidence of bronchial discomfort in this population. A handful of research have been done on humans' ingestion, but a substantial number of MPLs study is done on consumption by marine organisms (Blackburn and Green 2022), sea creatures (Haque et al. 2023), as well as other animals (Gunaalan et al. 2020). Consumable plant products like fruits offer additional evidence of MPLs that people consume. According to a study, apple was found to have the highest MPLs contamination. In a very similar way MPLs contamination leads to many complexities in the human body (Fig. 15.9).

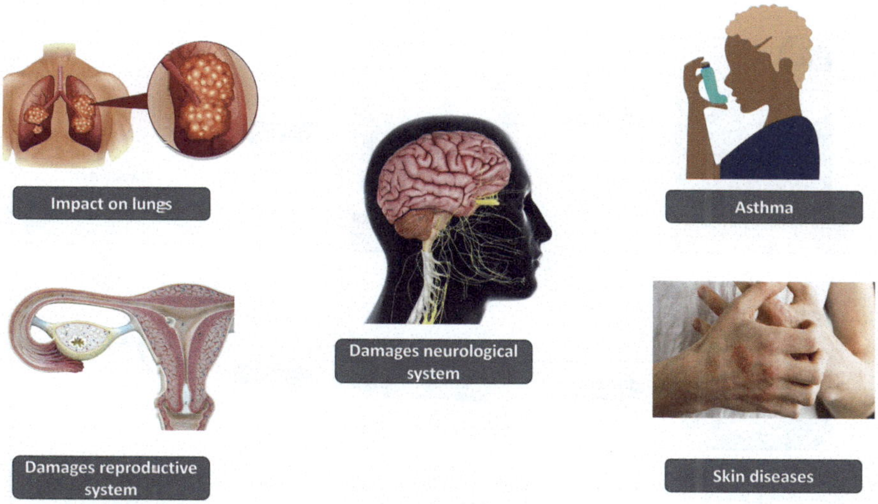

Fig. 15.9 Diseases that MPL consumption can trigger

15.5 Analysis

Identification and elimination of MPLs in marine ecosystem are more challenging than bigger plastics, often easier to get rid of through common or specific procedures. MPLs may remain on the surface or become stuck in deposits for an extended period because of their microscopic dimensions, and they may ultimately find a way up (Mendoza and Balcer 2019). MPLs have been found in the specialized cells of various creatures, including pisces, crustaceans, and the digestive tracts of sandworms that consumes MPLs deposits (Imhof et al. 2017a, b). The medical and ecological health risks of MPLs haven't been fully understood, so the dangers linked to their presence in the food supply remain inadequately described. Similar to larger plastics, microplastics eventually end up in the ocean primarily through water flows. Utilizing just one approach for analysis to completely and reliably detect MPLs of numerous dimensions, forms, and polymeric states is challenging (Law et al. 2014). Consequently, it has become common practice to combine two or more analytical methodologies. Screening often involves two stages first, physical identification of possible polymers using techniques like microscopy, and then chemical evaluation to validate the presence of polymers. Every approach, as well as other variants, has benefits as well as drawbacks. Figure 15.10 shows the methods employed for the screening and identification of MPLs from soil samples.

The four methods of analysis of MPLs in samples from the environment are the extraction, recognition, quantization, and exclusion (ultrafiltration) (Murphy et al. 2016). Traditionally, big MPLs in 1–5 mm size range can be identified visually using the unaided optical sense (Fig. 15.11).

Nevertheless, in order to properly conduct a scientific hazard evaluation, measurement at the lowest levels is necessary due to the ubiquitous occurrence of MPLs

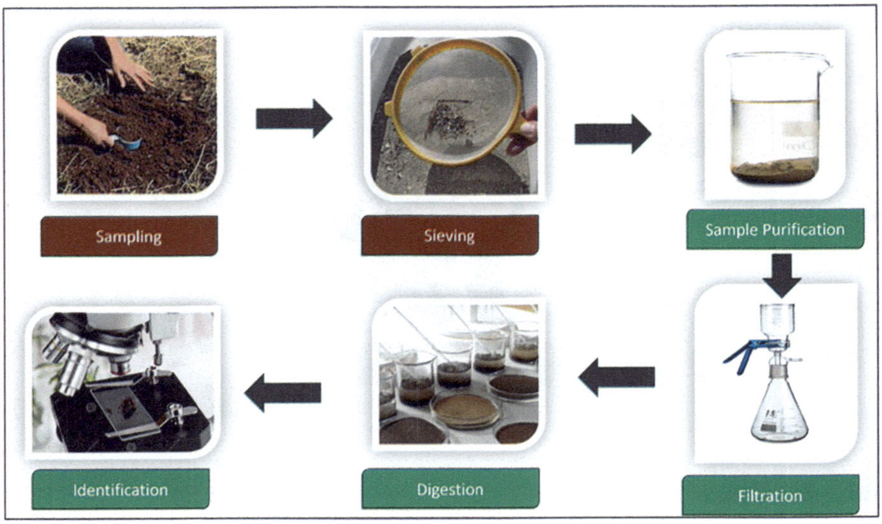

Fig. 15.10 Various phases of analysis and detection of microplastics primarily from soil sample

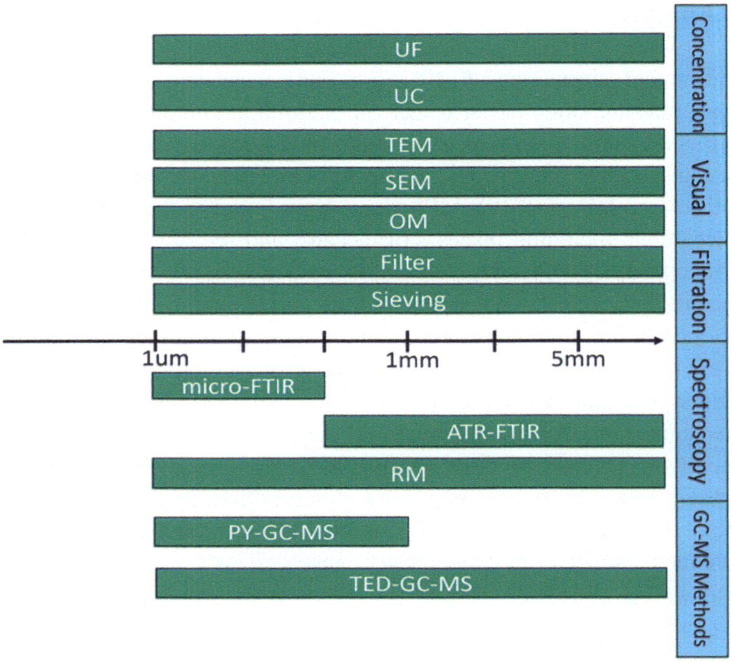

Fig. 15.11 Size categories for MPLs detection and methodologies for extraction or removal

in the surroundings and the detrimental physiological consequences of micro- and nano-sized plastics. Additionally, it has been difficult to produce a uniform categorization of MPLs data due to the vast variety of sizes and the complexity associated with their forms, colors, and polymeric forms. That renders comparing results considerably extra challenging. With advancements in analytical techniques, MPLs may now be extracted and sampled across an extensive range for ecological sources.

15.5.1 Optical Visualization

Owing to the very wide variety of sizes in MPLs, separating and detection were typically done in tandem on a platter using tweezers and human vision. Such optical approach may be used to identify certain tiny plastic fragments, and an unaided eye can distinguish colorful fragment pieces and epoxy granules having a dimension spectrum of 2–5 mm. There's a significant chance that tiny plastic components will be missed while separating collections from the sites having large concentrations of intervening dubious plastics that are about 1 mm, which also possess colors that seem comparable to those of intervening components. Optical screening and detection of big MPLs provide a straightforward, simple, and quick procedure.

15.5.2 Microscopy

When identifying MPLs having size in the range of microns, dissecting microscopy is one of the commonly utilized techniques. Microscopically intensified images offer comprehensive coating texture and structural details of objects, which are crucial for distinguishing unclear, plastic-like particles. While the majority of particles within this size range can typically be identified using microscopy, although it can be challenging to confidently classify as plastic particles in the sub-micron sized preferably less than 100 μm that lack color or a recognizable structure (Fig. 15.12).

Silt specimens that exhibit inadequate density differentiation of airy bits may impede the barely visible detection of MPLs. Imaging of fragments that resemble plastics that are incredibly detailed yet magnified are easily obtained using scanning electron microscopy (SEM). Distinguishing between MPLs and organic substances is made easier by pictures with excellent resolution of the particulate surfaces. Subsequent examination using energy-dispersive X-ray spectroscopy (EDS) reveals the fundamental makeup of the identical object. To distinguish C-based polymers from inanimate fragments, one can evaluate the fundamental makeup of the particles. The total quantity of specimens that might be processed by a SEM-EDS is limited due to its high cost and the significant time and effort required for the preparation of samples and inspection. Plastic colors can't be distinguished from one another in SEM. In several instances, polymer fragments were successfully identified using very sophisticated microscopic approaches. Polarized optical microscopy has been established to identify various polymers. A variety of factors is related to the production; crystalline structure can differ between various kinds of polymers as well as the same polymers. Polarized light propagation through polymers may be

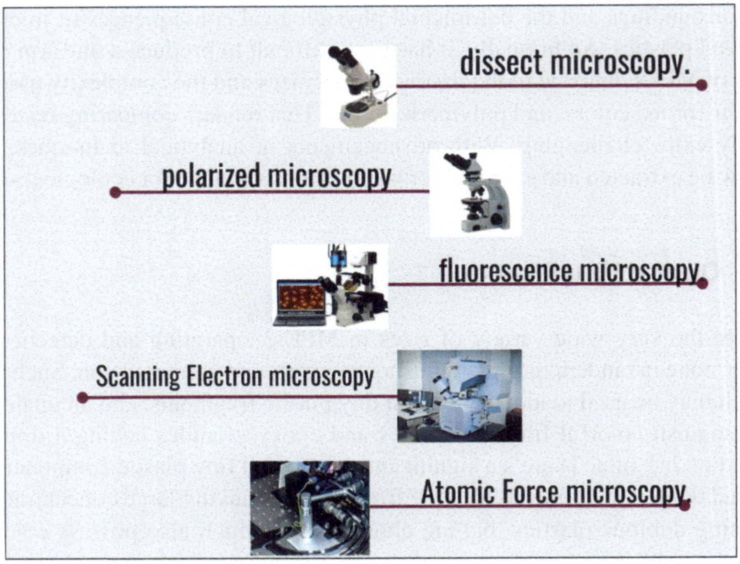

Fig. 15.12 Various types of microscopies that can be employed to visualize MPLs

evaluated based on its arrangement of crystals. The tiny plastic particles ought to be sufficiently thin to let enough polarized light penetrate. MPLs specimens on translucent filtering paper cannot be processed using this approach.

15.5.3 Fourier Transformed Infrared Spectroscopy

Spectroscopic analysis is a highly efficient and accurate method for positive identification of MPLs. Insight about those precise bonds that make up the particle is obtained by Fourier transformed infrared spectroscopy (Fig. 15.13). This approach makes it simple to identify C-derived polymers, and a distinct spectrum produced by varying bond combinations allows plastic materials to be distinguished among particulates. It's possible to identify particular polymeric states in addition to confirming the presence of plastic.

In Micro-FTIR, the object lens and Infrared sensor are switched to conduct microscopic inspection of particles before spectral validation. This is necessary for handling small MPLs. For FTIR MPLs evaluation, attenuated total reflectance (ATR) method is also accessible. Furthermore, from uneven MPLs surfaces, the ATR mode yields steady spectra, but in the reflectivity, the spectra are not so reliable. Theoretically, MPLs can be identified down to the size of the ATR probe's infrared beam aperture. MicroATR–FTIR provides a helpful way to find MPLs collections within the environment by first identifying polymeric pieces using

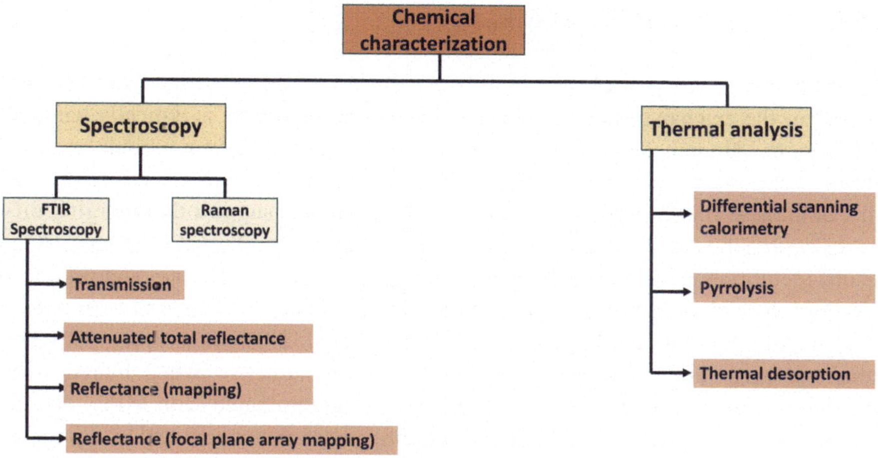

Fig. 15.13 Approaches followed for chemical characterization of MPLs

microscopes and then confirming their composition with spectrophotometer. Statistical analysis is one type of measurement using ATR-FTIR. Highly worn or delicate MPLs may be harmed by the applied pressure generated by the ATR probe, and small particles may be removed of the filtering membrane by adhering to or interacting electrostatically alongside the probing tip.

15.5.4 Raman Spectroscopy

MPLs are additionally being identified using Raman spectroscopy. Polymeric material has a particular spectra because of the arrangement of molecules and number of atoms in the material. In addition to identifying plastics, it offers characteristics of every specimen's polymeric makeup, just like FTIR. Intricate MPLs detection is often compromised by the disparate outcomes and spectrum obtained from a MPL using FTIR and Raman spectroscopy. MPLs with very small dimensions are capable of being identified because of the lower diameter of the beam compared to FTIR. A combination of Raman spectroscopy and confocal microscopy is able to be utilized to detect MPLs in Calanus specimens. The present challenges in MPLs detection may be resolved by developing novel techniques and instrumentation and linking with others or to regular tools that are already in use. Analytical techniques used to quantify plastic waste in ecological specimens have still not been established. There are numerous obstacles to overcome in the measurement of MPLs, including contaminants and excessively and underestimating.

15.6 Ways to Get Rid of MPLs

A significant quantity of plastic trash is usually produced as an inevitable consequence of the extensive usage of plastics. Unpleasantly, plastic is typically handled quite poorly. Nonetheless, efforts are being made to switch out nonbiodegradable plastics with biodegradable ones that are derived through either renewable or, ideally, nonrenewable resources (Zahid et al. 2024). Studies suggest that merging different approaches leads to increased cost-effectiveness or removal efficiency. As a result, membrane-based procedures are frequently paired with biological products preparation, such as membrane bioreactors (Talvitie et al. 2017) paired with physicochemical interventions (Bayo et al. 2020; Yang et al. 2019). Absorption, adsorption, and percolation strategies are therefore more effective when paired with biological degradation (Bayo et al. 2020). Water contaminated with MPLs can be remedied using a variety of techniques, including physicochemical and biochemical methods as well as biologically. All these approaches have some benefits and drawbacks (Lares et al. 2018). Although the deterioration process using photocatalysis is less efficient than the use of membranes, it is nevertheless an ecologically sound method (Fig. 15.14). Although membrane-based procedures often offer the greatest elimination efficiencies, a major disadvantage of this technology is to maintain the membranes.

15.6.1 Bioremediation

According to collard (2014), the most widely used method for cleaning up contaminated waterways is likely bioremediation. Utilizing optimal conditions for growth and ample nutrition, biological remediation breaks down pollutants through the utilization of the metabolic process of microorganisms. Microbes that are modified

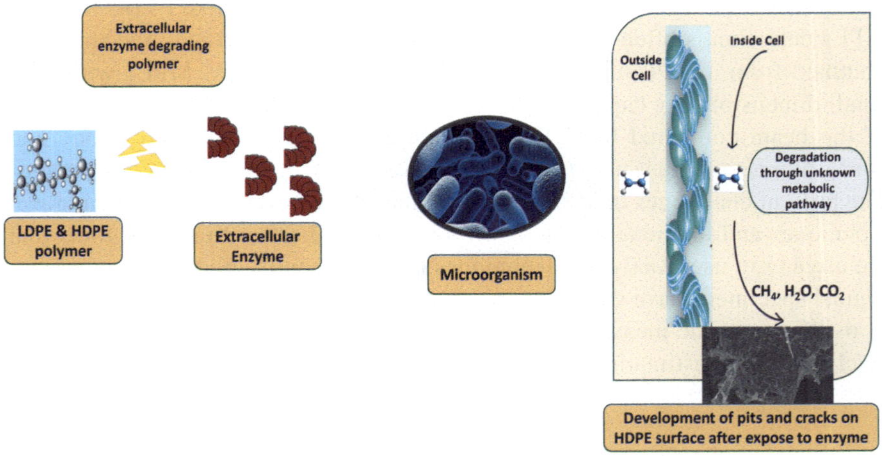

Fig. 15.14 MPLs' deterioration strategy for LDPE and HDPE

through genetics, extremely specific or nonspecific, are all capable of being employed in biological remediation. Microbes collected coming from soil, organic matter, or active effluent aren't specified. Targeted contaminants are broken down by specific enzymes found in some specific bacteria (Fouad et al. 2022). The process of creating genetically modified microorganisms (GMMs) involves manipulating their genomes to add or remove desired genes (Sun et al. 2023). Bioaugmentation and biostimulation are the most widely applied bioremediation strategies (Fig. 15.15). The introduction of microbes is essential to accelerate the deterioration of pollutant rate and is known as bioaugmentation while the process of changing the surroundings in order to encourage naturally occurring microorganisms that can perform bioremediation is known as biostimulation.

Bioenhancement by fungi is uncommon, as compared to bacteria, which is merely frequent in wastewater treatment systems. Specifically, the growth of bacteria is encouraged by a pH of roughly 6–7 during wastewater treatment (Fig. 15.16). As lower pH favors growth of fungi, those are not suggested to be introduced into treatment plants and that's why their number is very less as compared to bacteria in the sludge. Microalgae are frequently introduced as bioflocculants though a few of those can also biodegrade microplastics in treatment plant.

According to Barone et al. (2024), the utilized microbes transform contaminants into carbon dioxide, water, and other inorganic matter using their respective enzymatic mechanism. Two main biodegradation phases typically exist. Firstly, the polymer's molecular weight is reduced, facilitating the passage of biomolecules over the plasma membrane, enabling the targeting of subcellular enzymes, followed by metabolism of generated molecules with the help of citric acid cycle and β-oxidation of fatty acids (Folino et al. 2020). In the initial phase of biodegradation, the microbes colonize on the outer coating of the MPLs, which then secretes enzymes outside the

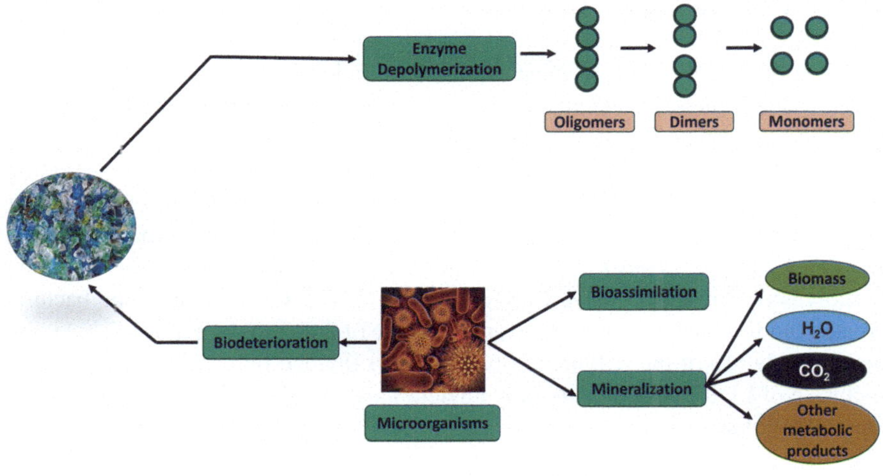

Fig. 15.15 Key phases in biodeterioration of MPLs

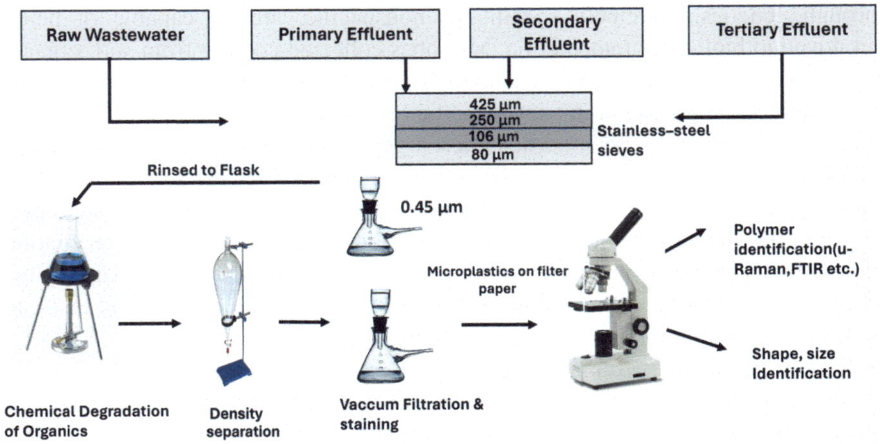

Fig. 15.16 Detection of MPLs from the wastewater sample

cell. This is greatly impacted by the hydrophobic nature of polymers in addition to the microbial membrane (Krasowska and Sigler 2014). These enzymes are primarily responsible for the immediate breakdown of polymeric forms following biological assimilation; molecules with low molecular mass are readily absorbed by the cells of microbes, where enzymes carry out the remaining breakdown process to produce carbon dioxide and water. The types of microbes that may break down MPLs vary according to the properties of the surroundings. Specifically, the physiological state of the applicable microbes is greatly influenced by the external conditions that determine the pace of decomposition.

15.6.2 Bacteriological Degradation

Because of the diversity of metabolism, bacterial cultures are supremely often employed in biological remediation (Pathak 2023). Many of them were previously identified as being capable of biodegrading polymers (Mehmood and Peng 2022) (Fig. 15.17). Applying film-forming bacteria is a smart option for biodegrading plastics. Sivan et al. (2006), for instance, looked into how *Rhodococcus ruber* breaks down polyethylene (PE) (Franco et al. 2021). Laccase, a type of enzyme secreted by *Rhodococcus ruber*, causes polyethylene to depolymerize by oxidizing its polymeric substance (Silva-Cavalcanti et al. 2017). Consequently, the bacteria created biofilm on the PE surface over the course of 2 months of exposure, breaking down nearly 8% of the polymer. Polyethylene is divided into two main categories inclusive of LDPE (low-density polyethylene) and HDPE (high-density polyethylene) (Yuan et al. 2020).

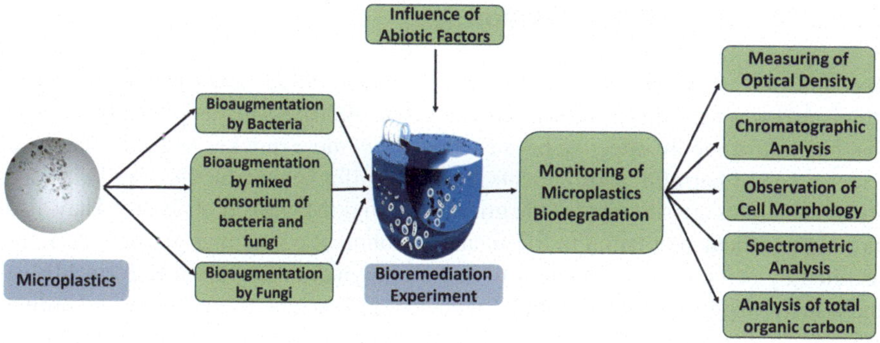

Fig. 15.17 The process of biodegradation and respective analysis approaches

As PE is perhaps the major pervasive form of contamination in the ecosystem, studies based on the strategies of various pathways followed by enzyme for degrading polymers have drawn attention. It has been observed that the enzymes laccase, alkane hydrolase, manganese peroxidase, and soybean peroxidase control the breakdown of PE in the surroundings. The majority of the polyethylene terephthalate found in the ecosystem originates from plastic waste leaching. The breakdown of polyethylene terephthalate was profoundly associated with two different kinds of enzymes. Polyethylene terephthalate and mono(2-hydroxyethyl) terephthalic acidase are always working in tandem. Polyethylene terephthalate is broken down into mono(2-hydroxyethyl) terephthalic acid by polyethylene terephthalatase. After mono(2-hydroxyethyl) terephthalic acid is absorbed, mono(2-hydroxyethyl) terephthalic acidase transforms it into terephthalic acid and 1,2-Dihydroxyethane. However, the workings of mono(2-hydroxyethyl) terephthalic acidase and polyethylene terephthalate are well known. Some other enzymes are also there that are capable of degrading polyethylene terephthalate like lipase and esterase. Most applications for polystyrene (PS) are in the food packaging sector. A great deal of research has been done on PS deterioration. It has been observed that many microbes can break down this contaminant. Styrene oxide isomerase, phenylacetyl coenzyme A ligase, styrene monooxygenase, and phenylacetaldehyde dehydrogenase are among the enzymes that have been linked to degrading polystyrene. The deterioration process results in the formation of acetyl-CoA, which is further used in the citric acid cycle (Kong et al. 2021). Additionally, serine hydrolase also has the ability to break down polystyrene, yet its exact process is unclear for now.

15.6.3 Mycological Degradation

The question of which is more effective in biodegrading microplastics between fungi and bacteria is up for debate (Oliveira et al. 2020). Some people really believe that bacteria are more effective because they grow more quickly and can adapt to a wider range of harmful environments. Fungi do offer some benefits. Firstly, a variety of substrates are suitable for the growth of fungi (Magnin et al. 2020). They also have an excellent adaptation to life in an environment containing plastics. Proteins secreted by cells of fungi enable it to exploit polymers that devote to form MPLs, and transforms it to potential sources of energy and carbon. Furthermore, according to some studies by Kumar et al. (2024), fungi are more efficient in biodegrading polymers with complicated organization; therefore, more research has been done on the biodeterioration of polyvinyl chloride (PVC) by fungi than by bacteria. Numerous generic enzymes found in fungi may trigger a wide range of functions (Russell and Webster 2021). They're thus ideal to facilitate the biodeterioration of MPLs, as the enzymes from fungi origin functions in a cellular detoxification, which is crucial to their adaptation (Olicón-Hernández et al. 2017). The cytochrome P450 family of enzymes, comprising transferases, which are in charge of conjugation activities, and epoxidases, which participate in oxidations, is what powers the whole thing (Ruthsatz et al. 2023) (Fig. 15.18).

There are two components in the external enzymatic complex inclusive of a hydrolyzing component, which generate hydrolases and is in charge of

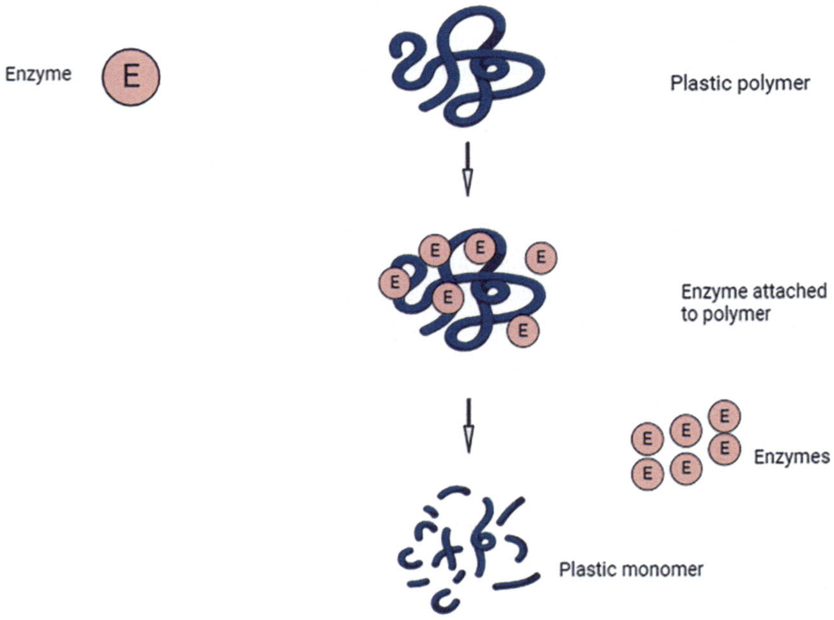

Fig. 15.18 Enzymatic degradation of polymers

polysaccharide breakdown. A nonspecific component, another one contributes to the oxidizing property of complicated materials. Most nonspecific oxidoreductases, such as diarylpropane oxygenase, peroxidase-M2, Reactive-Blue-5: hydrogen-peroxide oxidoreductase, reactive-black-5: hydrogen-peroxide oxidoreductase, and several nonspecific peroxygenases, are included in this nonspecific subcategory. Peroxidase-M2 is therefore effective at breaking down polyethylene, particularly when polysorbate 80, $MnSO_4$, and malonate buffer are present. Peroxidase-M2 is probably strongly associated with the production of the Mn(III) chelate complex (Zhang et al. 2021a, b). Furthermore, Aspergillus fungus is preferable to some bacteria in their ability to biodeteriorate LDPE. However, plasticizers are substances that can operate as a substrate to fungus by leaching from polyvinyl chloride (PVC) miniature particles toward the liquid phase. However, they can also improve fungal adherence by inducing microbial metabolism.

15.6.4 Phycological Degradation

For various experiments, algae have been a good choice among researchers for toxicology-related studies by examining the toxic effects of MPLs (Manzi et al. 2022). Algae can also be employed in the biodeterioration of MPLs. Algae comes under harmless microbes that are simple to work with (Moog et al. 2019). By secreting external polymeric molecules, they might inhabit over the MPLs, which may result in efficient breakdown (Chia et al. 2021). External polymeric molecules can be produced in response to the inclusion of certain polymers in general, particularly plastics (Song et al. 2020). Furthermore, different polymers promote their synthesis in different ways, leading to variations across respective constituents and amounts. This establishes the biofilms' variety in cohesiveness and, in turn, the resident microorganisms' biomass (Lagarde et al. 2016). Different biomass types and amounts of external polymeric compounds can result in different properties of film attachment to MPLs surfaces. MPLs can be biodegraded by a variety of algae species. Due to their ability to utilize LDPE as a supply of energy and carbon, *Phormidium lucidum* possesses a significant capability for the breakdown of MPLs (Vaid et al. 2021). A greater degree of efficiency has been noticed in the degradation of LDPE by certain microscopic algae, particularly the blue-green algae having the ability to perform photosynthesis (Pourebrahimi and Pirooz 2023). Plastic microparticles made of polypropylene and polyethylene can be degraded by blue-green algae called Spirulina species. Additional chemical compounds, such as carbonyl molecules, acidic substances, and OH groups (hydroxy), among others, are formed over the MPLs as a consequence of that biodeterioration (Devi et al. 2023). The backbones of polyethylene (PE), polypropylene, polystyrene (PS), and polyvinyl chloride (PVC) lack hydrolyzable functional molecules, thereby rendering biological degradation challenging. The combined effect of biological and abiotic variables appeared primarily responsible for the early disintegration of MPLs in the ecosystem and the resulting decrease in their molecular weights (Anand et al. 2023). The overall objective of Chemicophysical preprocessing is to break down chemical links in the molecular makeup of MPLs so that their polymers can biodegrade more easily (Van Cauwenberghe 2016). For best removal of MPLs, one can utilize the combination of more than one approach in order to properly eliminate the MPLs contamination.

References

Anand U, Dey S, Bontempi E, Ducoli S, Vethaak AD, Dey A, Federici S (2023) Biotechnological methods to remove microplastics: a review. Environ Chem Lett 21(3):1787–1810

Aragaw TA, Mekonnen BA (2021) Distribution and impact of microplastics in the aquatic systems: a review of ecotoxicological effects on biota. In: Microplastic pollution. Springer, pp 65–104

Barone GD, Rodríguez-Seijo A, Parati M, Johnston B, Erdem E, Cernava T, Radecka I (2024) Harnessing photosynthetic microorganisms for enhanced bioremediation of microplastics: A comprehensive review. Environmental Science and Ecotechnology, 100407.

Bayo J, Olmos S, López-Castellanos J (2020) Removal of microplastics from wastewater. Handbook of Microplastics in the Environment, 1–20.

Bellasi A, Binda G, Pozzi A, Galafassi S, Volta P, Bettinetti R (2020) Microplastic contamination in freshwater environments: a review, focusing on interactions with sediments and benthic organisms. Environments 7(4):30

Besseling E, Redondo-Hasselerharm P, Foekema EM, Koelmans AA (2019) Quantifying ecological risks of aquatic micro- and nanoplastic. Crit Rev Environ Sci Technol 49(1):32–80

Bilal M, Ali H, Ullah R, Hussain I, Khan BA (2020) Overview of microplastics threat in aquatic animals since 1960 to 2021. Int J Res Anal Rev 8:950–964

Blackburn K, Green D (2022) The potential effects of microplastics on human health: what is known and what is unknown. Ambio 51(3):518–530

Blair RM, Waldron S, Phoenix VR, Gauchotte-Lindsay C (2019) Microscopy and elemental analysis characterisation of microplastics in sediment of a freshwater urban river in Scotland, UK. Environ Sci Pollut Res 26:12491–12504

Bradney L, Wijesekara H, Palansooriya KN, Obadamudalige N, Bolan NS, Ok YS et al (2019) Particulate plastics as a vector for toxic trace-element uptake by aquatic and terrestrial organisms and human health risk. Environ Int 131:104937

Brennecke D, Duarte B, Paiva F, Caçador I, Canning-Clode J (2016) Microplastics as vector for heavy metal contamination from the marine environment. Estuar Coast Shelf Sci 178:189–195

Chia RW, Lee JY, Kim H, Jang J (2021) Microplastic pollution in soil and groundwater: a review. Environ Chem Lett 19(6):4211–4224

Cole M, Lindeque P, Halsband C, Galloway TS (2011) Microplastics as contaminants in the marine environment: a review. Mar Pollut Bull 62(12):2588–2597

Collard F, Das K, Parmentier E (2014) Microplastics caught in herring gill rakers: illustration by scanning electron microscopy. In VLIZ Young Marine Scientists' Day.

da Costa JP, Duarte AC, Rocha-Santos TA (2017) Microplastics—occurrence, fate and behaviour in the environment. In: Comprehensive analytical chemistry, vol 75. Elsevier, pp 1–24

da Costa JP, Paço A, Santos PS, Duarte AC, Rocha-Santos T (2018) Microplastics in soils: assessment, analytics and risks. Environ Chem 16(1):18–30

de Souza Machado AA, Kloas W, Zarfl C, Hempel S, Rillig MC (2018) Microplastics as an emerging threat to terrestrial ecosystems. Glob Chang Biol 24(4):1405–1416

Devi A, Hansa A, Gupta H, Syam K, Upadhyay M, Kaur M, Sharma R (2023) Microplastics as an emerging menace to environment: Insights into their uptake, prevalence, fate, and sustainable solutions. Environmental Research, 229, 115922.

Fendall LS, Sewell MA (2009) Contributing to marine pollution by washing your face: microplastics in facial cleansers. Mar Pollut Bull 58(8):1225–1228

Folino A, Karageorgiou A, Calabrò PS, Komilis D (2020) Biodegradation of wasted bioplastics in natural and industrial environments: A review. Sustainability, 12(15), 6030.

Fouad FA, Youssef DG, Shahat FM, Abd El-Ghany MN (2022) Role of Microorganisms in Biodegradation of Pollutants. In Handbook of Biodegradable Materials (pp. 1-40). Cham: Springer International Publishing.

Franco AA, Arellano JM, Albendín G, Rodríguez-Barroso R, Quiroga JM, Coello MD (2021) Microplastic pollution in wastewater treatment plants in the city of Cádiz: abundance, removal efficiency and presence in receiving water body. Sci Total Environ 776:145795

Free CM, Jensen OP, Mason SA, Eriksen M, Williamson NJ, Boldgiv B (2014) High-levels of microplastic pollution in a large, remote, mountain lake. Mar Pollut Bull 85(1):156–163

Frias JP, Nash R (2019) Microplastics: finding a consensus on the definition. Mar Pollut Bull 138:145–147

Galafassi S, Sabatino R, Sathicq MB, Eckert EM, Fontaneto D, Dalla Fontana G et al (2021) Contribution of microplastic particles to the spread of resistances and pathogenic bacteria in treated wastewaters. Water Res 201:117368

Geyer R, Jambeck JR, Law KL (2017) Production, use, and fate of all plastics ever made. Sci Adv 3(7):e1700782

Gilani IE, Sayadi S, Zouari N, Al-Ghouti MA (2023) Plastic waste impact and biotechnology: exploring polymer degradation, microbial role, and sustainable development implications. Bioresour Technol Rep 24:101606

Godswill AC, Godspel AC (2019) Physiological effects of plastic wastes on the endocrine system (Bisphenol A, Phthalates, Bisphenol S, PBDEs, TBBPA). Int J Bioinformatics Computat Biol 4(2):11–29

Groh KJ, Backhaus T, Carney-Almroth B, Geueke B, Inostroza PA, Lennquist A, Muncke J (2019. Overview of known plastic packaging-associated chemicals and their hazards. Science of the total environment, 651, 3253–3268.

Gruber ES, Stadlbauer V, Pichler V, Resch-Fauster K, Todorovic A, Meisel TC et al (2023) To waste or not to waste: questioning potential health risks of micro- and nanoplastics with a focus on their ingestion and potential carcinogenicity. Expo Health 15(1):33–51

Gunaalan K, Fabbri E, Capolupo M (2020) The hidden threat of plastic leachates: a critical review on their impacts on aquatic organisms. Water Res 184:116170

Haque MK, Uddin M, Kormoker T, Ahmed T, Zaman MRU, Rahman MS et al (2023) Occurrences, sources, fate and impacts of plastic on aquatic organisms and human health in global perspectives: what Bangladesh can do in future? Environ Geochem Health 45(8):5531–5556

He P, Chen L, Shao L, Zhang H, Lü F (2019) Municipal solid waste (MSW) landfill: a source of microplastics?—evidence of microplastics in landfill leachate. Water Res 159:38–45

Horton AA, Walton A, Spurgeon DJ, Lahive E, Svendsen C (2017) Microplastics in freshwater and terrestrial environments: evaluating the current understanding to identify the knowledge gaps and future research priorities. Sci Total Environ 586:127–141

Imhof HK, Rusek J, Thiel M, Wolinska J, Laforsch C (2017a) Do microplastic particles affect *Daphnia magna* at the morphological, life history and molecular level? PLoS One 12(11):e0187590

Imhof HK, Sigl R, Brauer E, Feyl S, Giesemann P, Klink S et al (2017b) Spatial and temporal variation of macro-, meso- and microplastic abundance on a remote coral island of the Maldives, Indian Ocean. Mar Pollut Bull 116(1–2):340–347

Issac MN, Kandasubramanian B (2021) Effect of microplastics in water and aquatic systems. Environ Sci Pollut Res 28:19544–19562

Ivar do Sul JA, Costa MF, Fillmann G (2014) Microplastics in the pelagic environment around oceanic islands of the Western Tropical Atlantic Ocean. Water Air Soil Pollut 225:1–13

Jambeck JR, Geyer R, Wilcox C, Siegler TR, Perryman M, Andrady A et al (2015) Plastic waste inputs from land into the ocean. Science 347(6223):768–771

Jiang Y (2021) Recent progress and research trends in microplastic pollution and the potential health risks. Bull Environ Contam Toxicol 107(4):575–576

Kim D, Chae Y, An YJ (2017) Mixture toxicity of nickel and microplastics with different functional groups on *Daphnia magna*. Environ Sci Technol 51(21):12852–12858

Kong Q, Li Y, Xu X, Cheng J, Sun Y, Zhang L et al (2021) Microplastics interfere with mixotrophic Ochromonas eliminating toxic Microcystis. Chemosphere 265:129030

Krasowska A, Sigler K (2014) How microorganisms use hydrophobicity and what does this mean for human needs?. Frontiers in cellular and infection microbiology, 4, 112.

Kumar M, Xiong X, He M, Tsang DC, Gupta J, Khan E et al (2020) Microplastics as pollutants in agricultural soils. Environ Pollut 265:114980

Kumar NM, Sharmila G, Muthukumaran C (2024) Microbial Degradation of Plastics: Current Perspectives and Challenges. Sustainable Production Innovations: Bioremediation and Other Biotechnologies, 233-272.Lagarde F, Olivier O, Zanella M, Daniel P, Hiard S, Caruso A (2016) Microplastic interactions with freshwater microalgae: hetero-aggregation and changes in plastic density appear strongly dependent on polymer type. Environ Pollut 215:331–339

Laranjo L, Arguel A, Neves AL, Gallagher AM, Kaplan R, Mortimer N et al (2015) The influence of social networking sites on health behavior change: a systematic review and meta-analysis. J Am Med Inform Assoc 22(1):243–256

Lares M, Ncibi MC, Sillanpää M, Sillanpää M (2018) Occurrence, identification and removal of microplastic particles and fibers in conventional activated sludge process and advanced MBR technology. Water research, 133, 236–246.

Law KL, Thompson RC (2014) Microplastics in the seas. Science, 345(6193), 144–145.

Lebreton L, Andrady A (2019) Future scenarios of global plastic waste generation and disposal. Palgrave Commun 5(1):1–11

Lei L, Wu S, Lu S, Liu M, Song Y, Fu Z et al (2018) Microplastic particles cause intestinal damage and other adverse effects in zebrafish *Danio rerio* and nematode *Caenorhabditis elegans*. Sci Total Environ 619:1–8

Magnin A, Pollet E, Phalip V, Avérous L (2020) Evaluation of biological degradation of polyurethanes. Biotechnol Adv 39:107457

Mahon AM, O'Connell B, Healy MG, O'Connor I, Officer R, Nash R, Morrison L (2017) Microplastics in sewage sludge: effects of treatment. Environ Sci Technol 51(2):810–818

Manzi HP, Zhang M, Salama ES (2022) Extensive investigation and beyond the removal of micro-polyvinyl chloride by microalgae to promote environmental health. Chemosphere 300:134530

Mao R, Hu Y, Zhang S, Wu R, Guo X (2020) Microplastics in the surface water of Wuliangsuhai Lake, northern China. Sci Total Environ 723:137820

Mason SA, Garneau D, Sutton R, Chu Y, Ehmann K, Barnes J et al (2016) Microplastic pollution is widely detected in US municipal wastewater treatment plant effluent. Environ Pollut 218:1045–1054

Massos A, Turner A (2017) Cadmium, lead and bromine in beached microplastics. Environ Pollut 227:139–145

Mearns AJ, Reish DJ, Bissell M, Morrison AM, Rempel-Hester MA, Arthur C et al (2018) Effects of pollution on marine organisms. Water Environ Res 90(10):1206–1300

Mehmood T, Peng L (2022) Polyethylene scaffold net and synthetic grass fragmentation: a source of microplastics in the atmosphere? J Hazard Mater 429:128391

Mendoza LR, Balcer M (2019) Association of hazardous compounds with microplastics in freshwater ecosystems. In: Microplastics in water and wastewater. IWA Publishing, London, pp 15–25

Mercy FT, Alam AR, Akbor MA (2023) Abundance and characteristics of microplastics in major urban lakes of Dhaka, Bangladesh. Heliyon 9(4):e14587

Mintenig SM, Löder MG, Primpke S, Gerdts G (2019) Low numbers of microplastics detected in drinking water from ground water sources. Sci Total Environ 648:631–635

Mistri M, Scoponi M, Sfriso AA, Munari C, Curiotto M, Sfriso A et al (2021) Microplastic contamination in protected areas of the Gulf of Venice. Water Air Soil Pollut 232:1–9

Moog D, Schmitt J, Senger J, Zarzycki J, Rexer KH, Linne U et al (2019) Using a marine microalga as a chassis for polyethylene terephthalate (PET) degradation. Microb Cell Factories 18:1–15

Murphy F, Ewins C, Carbonnier F, Quinn B (2016) Wastewater treatment works (WwTW) as a source of microplastics in the aquatic environment. Environ Sci Technol 50(11):5800–5808

Nanthini Devi K, Raju P, Santhanam P, Perumal P (2023) The microplastics: their occurrence and impacts in India—a review. Proc Natl Acad Sci India Sect A Phys Sci 93(2):205–210

Napper IE, Bakir A, Rowland SJ, Thompson RC (2015) Characterisation, quantity and sorptive properties of microplastics extracted from cosmetics. Mar Pollut Bull 99(1–2):178–185

Nizzetto L, Bussi G, Futter MN, Butterfield D, Whitehead PG (2016) A theoretical assessment of microplastic transport in river catchments and their retention by soils and river sediments. Environ Sci Processes Impacts 18(8):1050–1059

Olicón-Hernández DR, González-López J, Aranda E (2017) Overview on the biochemical potential of filamentous fungi to degrade pharmaceutical compounds. Front Microbiol 8:256025

Oliveira AR, Sardinha-Silva A, Andrews PL, Green D, Cooke GM, Hall S et al (2020) Microplastics presence in cultured and wild-caught cuttlefish, *Sepia officinalis*. Mar Pollut Bull 160:111553

Pathak VM (2023) Exploitation of bacterial strains for microplastics (LDPE) biodegradation. Chemosphere 316:137845

Pauna VH, Buonocore E, Renzi M, Russo GF, Franzese PP (2019) The issue of microplastics in marine ecosystems: a bibliometric network analysis. Mar Pollut Bull 149:110612

Pourebrahimi S, Pirooz M (2023) Microplastic pollution in the marine environment: a review. J Hazard Mater Adv 10:100327

Prata JC (2018) Microplastics in wastewater: state of the knowledge on sources, fate and solutions. Mar Pollut Bull 129(1):262–265

Revel M, Châtel A, Mouneyrac C (2018) Micro (nano) plastics: a threat to human health? Curr Opin Environ Sci Health 1:17–23

Rillig MC, de Souza Machado AA, Lehmann A, Klümper U (2018) Evolutionary implications of microplastics for soil biota. Environ Chem 16(1):3–7

Rillig MC, Leifheit E, Lehmann J (2021) Microplastic effects on carbon cycling processes in soils. PLoS Biol 19(3):e3001130

Rios-Mendoza LM, Ontiveros-Cuadras JF, Leon-Vargas D, Ruiz-Fernández AC, Rangel-García M, Pérez-Bernal LH, Sanchez-Cabeza JA (2021) Microplastic contamination and fluxes in a touristic area at the SE Gulf of California. Mar Pollut Bull 170:112638

Rochman CM, Brookson C, Bikker J, Djuric N, Earn A, Bucci K et al (2019) Rethinking microplastics as a diverse contaminant suite. Environ Toxicol Chem 38(4):703–711

Russell M, Webster L (2021) Microplastics in sea surface waters around Scotland. Mar Pollut Bull 166:112210

Ruthsatz K, Schwarz A, Gomez-Mestre I, Meyer R, Domscheit M, Bartels F et al (2023) Life in plastic, it's not fantastic: sublethal effects of polyethylene microplastics ingestion throughout amphibian metamorphosis. Sci Total Environ 885:163779

Schellenberger S, Jonsson C, Mellin P, Levenstam OA, Liagkouridis I, Ribbenstedt A et al (2019) Release of side-chain fluorinated polymer-containing microplastic fibers from functional textiles during washing and first estimates of perfluoroalkyl acid emissions. Environ Sci Technol 53(24):14329–14338

Schür C, Rist S, Baun A, Mayer P, Hartmann NB, Wagner M (2019) When fluorescence is not a particle: the tissue translocation of microplastics in *Daphnia magna* seems an artifact. Environ Toxicol Chem 38(7):1495–1503

Sengupta S, Dey S (2020) Microplastic menace in soil environment: source, impact and the way forward. Agric Food 2:845–848

Silva-Cavalcanti JS, Silva JDB, de França EJ, de Araújo MCB, Gusmão F (2017) Microplastics ingestion by a common tropical freshwater fishing resource. Environ Pollut 221:218–226

Sivan A, Szanto M, Pavlov V (2006) Biofilm development of the polyethylene-degrading bacterium Rhodococcus ruber. Applied microbiology and biotechnology, 72, 346–352.

Smiroldo G, Balestrieri A, Pini E, Tremolada P (2019) Anthropogenically altered trophic webs: alien catfish and microplastics in the diet of Eurasian otters. Mammal Res 64:165–174

Song C, Liu Z, Wang C, Li S, Kitamura Y (2020) Different interaction performance between microplastics and microalgae: the bio-elimination potential of *Chlorella* sp. L38 and *Phaeodactylum tricornutum* MASCC-0025. Sci Total Environ 723:138146

Sun XL, Xiang H, Xiong HQ, Fang YC, Wang Y (2023) Bioremediation of microplastics in freshwater environments: A systematic review of biofilm culture, degradation mechanisms, and analytical methods. Science of The Total Environment, 863, 160953.

Sutanto A, Widodo W, Rahayu ID, Sustiyana S, Nazizah F, Iswahyudi I, Bakhtiar A (2024) The impact of microplastics on yield and economic losses in selected agricultural food commodities. Environ Qual Manag. https://doi.org/10.1002/tqem.22188

Talvitie J, Mikola A, Koistinen A, Setälä O (2017) Solutions to microplastic pollution—removal of microplastics from wastewater effluent with advanced wastewater treatment technologies. Water Res 123:401–407

Thornton Hampton LM, Bouwmeester H, Brander SM, Coffin S, Cole M, Hermabessiere L et al (2022) Research recommendations to better understand the potential health impacts of microplastics to humans and aquatic ecosystems. Microplastics Nanoplastics 2(1):18

Trestrail C, Walpitagama M, Hedges C, Truskewycz A, Miranda A, Wlodkowic D et al (2020) Foaming at the mouth: ingestion of floral foam microplastics by aquatic animals. Sci Total Environ 705:135826

Vaid M, Sarma K, Gupta A (2021) Microplastic pollution in aquatic environments with special emphasis on riverine systems: current understanding and way forward. J Environ Manag 293:112860

Van Cauwenberghe L (2016) Occurrence, effects and risks of marine microplastics. Doctoral dissertation, Ghent University

Vandermeersch G, Van Cauwenberghe L, Janssen CR, Marques A, Granby K, Fait G et al (2015) A critical view on microplastic quantification in aquatic organisms. Environ Res 143:46–55

Veerasingam S, Vethamony P, Aboobacker VM, Giraldes AE, Dib S, Al-Khayat JA (2021) Factors influencing the vertical distribution of microplastics in the beach sediments around the Ras Rakan Island, Qatar. Environ Sci Pollut Res 28(26):34259–34268

Vethaak AD, Legler J (2021) Microplastics and human health. Science 371(6530):672–674

Vroom RJ, Koelmans AA, Besseling E, Halsband C (2017) Aging of microplastics promotes their ingestion by marine zooplankton. Environ Pollut 231:987–996

Wang J, Liu X, Li Y, Powell T, Wang X, Wang G, Zhang P (2019) Microplastics as contaminants in the soil environment: a mini-review. Sci Total Environ 691:848–857

Watts AJ, Lewis C, Goodhead RM, Beckett SJ, Moger J, Tyler CR, Galloway TS (2014) Uptake and retention of microplastics by the shore crab *Carcinus maenas*. Environ Sci Technol 48(15):8823–8830

Wright SL, Thompson RC, Galloway TS (2013) The physical impacts of microplastics on marine organisms: a review. Environ Pollut 178:483–492

Yang L, Li K, Cui S, Kang Y, An L, Lei K (2019) Removal of microplastics in municipal sewage from China's largest water reclamation plant. Water research, 155, 175–181.

Yuan J, Ma J, Sun Y, Zhou T, Zhao Y, Yu F (2020) Microbial degradation and other environmental aspects of microplastics/plastics. Sci Total Environ 715:136968

Zahid H, Afzal N, Arif MM, Zahid M, Nawab S, Qasim MM, Abidi SHI (2024) Microorganism-mediated biodegradation for effective management and/or removal of micro-plastics from the environment: a comprehensive review. Archives of Microbiology, 206(4), 198.

Zeng EY (ed) (2023) Microplastic contamination in aquatic environments: an emerging matter of environmental urgency. Elsevier

Zettler ER, Mincer TJ, Amaral-Zettler LA (2013) Life in the "plastisphere": microbial communities on plastic marine debris. Environ Sci Technol 47(13):7137–7146

Zhang K, Shi H, Peng J, Wang Y, Xiong X, Wu C, Lam PK (2018) Microplastic pollution in China's inland water systems: a review of findings, methods, characteristics, effects, and management. Sci Total Environ 630:1641–1653

Zhang Y, Gao T, Kang S, Allen S, Luo X, Allen D (2021a) Microplastics in glaciers of the Tibetan Plateau: evidence for the long-range transport of microplastics. Sci Total Environ 758:143634

Zhang X, Li Y, Ouyang D, Lei J, Tan Q, Xie L et al (2021b) Systematical review of interactions between microplastics and microorganisms in the soil environment. J Hazard Mater 418:126288

Zhou B, Wang J, Zhang H, Shi H, Fei Y, Huang S et al (2020) Microplastics in agricultural soils on the coastal plain of Hangzhou Bay, east China: multiple sources other than plastic mulching film. J Hazard Mater 388:121814

The manufacturer's authorised representative in the EU is Springer Nature Customer Service Centre GmbH, Europaplatz 3, 69115 Heidelberg, Germany. If you have any concerns regarding our products, please contact ProductSafety@springernature.com

Printed and bound by CPI Group (UK) Ltd, Croydon, CR0 4YY
01/10/2026
02225863-0009